Quantum Computing and Artificial Intelligence

Quantum Computing

Edited by
Pethuru Raj and Abhishek Kumar

Quantum Computing and Artificial Intelligence

Training Machine and Deep Learning Algorithms on Quantum Computers

Edited by
Pethuru Raj, Abhishek Kumar, Ashutosh Kumar Dubey,
Surbhi Bhatia and Oswalt Manoj S

DE GRUYTER

Editors

Pethuru Raj
Reliance Jio Platforms Ltd.
D 002, Prospect Princeton Apartment, C Block
Manipal County Road
560068 Bengaluru
Karnataka
India
peterindia@gmail.com

Abhishek Kumar
Chitkara University School of Engineering and Technology
Room 301
Pinjore-Nalagarh National Highway (NH-21A)
174103 Kalujhinda
Himachal Pradesh
India
abhishekkmr812@gmail.com

Ashutosh Kumar Dubey
Chitkara University Institute of Engineering and Technology
PH-III, Khajuri Kalan Road C-6
Tagore Nagar
462021 Bhopal
Madhya Pradesh
India
ashutoshdubey123@gmail.com

Surbhi Bhatia
University of Salford
School of Science, Engineering and Environment
Department of Data Science
34 Crescent
M5 4WT Salford
Great Britain
surbhibhatia1988@yahoo.com
S.Khan138@salford.ac.uk

Oswalt Manoj S
Department of Computer Science and Engineering
Sri Krishna College of Engineering and Technology
641008 Coimbatore
Tamil Nadu
India
oswaltmanojibm@gmail.com

ISBN 978-3-11-079125-9
e-ISBN (PDF) 978-3-11-079140-2
e-ISBN (EPUB) 978-3-11-079147-1
ISSN 2940-0112

Library of Congress Control Number: 2023934427

Bibliographic information published by the Deutsche Nationalbibliothek
The Deutsche Nationalbibliothek lists this publication in the Deutsche Nationalbibliografie; detailed bibliographic data are available on the internet at http://dnb.dnb.de.

Cover image: Vitalii Gulenok/iStock/Getty Images Plus
Typesetting: Integra Software Services Pvt. Ltd.
Printing and binding: CPI books GmbH, Leck

www.degruyter.com

Preface

Quantum artificial intelligence (QAI) is being proclaimed as the most powerful technology for the future. QAI is a strategically sound combination of two pathbreaking technologies: artificial intelligence (AI) and quantum computing. By running AI algorithms and models on quantum computers, it is possible for bringing in delectable revolutions and evolutions across industry verticals. However, there is a long way to traverse to build and use fault-tolerant quantum computers. AI can surmount many practical problems getting associated with quantum computers. The key use cases that can be accomplished by QAI include visual perception, speech recognition, decision-making, financial analysis, and language translation.

Running AI models on quantum computers is more advantageous than running the same on conventional and classical computing. As illustrated in the chapters of this book, quantum computing carries special capabilities in the form of superposition and entanglement. A fault-tolerant quantum computer that naturally gets empowered with superposition and entanglement could do complex AI tasks quickly; that is, fault-tolerant quantum computers can finish complicated and challenging tasks in less time consuming less power. With less power consumption, quantum computer tasked to tackle sophisticated AI problems has turned out to be environment-friendly. Researchers and quantum computing service providers are exploring different complex problems in sync up with AI power.

AI is emerging as one of the key industry trends. There are AI-driven voice assistants such as Alexa and Siri. There are many natural language processing (NLP)-based chatbots. Autonomous vehicles immensely leverage the computer vision (CV) power. In the recent past, the world is excited with OpenAI's GPT-3, which can write prose like humans, and DeepMind by Google which can beat human chess experts. IBM Watson is creating waves across industry verticals. All these clearly illustrate that AI is fast becoming mature enough to resolve real-life problems. Thus, AI becomes penetrative, pervasive, and persuasive too.

QAI, when it gets fully perfected, can be a game-changer for a number of industry verticals. In fact, AI systems powered up by quantum computers can be a huge blessing for every enterprising business. CV and NLP are complex problems mandating for huge amount of computational resources. Herein, the unique power of quantum computing comes to the rescue. Autonomous vehicles will get a huge improvement with the QAI capability. Going forward, QAI is being explored by drug discovery companies as quantum computer can do large-scale data processing really fast. New materials can be identified through the combination power of QAI.

In short, quantum computing is pitched and presented as a new frontier of computer science that will facilitate high-speed AI-based data processing. Such a unique capability is demanded for tackling complex problems. Quantum computers are a kind of supercomputers based on the principles of quantum mechanics. Therefore, big data processing gets simplified and speeded up significantly.

https://doi.org/10.1515/9783110791402-202

How can quantum computers benefit from AI?

Based on a number of decision-enablement features, AI can identify which problems are precisely fit to be tackled through the power of quantum computing. Also several other improvements can be obtained by quantum computing through the immense contributions of AI. All kinds of quantum computer data can be collected, cleansed, and crunched by AI algorithms in order to emit out actionable insights that in turn help quantum computer to do more with less.

Thus, the potential of quantum computing is limitless. But its intimate integration with AI can produce a plethora of state-of-the-art applications. This book has chapters vividly illustrating quantum computing, AI, and how their convergence can be path-breaking in envisaging and realizing breakthrough systems for the ensuing era of knowledge.

Contents

List of contributing authors

Chapter 1
M. Rohini
Sri Krishna College of Engineering and Technology
Coimbatore
India
rohinim@skcet.ac.in

R. Gowthamani
Sri Krishna College of Engineering and Technology
Coimbatore
India
gowthamanihit@gmail.com

M. Indira Priyadharshini
Sri Ramakrishna Engineering College
Coimbatore
India
indirapriyadharshini.m@srec.ac.in

D. Surendran
Karpagam College of Engineering
Coimbatore
India
d.surendran@gmail.com

Chapter 2
Shubham Sharma
Consultant Technology-Virtusa
India
shubhamsharma151998@gmail.com

Pronika Chawla
Assist. Prof.
Manav Rachna International Institute of Research and Studies
India
pronika.fet@gmail.com

Sanu Kumar
B.Tech Computer Science and Engineering
Manav Rachna International Institute of Research and Studies
India
anukumar763390@gmail.com

Mukul Nair
B.Tech Computer Science and Engineering
Manav Rachna International Institute of Research and Studies
India
m3mj2002@gmail.com

Chapter 3
Ananth J. P.
Prof., Sri Krishna College of Engineering and Technology
Kuniamuthur, Coimbatore
India
jude.ananth@gmail.com

G. Raghuraman
Assoc. Prof.
Sri Sivasubramaniya Nadar College of Engineering
Chennai
India
raghuramang@ssn.edu.in

Premnath S P
Assist. Prof., Sri Krishna College of Engineering and Technology
Kuniamuthur, Coimbatore
India
premnathsp@yahoo.com

A. Francis Alexander Raghu
Research Scholar
Sri Krishna College of Engineering and Technology
Kuniamuthur, Coimbatore
India
francismtech@gmail.com

Chapter 4
Sourabh Sharma
Department of Computer Science and Engineering
The NorthCap University
Haryana
India
sourabh.714sharma@gmail.com

https://doi.org/10.1515/9783110791402-204

Poonam Chaudhary
Department of Computer Science
and Engineering
The NorthCap University
Haryana
India
poonamchaudhary@ncuindia.edu

Chapter 5
S. Stewart Kirubakaran
Department of Computer Science
and Engineering
Karunya Institute of Technology
and Sciences
Karunya Nagar, Coimbatore
Tamil Nadu
India
stewartkirubakaran@gmail.com

R. Senthil Kumar
Assist. Prof., Department of
Computer Science and Engineering
Saveetha School of Engineering
Saveetha Institute of Medical
and Technical Sciences
Tamil Nadu
India
rsenthilmecse@gmail.com

A.M. Anusha Bamini
Department of Computer Science
and Engineering
Karunya Institute of Technology
and Sciences
Karunya Nagar
Coimbatore, Tamil Nadu
India
anushabamini@gmail.com

Chitra.R
Department of Computer Science
and Engineering
Karunya Institute of Technology
and Sciences
Karunya Nagar
Coimbatore, Tamil Nadu
India
chitrajegan5@gmail.com

Chapter 6
Anusha Bamini A.M
Department of Computer Science
and Engineering
Karunya Institute of
Technology and Sciences
Coimbatore
India
anushabamini@gmail.com

Chitra. R
Department of Computer Science
and Engineering
Karunya Institute of Technology
and Sciences
Karunya Nagar
Coimbatore, Tamil Nadu
India
chitrajegan5@gmail.com

Jeba Priya S
Department of Computer Science
and Engineering
Karunya Institute of Technology
and Sciences
Coimbatore
India
jebapriya@karunya.edu

Chapter 7
Sruthi Anand
Assist. Prof., Department of
Information Technology
Sri Krishna College of
Engineering and Technology
Coimbatore
India
sruthianand@skcet.ac.in

Granty Regina Elwin
Assoc. Prof., Department of
Information Technology
Sri Krishna College of
Engineering and Technology
Coimbatore
India
grantyreginaelwinj@skcet.ac.in

R Priyadharsini
Assist. Prof., Department of Science and Humanities
Sri Krishna College of Engineering and Technology
Coimbatore
India
priyadharsini@skcet.ac.in

K Ranjeetha Priya
Assist. Prof., Department of Computer Science and Engineering
Sri Ramakrishna Engineering College
Coimbatore
India
ranjeethapriya.k@srec.ac.in

D Prabha
Prof., Department of Computer Science and Engineering
Sri Krishna College of Engineering and Technology
Coimbatore
India
prabha@skcet.ac.in

Chapter 8

G Ignisha Rajathi
Assoc. Prof., Computer Science and Business Systems
Sri Krishna College of Engineering and Technology
Coimbatore
India
ignishaelton@gmail.com

R. Vedhapriyavadhana
Assist. Prof./Senior Grade, School of Computer Science Engineering
Vellore Institute of Technology
Chennai
India
vedhapriya.vadhana@gmail.com

R. Niranjana
Assist. Prof., Electronics and Communication Engineering
Sri Krishna College of Engineering and Technology
Coimbatore
India
niranbme@gmail.com

N. Pooranam
Assist. Prof., Computer Science and Engineering
Sri Krishna College of Engineering and Technology
Coimbatore
India
pooranam.n@gmail.com

R. Johny Elton
Indsoft Technologies
Tirunelveli, Tamil Nadu
India
erjohnyelton@gmail.com

Chapter 9

N. Pooranam
Assist. Prof., Department of Computer Science and Engineering
Sri Krishna College of Engineering and Technology
Coimbatore
India
pooranam.n@gmail.com

Dr. D. Surendran
Prof., Department of Computer Science and Engineering
Karpagam College of Engineering
Coimbatore
India
d.surendran@gmail.com

Dr. N. Karthikeyan
Assist. Prof., School of Computer
Madanapalle Institute
of Technology & Science
Madanapalle, Andhra Pradesh
India
narenkarthikeyan.mecse@gmail.com

Dr. G. Ignisha Rajathi
Assist. Prof., Department of Computer Science
and Engineering
Sri Krishna College of Engineering
and Technology
Coimbatore
India
ignishaelton@gmail.com

Chapter 10

M M. Rohini
Sri Krishna College of Engineering
and Technology
Coimbatore
India
rohinim@skcet.ac.in

D. Surendran
Karpagam College of Engineering
Coimbatore
India
d.surendran@gmail.com

Oswalt Manoj S
Sri Krishna College of Engineering
and Technology
oswaltmanojibm@gmail.com

Chapter 11

B. Kiruba
Assist. Prof., Department of
Artificial Intelligence and
Data Science
Sri Krishna College of Engineering
and Technology
Coimbatore
India
kirubamugesh@gmail.com

M. Manicka Raja
Assist. Prof., Department of Computer Science
and Business Systems
Sri Krishna College of Engineering
and Technology
Coimbatore
India
manickaraja89@gmail.com

J. Jasmine Selvakumari Jeya
Senior Associate Professor,
School of Computing Science
and Engineering
VIT Bhopal University
Madhya Pradesh
India
wjasminejeya@gmail.com

C. Angel
Assist. Prof., Department of Computer Science
and Engineering
DMI Engineering College
Kanyakumari
angel.dhas19@gmail.com

Chapter 12

R. Vedhapriyavadhana
Assist. Prof./Senior Grade, School of
Computer Science Engineering
Vellore Institute of Technology
Chennai
vedhapriya.vadhana@gmail.com

G. Ignisha Rajathi
Assoc. Prof., Computer Science
and Business Systems
Sri Krishna College of Engineerin
and Technology
Coimbatore
India
ignishaelton@gmail.com

R. Niranjana
Assist. Prof., Electronics and
Communication Engineering
Sri Krishna College of Engineering
and Technology
Coimbatore
India
niranbme@gmail.com

N. Pooranam
Assist. Prof., Computer Science
and Engineering
Sri Krishna College of Engineering
and Technology
Coimbatore
India
pooranam.n@gmail.com

Chapter 13

Nisha Soms
Assoc. Prof., Department of Computer Science
and Engineering
KPR Institute of Engineering and Technology
nishasoms@gmail.com

S. Kalyani
Assist.Prof. S&H
Sri Ramakrishna Institute of Technology
Coimbatore
India
kalyani.sh@srit.org

S. Nagarani
Professor &Head, S&H
Sri Ramakrishna Institute of Technology
Coimbatore
India
sureshnagarani@gmail.com

M. Rohini
Assist. Prof, Department of Computer Science
and Engineering
Sri Krishna College of
Engineering and Technology
Coimbatore
India
rohinibtech@gmail.com

Oswalt Manoj S
Department of Computer Science
and Engineering
Sri Krishna College of
Engineering and Technology
Coimbatore
India
oswaltmanojibm@gmail.com

Chapter 14

K. Sasi Kala Rani
Prof. & Head, Department of Computer Science
and Engineering
Sri Krishna College of
Engineering and Technology
Coimbatore
India
sasikalaranik@skcet.ac.in

Priyadharsheni J M
Assist. Prof.
Robotics and Automation
Sri Ramakrishna Engineering College
Coimbatore
India
priya86.btech@gmail.com

B. Karthikeyan
Assist. Prof, Department of Computer Science
and Engineering
KPR Institute of Engineering and Technology
karthikeyanb@krpiet.ac.in

G.S. Pugalendhi
Assist. Prof.,AI&DS
Sri Krishna College of Engineering
and Technology
Coimbatore
India
pugalendhigs@skcet.ac.in

M. Rohini, R. Gowthamani, M. Indira Priyadharshini, D. Surendran

Chapter 1 Digital transformation technology and tools: shaping the future of primary health care

Abstract: Artificial Intelligence enablement is often incorporated to facilitate and smoothen business operations that require large-scale human activities in an effortless manner that relatively interacts with human intelligence. Artificial intelligence augmentation in business functions implies quality enhancement, process streamlining, ensuring security and privacy, detection of defects, bugs and errors that extend to design a strong system approach for enabling the completion of tasks more efficiently, enhances workforce productivity, improves customer engagement and experience, drives to lower operating cost and thus impact the organization's role. The organizations which incorporate and focus strongly on Artificial Intelligence enablement practices are found to be successful in attracting huge investments and talented human resource, in achieving corporate missions, in satisfying the clients, attaining competitive advantage over the others, and importantly, helping the organization in maximizing its profits. It is found that organizations incorporating Machine Learning and Artificial Intelligence have gained better financial returns from their Artificial Intelligence investments through the study of artificial intelligence in the enterprises. This work automatically detects the abnormalities from medical images in the computer-aided diagnosis context. Multimodal Medical image analysis is a challenging problem. Medical images are different in texture and shape. The current research work is an attempt to deal with problems in Multimodal Medical image analysis and provide the solution for Medical image analysis. Image Acquisition procedures refer to any or all Medical image capturing systems. Medical images differ from pictures procured from visualizations in that the delineated physical parameters in restorative pictures are generally distant for investigation. There are different strategies for making medical images accessible in the restorative business sector like r-ray, fluoroscopy and angiography, DSA, X-ray computed tomography (CT), CT angiography, magnetic resonance imaging (MRI), MR angiography, functional MRI, perfusion MRI, diffusion MRI, SPECT, PET, and so on discussed in this work.

M. Rohini, Sri Krishna College of Engineering and Technology, e-mail: rohinim@skcet.ac.in
R. Gowthamani, Sri Krishna College of Engineering and Technology, e-mail: gowthamanihit@gmail.com
M. Indira Priyadharshini, Sri Ramakrishna Engineering College, e-mail: indirapriyadharshini.m@srec.ac.in
D. Surendran, Karpagam College of Engineering, e-mail: d.surendran@gmail.com

https://doi.org/10.1515/9783110791402-001

1.1 Introduction

Medical imaging is defined as "the discipline and field of healthcare involving the development and application of imaging technologies and procedures to offer biochemical analyses of tissues and organs" (WebMD). Medical imaging has turned into one of the principal essential visualization and interpretation methods in science and medication over the previous decade. Medical images are, today, an essential and amazingly popular source of information used by medical practitioners for diagnosing and for arranging relevant treatment. Hence medical imaging in light of the fact that the strategy of physical examinations turns in a standout amongst the most vital instruments inside the entire biomedical designing, and conjointly, the business sector of imaging gadgets might be discovered on the grounds of precise quick creating parts of med-frameworks. Aided by the mixed bag of the latest medicinal imaging gadgets, therapeutic experts can gain totally distinctive sorts of images that present interesting organs and examine tissues in numerous different structures. Among the big advances in current pharmaceuticals, therapeutic imaging has made real headway. The adaptability to obtain images of the human body has a few accommodating clinical requisitions.

Totally distinctive modalities of restorative imaging have developed through the years, each with its advantages and downsides. For any patient who needs his or her body visualized, specialists pick the principal appropriate imaging studies to convey the best alternative [1]. Modern medical imaging developments offer novel insights into the anatomy, physiology, metabolism, hemodynamics, and pathology of the human body. Medical imaging covers a wide range of areas, including image acquisition and formation, image processing and advancement, edge detection, filing, simulation, visualization, and encoding. It also includes areas like computer-aided diagnosis, computer-aided surgery, and online medical imaging [2]. Different medical screening tests need to be taken into account.

The approach that delivers these images additionally assumes an exceptionally essential part in recognizing what to apply to show signs of improvement. To this end, the computerized characteristics of a few modalities like CT, MRI, X-ray, ultrasound (US), microscopic pictures by utilizing fuzzy logic, and some earlier information like functional features called multimodal medical image analysis is proposed. Since presently there are no broadly acknowledged routines, automatic and reliable techniques for medical image analysis are an incredible need and investment.

1.2 Smart applications toward digital transformation

Image acquisition procedures refer to any or all medical image capturing systems. The distinguishing physical parameters in restorative images are typically remote for exploration, which makes them different from photographs. The medical imaging industry in the field of quality management offers a variety of methods for accessing health images, including X-rays, radiology, and angiography, DSA, X-ray CT, CT angiography, MRI, MR angiography, functional MRI, perfusion MRI, diffusion MRI, SPECT, PET, and others. In current research work, five types of medical images are used. They are X-ray, CT, MRI, US, and MS.

1.2.1 Image preprocessing

Image preprocessing subsumes all image-to-image transforms that are done throughout Image Acquisition. The obvious reasons for the need of image preprocessing are enhancement of image quality to fulfil the necessities of medical practitioner, noise reduction, contrast enhancement, misclassification of missing pixels, optimal preparation of information for postprocessing, and elimination of acquisition-specific artifacts. The preprocessing techniques are used here to remove the artifacts from acquired images.

Image segmentation is the process of dividing or collating a picture into diverse parts. These segments usually compare to things that people will effortlessly separate and take a glance at, as individual articles. Machines do not have any method for keenly perceiving items, and afterward, numerous different systems are created, so as to fragment clinical images. The division strategy is based on various peculiarities found inside the picture. As a result of the misclassification that is visible in a few parts of image process, fuzzy process is appropriate. Feature Extraction in machine vision and image processing refers to procedures that go for processing deliberations of image information and making local at each image point, whether there is an image feature of a given sort by then or not.

Several significant steps in image analysis include Feature Extraction. Features are the contents or properties or significant characteristics of the image using which one can classify the image into its category. For the same purpose, features in numeric form are extracted from images using different techniques. The classification model is designed from the extracted features. The performance and accuracy rate of the classifier also depends on features.

Image Classification Object (image or parts of image) is a real-world entity and classification is the technique used to describe a group or classify the object. The real-world problems are very critical and require an expert as decision maker to make certain decisions. This decision is taken by using two different methods based on the availability of information (image features). These steps include training the clinical and follow-up sets with classification algorithms. Supervised learning techniques classify the objects using prior knowledge of objects, whereas unsupervised learning techniques do not need prior knowledge and form clusters of input objects.

1.2.1.1 Computer-assisted digital diagnosis system

The term “computer diagnosis” or “automated diagnosis” was first used in the field of radiology in 1960, because it was believed that since computers and machines are more adept at performing some tasks than humans, they could eventually replace the radiologists in the detection of abnormalities. The idea of computer-aided diagnosis rose in the 1980s, presuming that radiologists could use the computer output but that it would not replace them. Digital medical images are quantitatively examined by computers as is the standard practice in CAD and Automated Computer Diagnosis. As a result, Machine Learning must be designed for both CAD and computer diagnostics. The way the machine information is used for the diagnostic is the key difference between CAD and computer assessment. The final choices are made by physicians using the computer output as “feedback” in CAD. However, a rather greater standard of computer coefficient of performance is needed for automated software diagnosis. Therefore, in order to apply automated computer diagnosis, high sensitivity and high specificity are needed [3, 4]. The architecture for an intelligent clinical disease classification steps is shown in Figure 1.1. The prognosis precision can be improved by including the intelligent agents in clinical sets and most weighted random features of disease, particularly in circumstances involving novice clinicians [5].

CAD is an analysis performed by a radiologist, who uses the results of a computerized examination of medical pictures as a second assumption in determining where damages are present and in making choices. The last conclusion is made by the radiologist. A Computer-aided Diagnostic System is a computer program that might be used to complete a proper medical diagnosis. There is an increasing burden at the hospitals because of population increase. Consequently, the tasks of specialists and clinicians are getting to be harder day by day. The application of cutting-edge digital technologies can simplify this work [6]. Physicians are increasingly using computer-aided detection and diagnosis (CAD) technologies to help identification and interpretation of diseases.

Computer-aided Detection and Diagnosis (CAD) systems highlight portions of an image that could reveal certain anomalies and direct physicians’ focus there, while comprehending the image.

Clinicians use computer-aided diagnosis tools as a decision support when making diagnoses, because they provide an assessment of a condition using image-based information alone or in conjunction with other pertinent clinical information [7].

Various CAD systems such as conventional projection radiography, computed tomography (CT), and MRI are being developed for the identification and/or categorization of various anomalies in medical imaging.

1.2.1.2 Need for CAD

In reality, detection and prevention of disease faces several challenges like limited medical staff, geographical reach, awareness of the disease, and unavailability of medical practitioners. Several challenges are explained as follows. (a) Limited medical staff and geographical reach: Computerized framework structure can decrease the troublesome work included in quantitative diagnosing, inside out. Mostly, tomography centers are secured in urban zones and thus, rural zone patients have to experience the ill effects of logistics issues and invest additional time for analysis and related medication. This builds the need for portable computer-aided frameworks. (b) Disease awareness and social obstructions: Awareness of a disease is a critical element and it can be avoided at the early stage. (c) Accessibility of expert: Effective diagnosis from medical images requires expertise and experience in studying human anatomy and disease. Availability of such medical practitioners in remote areas is a challenge.

1.2.2 Applications of CAD computers

Nodules in thyroid ultrasound images are automatically classified using assisted diagnosis systems. Two of the most often used imaging techniques for thyroid diagnosis are US and CT. Malignant thyroid nodules frequently blend with surrounding tissues and have diverse histological components, making it challenging to determine the shape of the nodule. Therefore, CAD is essential for identifying and classifying such thyroid nodules [9]. High-resolution brain MRI is necessary for the classification of brain tumors. MRI, positron emission tomography (PET) and CT scan are used for the majority of medical imaging studies and detection. Data from various successive pulses are delivered in a multidimensional manner by MRI. CAD is intended to detect brain tumors in MRI images automatically. CAD offers useful perspective and performance accuracy for clinical image processing.

Early detection of liver cancer is a challenging task due to cell structure. Sometimes, even for experienced practitioners, interpretation is difficult and time-consuming. CAD is used as a solution to this early cancer detection problem [11]. Mammography is a screening tool for the early detection of breast cancer. There are no ways to prevent the cancer, but if the cancer is detected in earlier stage there are better chances of proper treatment. The efficient identification of breast cancer is a much-needed improvement. Nowadays, CAD has been developed for breast cancer, including detection of masses, calcification, architectural distortion, and bilateral asymmetry in mammograms [12]. The success of a particular CAD system can be measured in terms of accuracy of diagnosis, speed, and automation level [5].

1.2.2.1 Industry-grade CAD tools

Digital tomography with ImageChecker CAD: Hologic-produced products support Picture Archiving and Communication System (PACS) or Digital Imaging and Communications in Medicine (DICOM) output. The majority of direct capture full-field digital mammography (FFDM) detector images can be processed by ImageChecker CAD, which can then display the results on a variety of workstation setups.

The chosen viewing solution determines how digital CAD marks will appear on screen.

ImageChecker CAD for mammography on screen film: The computer-aided detection (CAD) application, R2 ImageChecker from Hologic, improves screen-film mammography inspection. Radiologists can benefit from CAD review in a screen-film setting by using ImageChecker CAD software on analogue mammograms. The American Food and Drug Administration (FDA) granted its initial approval to a CAD system created by R2 Technology for mammography in 1998 [13].

SecondVisualizationTM: A product manufactured by its iCAD, July 2004, combines the CADstream system for CAD analysis of breast MR imaging with the early breast cancer detection capabilities of the iCAD Second Look line of film-based mammogram devices. Radiologists can simultaneously access radiography and MR CAD findings, while evaluating the relevant patient film with the results of the CADstream analysis being made available on the Second Look review station.

The CAD system for interpreting breast MRI results is called CADstream TM (Confirma).

RapidScreenTM analyses digital chest radiographs to detect lung cancer in its early stages [14].

MedSignificant: CAD analysis of the data for the heart, colon, and lungs.

ColonCAD: CT colonography and LungCAD for CT thorax scans are two of the company's CAD offerings. Clinicians using CT scans can discover, measure, and analyze suspected disease in the colon and lungs, with the aid of Medicsight's CAD solutions [15].

MeVisual: A CAD program for hepatic treatment scheduling has acquired FDA clearance (510(k) clearance). MeVis creates cutting-edge software for the analysis and processing of image data that is sold to producers of medical equipment and medical IT platforms. Important clinical disorders such as breast, lung, liver, prostate, and intestinal cancer as well as neurological diseases are the focus of image-based early detection and diagnostics [16]. Around the world, there are numerous CAD projects. Several efforts from China and Japan are discussed in [5].

The project Intelligent Assistance in Diagnosis of Multi-dimensional Medical Images was funded by a grant-in-aid for scientific research on priority areas from MEXT, Japan (October 2003 to March 2007). This task included computational modelling of organs, intelligent CAD, and simultaneous segmentation of numerous organs in multidimensional medical images based on local pathological morphology. It also included intelligent CAD based on normal structure recognition of the human body, intelligent CAD based on quantitative analysis of spatiotemporal change in medical image sequencing, the development of a navigation-based diagnosis system, and more. Multimodal, Intelligent CAD based on quantitative analysis of spatiotemporal change in medical image sequencing, based on typical structural recognition of the human body, and creation of a diagnosis system based on navigation, including multimodal are all integrated.

In this work, the mentioned techniques are used to extract features from medical images. Texture, shape, and simple statistical features are used for medical image classification according to modality. Each type of feature is extracted separately and combined with another type. These features are stored in a feature database and used to design the classification model. Features are extracted at two levels from the image. Medical Images are used to visualize different internal structures and recommended as per the need of diagnosis. Medical images vary in tissue appearance structures. Hence first, features are extracted for identification of the image modality. Secondly, features are extracted from image clusters for cluster labeling using GLCM texture features and first-order statistical features.

1.2.3 General feature extraction approach

- Training dataset is of size 250 images (50 images of each modality CT, MRI, X-ray, US, MS).
- Read Image from Training dataset.
- Apply median filtering for noise removal.
- Extract features from image using one of the feature extraction techniques or a combination of feature extraction techniques. Repeat steps 2–4 for all 250 images.
- Compute the mean and standard deviation of each feature set.
- Store the results into Feature Database. Some of the approaches for medical image classification according to modality are mentioned here.

1.2.3.1 Statistical feature-based approach

Gray Level Co-occurrence Matrix (GLCM), along with first order statistical features-based approach, is shown in Figure 1.1. In contrast, correlation, entropy, homogeneity, local homogeneity, mean, and standard deviation (07 features) of each image are computed using the abovementioned techniques. The mean and standard deviation of each feature set is computed for each modality. For each modality, 14 features are generated. A total of 70 features are generated. The contrast of each image is computed using Harlik texture feature extraction method. Mean and standard deviation of the image are computed.

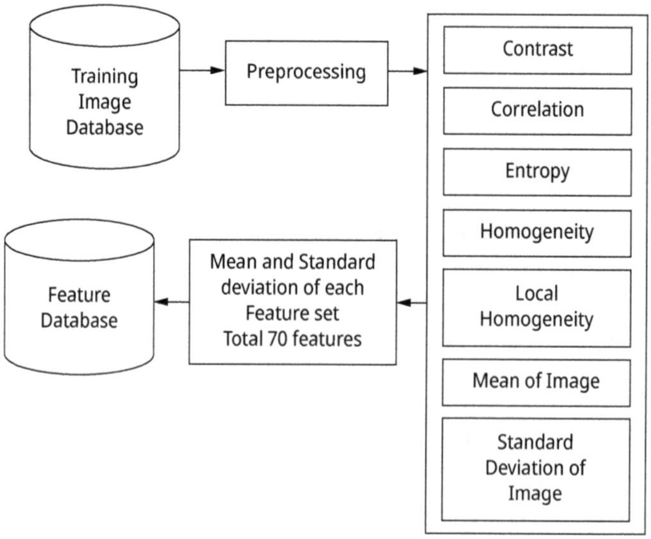

Figure 1.1: Statistical feature-based approach.

1.2.3.2 Extraction of first-order statistical features

Figure 1.2 shows the first-order statistical feature-based feature extraction. The mean and standard deviation of each feature set is computed for each modality. For each modality, 06 features are generated. A total of 30 features are generated.

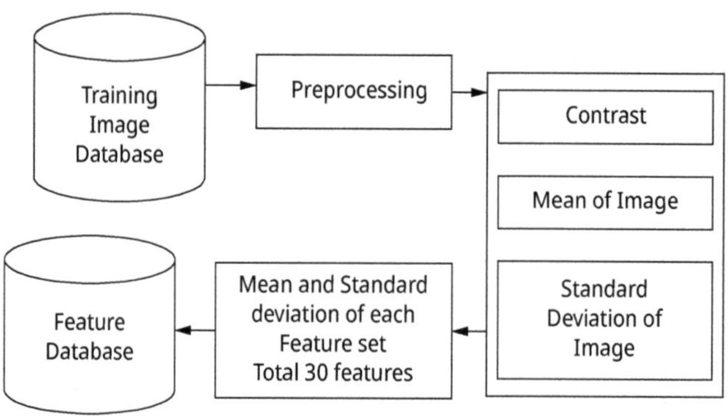

Figure 1.2: First-order statistical feature-based feature extraction.

1.2.3.3 Discrete wavelet transform-based feature extraction

Figure 1.3 explains discrete wavelet transform (DWT)-based feature extraction. DWT (Haar wavelet transform is used) gives approximation coefficient and horizontal, diagonal, and vertical coefficients of the image (total 08 features). Mean and standard deviation of each coefficient are extracted. Mean and standard deviation of these features are computed to reduce the size of feature set (02 features of each modality).

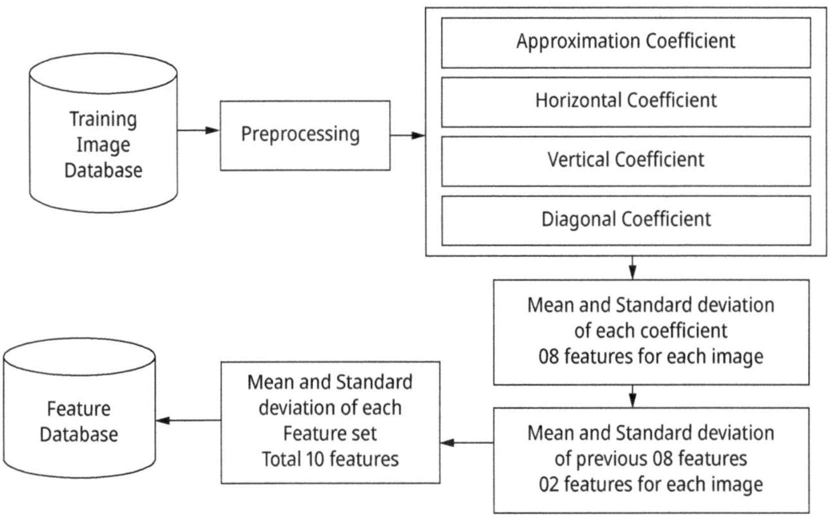

Figure 1.3: DWT-based feature extraction.

1.2.3.4 Discrete cosine transform-based feature extraction

Discrete cosine transform (DCT) gives a similar size resultant matrix of the input image. Mean and standard deviation of each resulting image are computed. Mean and standard deviation of these features of each modality are computed to reduce the size of feature set (02 features of each modality). A total of 10 features are generated. DCT-based feature extraction is shown in Figure 1.4.

1.2.3.5 Invariant moment-based feature extraction

Invariant moment techniques extract 07 moment features. Absolute log of each feature is computed. Mean and standard deviation of each moment set is computed. A total of 14 features for each modality, i.e., a total of 70 features are computed for five modalities. Figure 1.5 shows the invariant moment-based feature extraction.

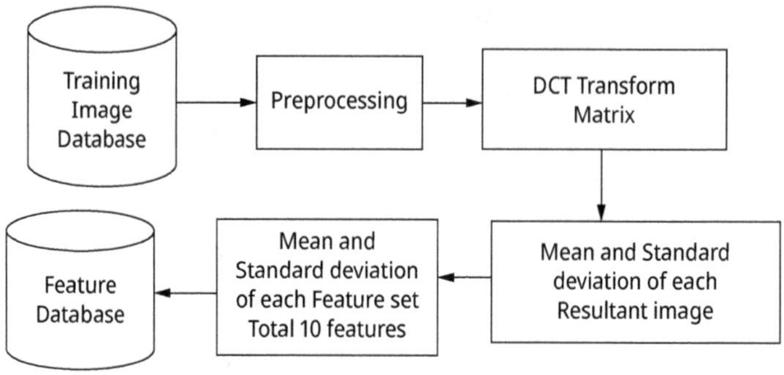

Figure 1.4: DCT-based feature extraction.

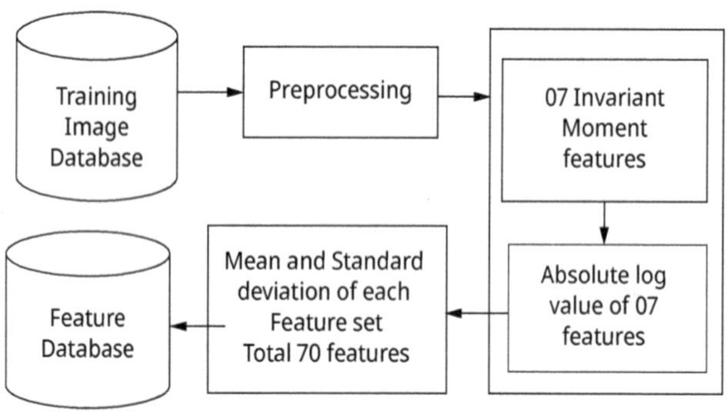

Figure 1.5: Invariant moment-based feature extraction.

1.2.3.6 Invariant moment + GLCM texture features + first-order statistical-based feature extraction

Invariant moment techniques extract 07 moment features. The absolute log of each feature is computed. Contrast, correlation, entropy, homogeneity, local homogeneity, mean, and standard deviation (07 features) of each image are computed using the abovementioned techniques. Mean and standard deviation of each feature set is computed for each modality. For each modality, 28 features are generated. A total of 140 features are generated.

1.2.3.7 System based on fuzzy rules

To create fuzzy frameworks, we can employ fuzzy logic. A fuzzy system is a static or dynamic system that uses fuzzy sets. The most typical fuzzy systems use if-then logic. These are referred to as rule-based systems, or fuzzy models. All methods that comprehend, represent, and process pictures, their segments, and features as Fuzzy Sets are together referred to as fuzzy image processing. The chosen fuzzy approach and the issue at hand determine the representations and processing [19]. Figure 1.6 depicts the general organization of fuzzy image processing.

Fuzzy image processing plays a key role in conveying data ambiguity. In general, an if-then rule is expressed as: IF an antecedent proposition then a subsequent proposition. A fuzzy statement of the form "x is A" – where x is a linguistic variable and A is a linguistic constant – is always the antecedent proposition. (For instance, if the temperature is high, then). However, the framework we choose will determine how the subsequent proposition is structured. Both the antecedent and the consequent are fuzzy propositions in a linguistic fuzzy model. The linguistic fuzzy model is expanded upon in the fuzzy relational model. A fuzzy antecedent may now be related to several fuzzy propositions simultaneously. The consequent in the Takagi-Sugeno (TS) fuzzy model is an exact function.

1.2.3.8 Fuzzy c-means clustering

In order to solve issues with Pattern Recognition, Computer Vision, and CAD, among other things, fuzzy c-means clustering is crucial. This method is grounded in fuzzy logic. Dunn created fuzzy c-means (FCM) clustering in 1973. Based on their membership value, which indicates the degree to which data belong to distinct clusters, FCM clustering enables data to belong to more than one cluster. The centroid of a cluster in FCM is calculated as the average of all points weighted by how much they belong to the cluster. The mathematical morphology algorithms have numerous solutions that are based on fuzzy logic and fuzzy set theory.

They have improved the manner that grayscale photos are processed with the use of fuzzy morphological operators. Theory of Fuzzy Sets and Fuzzy Logic Fuzzy Set and Fuzzy Logic theory are new research areas for defining new algorithms [18] and solutions in a mathematical morphology environment [20]. According to [21], fuzzy image processing techniques have a number of advantages. Fuzzy approaches are effective methods for processing human information in the form of fuzzy if-then rules. Techniques that use fuzzy thinking can effectively handle ambiguity and vagueness. Since fuzzy logic can accept imperfect data, it is excellent for highly complex or somewhat nonlinear processes. As a result, it makes it easier to avoid inconsistent behavior.

1.2.3.9 Fuzzy logic in medical diagnosis

Quantitative analysis, in addition to qualitative evaluation, is important in identification of disease in its earlier stage. Initial diagnosis is often aided by medical imaging. Problems with medical diagnosis and prognosis are excellent instances of decision-making under uncertainty. Dealing with uncertainties is a challenge that Pattern Recognition faces frequently, and the use of Fuzzy Set theory has led to the development of several novel techniques for Pattern Recognition in medical diagnostics. Traditional diagnostic techniques for medical imaging can be improved by using fuzzy pattern recognition techniques [22]. Radiologists often have a great amount of knowledge about characteristics of lesion in Medical Image. They use this knowledge to identify lesions from patient's medical image. Hence, it can be beneficial to encapsulate this knowledge in the computer system for Medical Image Processing. And, one important technique for representing experts' knowledge regarding pattern recognition is Fuzzy if-then rules. Fuzzy rule-based systems are used to segment an image, to recognize the object in the image, or to analyze the relationships between the image objects [23].

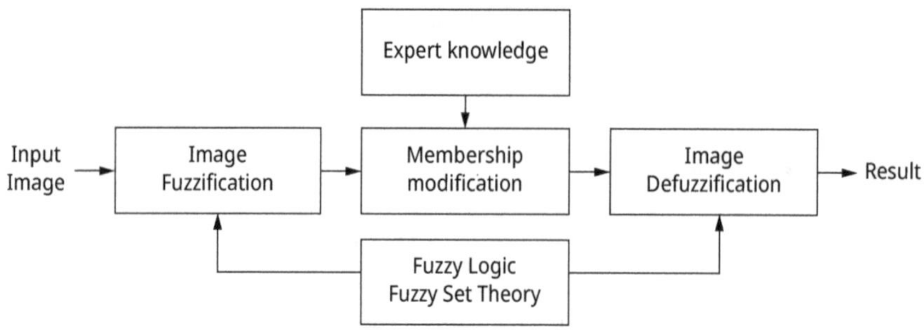

Figure 1.6: Fuzzy Image Processing.

They automatically detect and localize region of interest in clinically acquired medical images (CT, MRI, X-ray, US, and microscopic) to assist medical practitioners. This study ought to give a working answer for outlining the capability of fuzzy frameworks for tackling particular Computer Vision issues in the clinical Medical Imaging area.

Due to the rate of increase in Medical Imaging, there is need for computer-based systems to assist radiologists. To reduce the burden of radiologists, CADD are needed. Though radiologists do work on different types of medical images, multimodal systems can produce a second opinion to radiologists. X-rays, CT scans, and MRI are examples of clinical computed tomography that are used by radiologists to diagnose and decide how to treat patient health issues. Radiologists provide image-guided operations, but they typically do not deal with a patient's general needs. Instead, a radiologist uses the resources at his or her disposal to diagnose a patient and then provides the clinical investigations to the patient's doctor.

1.3 Image preprocessing

Image preprocessing subsumes image-to-image transforms that are done during image acquisition. One can preprocess the image for improvement in image quality, reduction of noise, contrast enhancement, and to remove the special artifact produced during image acquisition. These different image preprocessing techniques can be used to eliminate image artifact acquired during image acquisition procedures according to the need of physicians. Medical images also suffer from various artifacts during image acquisition. Hence, image preprocessing is a crucial step in Medical Image Processing. Image preprocessing phase is carried out to bring out specific features of an image and the resulting image provides more details than the original image for post processing. Image preprocessing techniques can be classified into image restoration and image enhancement. Image restoration process reconstructs an image that has been degraded by degradation phenomenon.

Noise removal and deblurring are image restoration techniques, whereas image enhancement process improves an image in some predefined sense. Contrast Stretching is an example of image enhancement technique. Image Preprocessing in multimodal medical image processing is a vital task. There is no fixed noise removal and image enhancement Technique for medical image processing. Medical images are obtained from different modalities. Each modality suffers from different artifacts. In this work, a study of different preprocessing techniques is conducted and these techniques are applied on multimodal medical images. From a comparison of obtained results, noise removal technique and image enhancement technique are chosen for current work.

1.3.1 Medical image acquisition

Medical imaging is a process that uses specialized equipment and methods to provide images or pertinent information about the inside biological structures and functioning of the body, in a clinical setting [25]. Images of a particular organ or body part are taken for clinical inspection in order to determine the diagnosis or pathology. The imaging method was specifically chosen since it is notable for producing images that describe information that is pertinent to diagnosis. This knowledge is generally used by efficient analytic techniques as a component of their domain knowledge to compensate for the unavailability of the measured attribute.

To be relevant, a characteristic must fulfil three requirements: it should be measured by an imaging equipment and displayed as an image; it must enter the body without undue interference; and it must serve the aim of asking the respondent a medically pertinent inquiry [2]. Medical projection and tomographic images are the two forms of images. For a 3D body, projection creates a single picture that acts as the body's "shadow" in a certain direction (integration through the body), such as an X-ray image, unlike tomographic pictures, which produce a succession of images from each slice of a 3D object in a certain direction (axial, coronal, and sagittal). Each slice's image is created by first obtaining projections in various directions, and then reconstruction is done from those images [3].

To successfully gather the necessary data for a pathological examination, it is critical to choose the right medical imaging modality [1]. An energy source such an X-ray, CT scan, US, PET scan, or an MRI is needed for the basic process of picture production, in order to learn more about the item that is being represented by the image. Object tissue or organs interact with some type of radiation, such as optical light, X-rays, gamma rays, Radio Frequency (RF), or acoustic waves, to reveal information about the characteristic quality of the object [1]. The imaging modalities used for this study – CT, MRI, X-ray, US, and microscopic imaging – as well as their underlying principles are briefly covered in this work.

1.3.2 X-ray imaging

Electromagnetic Spectrum with their frequency is mentioned in figure 1.7. The X-rays pass through most objects, as well as the body. Once it is rigorously aimed toward the part of the body being examined, an X-ray machine produces a tiny, low burst of radiation that passes through the body, recording a picture on photographic paper or a special detector. Completely different parts of the body absorb the X-rays in varied degrees. Dense bone absorbs an abundance of the radiation, whereas soft tissue, like muscle, fat, and organs, enable a lot of the X-rays to pass through them.

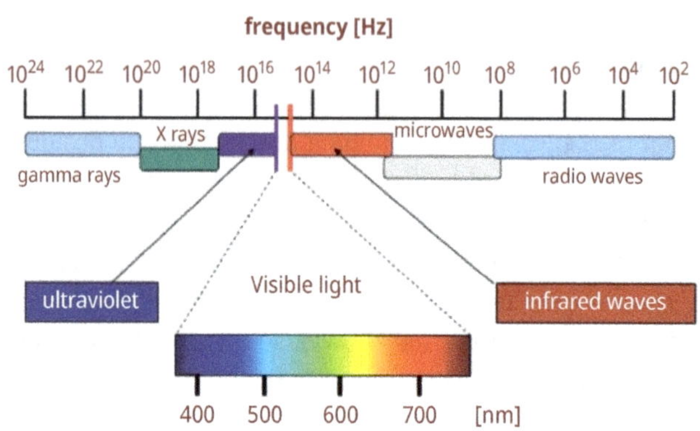

Figure 1.7: Electromagnetic spectrum.

The low X-ray linear attenuation coefficient is imaged as black; bone, because of its much higher X-ray linear attenuation coefficient is imaged white; whereas soft tissue is often presented in between these two extremes as shades of gray [4, 6]. The diagnostic range of X-rays is between 0.5A and 0.01 A. A wavelength, which corresponds to the photon energy of approximately 20 keV to 1.0 MeV. X-ray medical imaging uses an external ionized radiation source.

1.3.3 CT tomography

The structures are projected on top of one another. In typical X-ray imaging, the complete thickness of the body is projected on a film; structures overlap and are troublesome to differentiate and there is loss of data regarding depth. High attenuating objects like bone could hide other objects. In bone pictures, for example, the bone would hide most of the details of the biomarker with soft tissue [2]. CT or X-ray CT produce a group of 2D cross-sectional images rather than one 2Dimage. CT scanning combines the use of sophisticated computers and X-ray equipment to produce several images or photos of the inside of the body. During a CT scan, the subject is maneuvered across a revolving gantry that has an X-ray tube on one side and a collection of detectors on the other. A computer analyses data from the detectors and displays the results as a grayscale image [7].

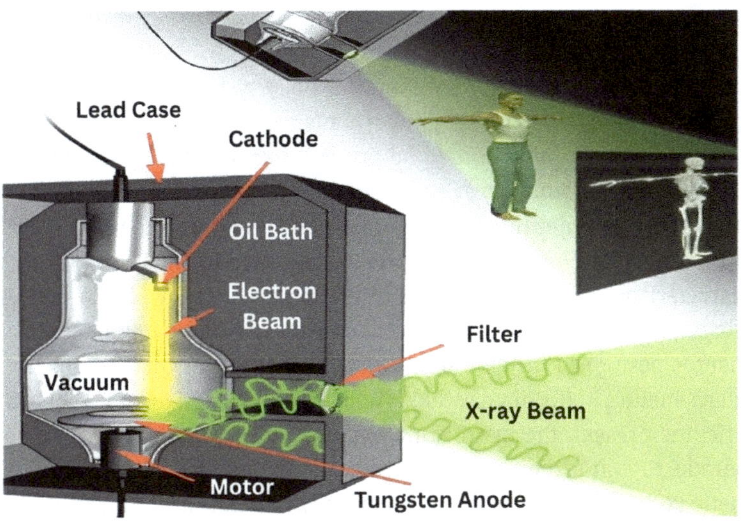

Figure 1.8: CT imaging procedure.

The cross-sectional images of the study region can subsequently be reproduced, copied to a CD, or viewed on a computer monitor. When compared to standard X-ray examinations, CT scans of internal organs, bones, soft tissue, and blood arteries are clearer and disclose more information [8]. An X-ray source revolves around a single slice or multiple slice detector ring in a CT scanner to capture projections. There are continual detections during rotation. It can be compared to a machine that produces one or more lines of X-ray images from various angles. Measurements are produced significantly more quickly than comprehensive X-ray images from various angles. Figure 1.8 depicts the CT imaging process.

"The variations between the various soft tissues are modest, whereas the attenuation of air, fat, water, and bone is dramatically different" [2]. Radiologists and other medical professionals can identify internal structures in CT images and observe their size, shape, density, and texture. Due to the extent, precise position, and other pertinent details of the information, it may be used to establish whether a medical issue exists. The photos can also demonstrate whether there are no abnormalities. Lymphoma, neuroblastoma, kidney tumors, congenital heart, kidney, and blood vessel malformations, cystic fibrosis, complications of acute appendicitis, complications of pneumonia, inflammatory bowel disease, and serious traumas are more frequently evaluated by CT imaging [4].The electronic measurement of the relative density of a region of interest, expressed as attenuation value for water is 0, for muscle it is 40, for bone 500, for fat 120, and for air 1,000. Substances less dense than water, including fat and air, have negative values; substances of greater density have positive values.

1.3.4 MRI imaging

Organs, soft tissues, bone, and almost all other interior body systems may all be seen in great detail, thanks to MRI, which uses a powerful magnetic field, radio frequency pulses, and a computer. The visuals can then be printed, transcribed, or transferred to a CD, viewed on a computer monitor, or sent digitally. Unlike X-rays, MRI does not employ radioactivity [10].The neuroimaging process and several MRI images are shown in Figure 1.9. The operation of MRI scanners was detailed by David A. Lisle [7]. The magnetic characteristics of hydrogen atoms in motion are used by MRI to create images. The deployment of a powerful, external magnetic field is the initial stage in an MRI. The patient is positioned inside a sizable, strong magnet for this reason. The majority of modern medical MRI devices have field strengths of 1.5 or 3. When a strong external field is present, the hydrogen atoms within the patient align either parallel or antiparallel to it. A higher percentage aligns in the parallel direction, causing the net magnetic vector and alignment net vector to both points in the direction of the external field. The term "longitudinal magnetization" refers to this.

In a right angle to the initial electromagnetic force, a secondary energy is applied. Due to the fact that the frequency at which this second magnetic field is applied is located in the same region of the electromagnetic spectrum as radio frequency, it is also referred to as a radiofrequency pulse (RF pulse).The RF pulse is delivered by an electromagnetic coil known as the RF coil. The RF pulse causes the hydrogen atoms' net polarization vector to shift in the orientation of the plane, or a plane that is perpendicular to the original, strong external field. The RF coil experiences an electrical current due to the net magnetization vector component in the transverse plane. The development of an image is based on this current, also referred to as the MR signal.

An MR Image is created via computer analysis of the complicated MR data coming from the RF receiver coils [7]. There are two types of MRI images. The repose period affects T1 and T2 MRI. Repetitive duration, TR, is the interval among two parameters. A proton-density-weighted picture will be created by a short TE of 20 msec and a long TR of 2,000 msec. Given that T1 relaxation typically takes longer than 200–600 msec, using a shorter repetition time (TR = 300–600 msec) will result in a T1-weighted picture. A T2-weighted picture is created with a long TE (>60 msec) and long TR (2,000 msec).Tumors of the chest, abdomen, or pelvis, liver illnesses like cirrhosis, and other conditions are some of the conditions that doctors use an MR examination; to assist diagnose or track treatment for anomalies of the bile ducts and pancreas, inflammatory bowel diseases including Crohn's disease and ulcerative colitis, congenital heart disease, blood vessel abnormalities, vessel inflammation (vasculitis), and a fetus inside the womb of a pregnant woman are just a few other examples [4].

1.3.5 Ultrasound imaging

Ultra-high-frequency audio waves are used in US scanning to produce cross-sectional images of the body. The ultrasonic craft's core element is a piezoelectric crystal. The photovoltaic material is activated when an electronic impulse excites a crystal, causing it to create ultra-high frequency sound waves. The body's varied tissues act as mirrors for sound waves, reflecting them back to the crystal. Once more, using the piezoelectric effect, these echoed sound waves (echoes) interact with the piezoelectric crystal in the ultrasonic probe to generate an electric signal. A cross-sectional image is created by a computer's analysis of this electric signal. The degree to which solid organs, fluid-filled structures, and tissue interfaces reflect sound waves is referred to as their echogenicity. Hyperechoic tissues reflect more sound than anechoic tissues do. Hyperechoic tissues appear white or light blue in an ultrasound image, while echogenic cells seem dark gray. Pure plasma appears black on US pictures and its tracks (tends to reflect almost no sound). Furthermore, because nearly all sound waves pass through a fluid-filled space, tissues farther away pick up more sound waves and appear lighter, as a result. Acoustic enhancement is the term for this effect, which manifests in tissues near the gallbladder, urinary bladder, and small cysts. Gallstones, kidney stones, breast cancer, and bowels that contain gas have the opposite effect, known as "acoustic shadowing" [7]. Skin looks "bright and silky" (echogenic, hyperechoic, highly reflective). Subcutaneous fat is normally black but can be bright (hypoechoic).When seen in cross section, muscle is also black. Long sections of sound are reflected back by the muscle fibers, making it possible to observe the internal organization of the muscle. Blood, effusions, and cysts are all typically black (anechoic) fluids; however, thicker fluids like pus can be brilliant or dark. Tendons are usually bright, though this varies depending on how they are positioned in relation to the probe. When looking at the shoulder, nerves are typically not visible, but their appearance is comparable to that of tendons. Due to the stark contrast in acoustic impedance between bone and soft tissue, bone stands out as an especially bright line.

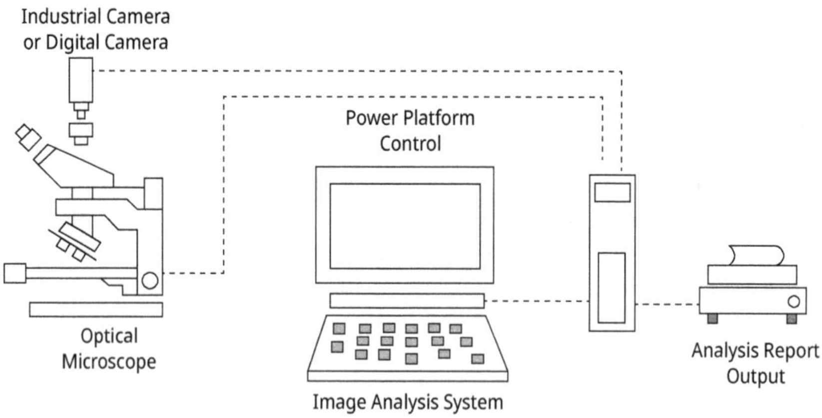

Figure 1.9: Modern digital microscope imaging system.

1.3.6 Microscopic imaging

The majority of microscopes used in medical imaging are light-optical devices that can examine living structures between the diameters of 0.1 m and 1 mm. Through a set of lenses, magnification is produced. With more spatial detail, photosensitivity declines. Pathology in a tissue sample is often identified at the cellular level using microscopic images. Color photos with a good signal/noise ratio and contrast are occasionally used in photographs. Microscopic images are used for cell distribution analysis, cell shape analysis, cell counting, and structural analysis. As seen in Figure 1.9, a digital microscope imaging system integrates the conventional optical microscope, digital multimedia, and digital processing technologies.

Typically, this system consists of three parts:
- Data capture component
- Digital image processing and application
- Microscope ocular module

The initial optical module deduces the role of the microscopic imaging from these modules.

The second data acquisition module uses digital cameras with CMOS, CCD, and other sensors to take images, which are then saved in digital format on computer storage devices. To best enhance the image quality, the third software control module manages the image capture, processing, and measurement in real time. These digital photos are being seen in real time on computer monitors. The picture details can be captured and displayed by digital microscopic imaging systems with greater sensitivity after employing a variety of digital image processing techniques. For their experimental work, research scientists and physicians can use various image processing and analysis techniques on these digital images [10]. K.D. Toennies [2] claims that the term "Medical Image" encompasses a wide range of various acquisition methodologies and signals to be captured. Therefore, every time, effort is needed to comprehend the visuals. The use of automatic or semiautomatic technologies can help tackle these new analysis tasks.

1.4 Medical image categorization

Hence, the researcher coined the name "artificial intelligence alacrity factor," where there is a high level of growth involved to carry out their work in a deliberate manner. The factors like artificial intelligence (AI)-enabled clinical decision support system software reminds healthcare professionals about the procedures and tests, AI-enabled solutions improve clinical workflows through digital healthcare applications, AI-enabled solutions provide data driven clinical decision support to doctors and 213 healthcare professionals, AI-enabled intelligent systems can detect and reduce the errors in the process of diagnosis and provide sabbaticals. These mainly focus on artificial intelligence systems in reducing the defects and removing the malware with regard to day-to-day operations in the technological and digital space [19]. Hence, the researcher coined the name artificial intelligence as the reducer of defects density rate factor.

Here, this practice helps the management of digital healthcare organizations to enrich and secure their system from threats and reduce the errors, thus improving the organizational workforce productivity. AI enablement helps healthcare professionals to view medical imaging and test results on digital gadgets at their convenience; AI-enabled algorithmic symptoms checker can diagnose and treat mild illness. These variables perform the characteristics of diagnosing the illness with the help of deep learning techniques using devices that are technology-enabled. Hence artificial intelligence can be called a device factor. AI-enabled electronic medical records and AI-enabled solutions streamline the process, help the healthcare staff process by managing electronic health records faster and more efficiently, AI-enabled solutions improve and streamline the process of insurance verification eligibility before taking treatment, and AI-enabled solutions improve surgical precision.

These statements are highly related with the process streamlining of day-to-day healthcare operations of the digital healthcare providers [20]. Hence, the researcher coined the name artificial intelligence as process streamliner factor. AI-enabled clinical decision support system software to remind healthcare professionals on reduction of medication errors, AI-enabled device alerts on drug reactions like dosage errors and drug interactions using Adverse Drug Reaction (ADR) system, AI-enabled solutions improve in managing better patient flow, AI-enabled solutions help in alerting doctors in developing treatment protocol, and AI-enabled microscopes can scan harmful bacteria in blood samples at a faster rate than manual scanning. AI-enabled intelligent systems work before any the challenges arise in terms of treatment and diagnoses; it will notify the healthcare practitioners and help management to incorporate 214 such systems in place that also save lives of patients and indirectly improves the workforce productivity, most of which are problem-specific and have been shown to be complementary in medical imaging, leading to the solution for high-level problems on image recognition.

1.5 Medical image segmentation

Segmentation is an image analysis process that partitions the selected image into various regions that show similar attributes. It is a technique in which image features, edges, and feature attributes are acquired from structural and functional image modalities that can provide necessary data and input vectors about each organ sample [21]. During earlier days, image acquisition and filtering were performed by automation techniques with several layer steps. With the evolution of deep learning techniques, knowledge acquisition and image processing tasks are accomplished for each layer with pixel partitioning and labelling, where the images with similar properties are analyzed. The medical image analysis involves organization of functional structures of organs, blood samples, or genetic structures with segmentation activities. This segmentation of clinical subdomains has rendered the best labelling results in decision-making phase through multiple image modalities. Figure 1.10 shows the types of neural networks.

In older days, Convolutional Neural Network (CNN) [17] mainly supported segmentation. Later, image segmentation could be achieved successfully using a fully convolutional neural network (FCN); it will segment an image in a proper manner. FCN is an optimized technique of CNN, which classifies the big segmentation task into an easier one in network upsampling and a pixel wise loss. The medical image segmentation can be classified as 2D method or 3D method, based on the image dimension [22]. The most powerful fully connected network architecture for segmenting images is U-Net learning architecture. It consists of an input equipment path, the image classification path, and decision-making path [23]. Figure 1.11 shows the process of CNN.

The typical neural network architecture follows the sampling of data from different layers and the amount of feature extracted from input and applied to normalization of model parameters. With each error identification phase, the amount of feature channels also doubles. Each training path consists of oversampling of the characteristic pattern followed by a deconvolution, which halves the number of characteristic

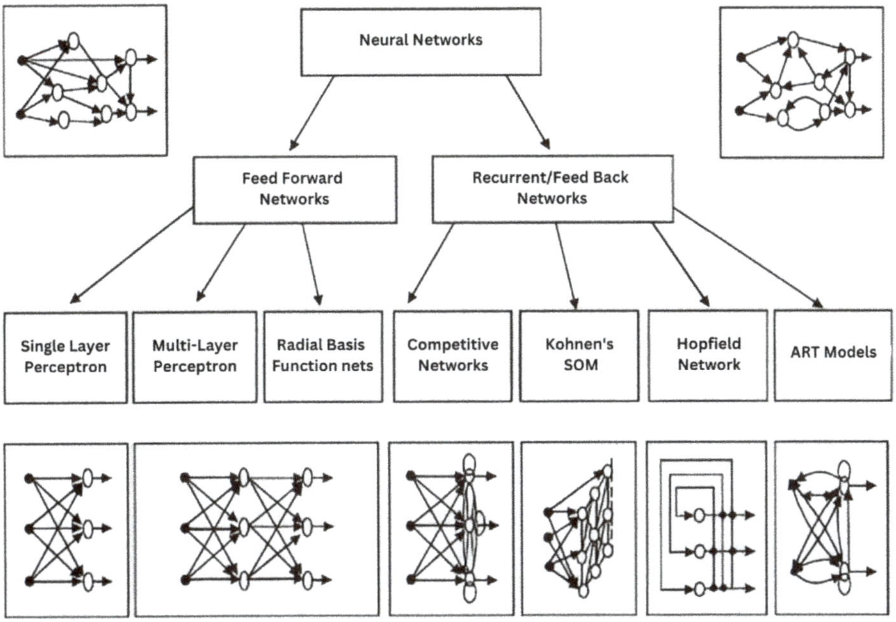

Figure 1.10: Types of neural networks.

channels. A concatenation is also applied with the characteristic map cut out accordingly by the contraction path. Variants of architectures based on UNet have been proposed. Another framework called nnU-Net supports segmentation of medical images, tried in fingerprint dataset to optimize the segmentation process. This nnU-Net has worked well across various segmentation images like tumors, different organ structures, and different images from MRT, CTC, etc. [24].

Image segmentation is an important technique that practices image segregation and distribution of center of locality among various image test cases, sharing the locality of reference among qualitative and quantitative features for edge detection of objects. For 50 years, these segmentation concepts have been used in the medical field. The main challenge of deep learning is medical imaging. The most widely used machine learning

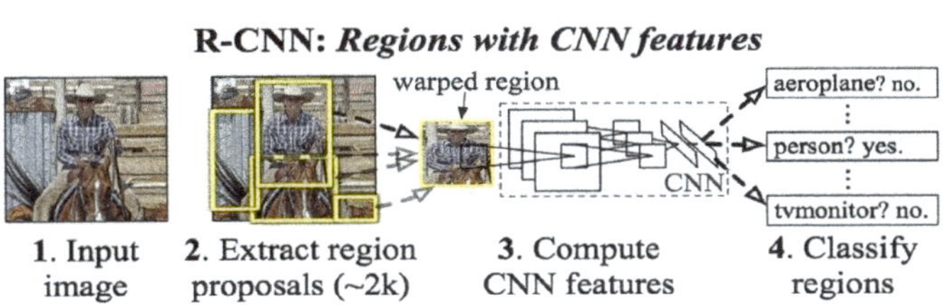

Figure 1.11: Convolutional Neural Network.

segmentation technique is the multispectral tissue classification report. Statistical pattern recognition is a leading technology used in the medical field for segmentation [25].

1.6 Image registration

This technique includes the image fusion, image matching, and image warping techniques, which is the process of arranging or aligning two or more images, based on the appearance. It was an important preprocessing step applied in various medical applications. The goal is applying the significant follow-up and implementation within several dimension images at various times, by diversified patients and image modalities and from different views. Registration of medical images aims to obtain the gradient descent for learning algorithm with minimum error and loss rate. The registration of medical images is a vast subject that can be collectively organized from different perspectives. From the input image acquisition point of view, the registration can be divided into multimodal, clinical, and genetic-based image features. Thus, various images acquired are implemented with several machine learning and deep learning hypothesis. Each image set consists of object identification with the benchmark datasets. They train the model with model parameters that are acquired for 2D and 3D image modalities with minimum loss ratio. These combine moving and field images into a single image as deformation field.

1.7 AI in image quantification

The factors like such as AI-enabled bar coding medications and automatic dispensing, and AI enablement help radiologists in imaging scans and automatically analyze various clinical findings and diagnoses more effectively [26]. These works mainly speak on how AI-enabled intelligent systems can regulate and notify the doctors and patients on timely medications to prevent disease and also channelize the clinical dispensaries at the pharmacy division of the hospital for the clear and smooth flow of processes that are mechanized and also provide healthcare professionals by automatic screening of patients' medical imaging. Thus, it is able to identify the signs and symptoms of mild diseases that are already mechanized in the AI systems by training them with a good amount (215) of data. It helps the healthcare professionals increase their productivity and improve the professional growth and commitment to work and also improve an organization's growth, profitability, and attainment competitive advantage.

1.8 Analysis and assessment of technology

The most effective technique to assess the success of any application sequences using a novel test data set that contains material that was not observed during training. The test set is used to evaluate the genome learning model and plot the results as a confusion matrix. Almost all test data variants should be classified accurately. The probability of 0 to N feature vectors that classifies the disease gives the correlation measure between various structural and demographic AD features, such as $\mu x, \mu y$ is represented in eq. (1.1).

$$C = \sum_{i,j=0}^{N} \frac{[1-\mu x][1-\mu y]}{\sqrt{\mu x.\mu y}} \tag{1.1}$$

The Mean Absolute Percentage Error (MAPE) is calculated as the sum of negative gene classes obtained for every training period Ti, as given in eq. (1.2), where the Mean absolute error is not confined to the training period. The square root of the average sum of negative gene classes obtained for every training period scaled to the total training periods is denoted as:

$$MAEAPOE = \frac{1}{N}\sum_{(i=1)}^{xi} |Error_{APOE}| \tag{1.2}$$

$$MAPEAPOE = \frac{1}{N}\sum_{(i=1)}^{xi} \frac{|Error_{APOE}|}{T_i} \tag{1.3}$$

$$RMSEAPOE = \sqrt{\frac{1}{N}\sum_{i=1}^{xi} \frac{|Error_{APOE}|}{T_i}} \tag{1.4}$$

Support vector classifiers had an accuracy of more than 85% for N-fold cross-validation, 0.78 F1 weighted scores, confusion matrix and low mean absolute error (MAE), and root mean square error (RMSE) as shown in eqs. (1.2) and (1.3).

1.9 Conclusion

This work briefs the introduction of X-ray Imaging, CT scan Imaging, MRI Imaging, US Imaging and Microscopic Imaging. From this, the tissue appearance information is obtained, which will be used for determination of number of clusters in the medical image. For CT, X-ray, US, and Microscopic images, the number of clusters is 3, whereas for MRI image, the number of clusters is 4. Also, in T1 and T2, MRI images have different tissue characteristics, and hence, the MRI T1 and MRI T2 image classification is needed. One of the common artifacts identified is noise; hence noise removal processing is needed for all types of images. Also the contrast improvement is needed in case of X-ray, US, and Microscopic images. These statements reflect on the competency and potentiality of AI in transforming healthcare demarcations; all those abovementioned factors are highly correlating with competency that leads to improving the organizational workforce productivity. Hence, the researcher names this segment as AI as a competency factor. The above factors are grouped as AI competency in the digital healthcare space that leads to increase in workforce productivity [27]. These statements show how AI helps healthcare professionals in enhancing their confidence, improving interest and boosting the morale of the employees, and improving their professional growth at the workplace for the long run. Hence, these variables can be correlated as AI vigilance factor. The second part of this research is Medical Image Analysis in which the medical image is further processed.

For CT image analysis, firstly the CT scan image of brain is segmented into three clusters: skull, gray matter (GM), and background (BG). The mean of skull cluster is used to classify the CT scan image into normal or abnormal class. F mean (mean of features of all clusters), C mean (mean of contrast of all clusters) and standard deviation (all features of all clusters) extracted features of segmented tissues clusters, namely skull, brain matter (gray matter) and Background and Air class. An abnormal image is post processed to extract the abnormality region using morphology. The segmentation accuracy for skull cluster is Conclusion and Future Work 163 72%, for GM 64%, and BG 58%. The recognition rate for normal and abnormal image classification is 84%.

This statement mainly explains how AI is designed to meet the perspective of patient queries and help them seek an appointment with a doctor any time, making it more agile and lean; and sometimes intelligent systems are developed in such a manner so that they can react and handle the minor queries of patients on its own, which means it helps management to incorporate an effective communication platform and create a deeper sense of understanding for the doctor on the patient health improvements, and thus impacts on the betterment of patient experience as well; and this impacts on the positive developments on the organizational workforce productivity. Since it strongly relates to patientcare, AI can be called a patient-centric system factor.

AI-enabled solutions have reduced the workload of paramedical healthcare professionals and AI-enabled intelligent system can perform assisted minimal invasive surgeries and assist the healthcare practitioners in the treatment protocol. These statements mainly explain how AI supports healthcare professionals, be it in medical diagnoses or minimizing their extra efforts at the workplace, which, in turn, can be automated and leveraged better and at a faster phase. Hence, the researcher has coined the phrase, AI, an assistant to healthcare practitioner's factor.

References

[1] A. Vezhnevets, V. Ferrari, A. Gonzalez-Garcia, “An active search strategy for efficient object class detection”, CVPR IEEE conference, 2015.

[2] A. A. Efros, A. Gupta, et al., “Unsupervised discovery of mid-level discriminative patches”, ECCV, 2012, Springer.

[3] C. L. Zitnick, P. Dollar, “Edge boxes locating object proposals from edges”, E.C.C.V, Springer, September 2014.

[4] D. Silver, K. Kavukcuoglu, et al. “Playing atari with deep reinforcement learning” arXiv preprint arXiv-1312-5602, 2013.

[5] D. Wierstra, D. Silver, et al.“Playing Atari with deep reinforcement learning” arXiv preprint arXiv 1312.-5602, pub 2013.

[6] R. Girshick, J. Donahue, T. Darrell, J. Rich, et al., “Feature hierarchies for accurate object detection and semantic segmentation,” Proceedings of the IEEE conference on computer vision and pattern recognition, 2014, pp. 580–587.

[7] H. Harzallah, F. Jurie, C. Schmid, “Combining efficient object localization and image classification,” Proc. IEEE 12th Int. Conf. Comput. Vis, September/October 2009, pp. 237–244.

[8] H. Kagaya, M. Ogawa, “Food detection and recognition using convolutional neural network,” Proc. 22nd ACM Int. Conf. Multimedia, 2014, pp. 1085–1088.

[9] J. Redmon, S. Divvala, R. Girshick, A. Farhadi, “You only look once: Unified, real-time object detection,” Proceedings of the IEEE conference on computer vision and pattern recognition, 2016, pp. 779–788.

[10] J. Girshick, A. Farhadi, “Yolo real-time object detection,” IEEE conference on computer vision and pattern recognition, 2016.

[11] J. C. Caicedo, S. Lazebnik, “Active object localization with deep reinforcement learning,” Proceedings of the IEEE Inter. Conference on Computer Vision, 2015, pp. 2488–2496.

[12] M. D. Zeiler, Fergus, et al., “Visualizing and understanding convolutional networks,” Euro conf. on Computer Vision, 2014, pp. 818–833, Springer.

[13] F. Matthew, et al., “Understanding convolutional networks,” European conference on computer vision, 2014, pp. 820–823, Springer.

[14] M. B. Buenoa, X. G.-I. Nietob, F. Marquesb, J. Torresa, Hierarchical object detection with deep reinforcement learning, Deep Learn. Image Process. Appl. 31, 2017, 164.

[15] R. Goel, V. Kumar, S. Srivastava, A. K. Sinha, A review of feature extraction techniques for image analysis, ICACTRP 6(2), February 2017.

[16] S. Ren, K. H, R. Girshick, Real-time object detection with region proposal networks, Neural Inf. Process. Syst. 2015, 91–99.

[17] R. Girshick, “Convolutional neural net,” Proceedings of the IEEE international conference on computer vision, 2015, pp. 1440–1448.

[18] S. Albers, et al. Energy-efficient algorithms, Commun. ACM 53(5), 2010, 86–96.

[19] W. Liu, C. Szegedy, S. Reed, F. Cheng-Yang, A. C. Berg, et al., “SSD Single shot multibox detector,” European conference on computer vision, 2016, pp. 21–37, Springer.

[20] S. Oswalt Manoj, J. P. Ananth, M. Rohini, B. Dhanka, N. Pooranam, S. R. Arumugam, FWS-DL: Forecasting Wind Speed Based on Deep Learning Algorithms, in: Woodhead Publishing Series in Energy, Artificial Intelligence for Renewable Energy Systems, Woodhead Publishing, 2022, pp. 353–374. ISBN 9780323903967. https://doi.org/10.1016/B978-0-323-90396-7.00007-9.

[21] W. Erhan, C. Szegedy, S. Reed, F. Cheng-Yang, A. C. Berg, “Ssd: Single shot multibox detector,” European conference on computer vision, 2016, pp. 21–37, Springer.

[22] X. Wang, Y. Lin, et al. Generic object detection with dense neural patterns and region lets, arXiv April 2014, 4309–4316.

[23] Zitnick, et al., "Locating object proposals from edges," European Conference on Computer Vision, 2014, pp. 391–405, Springer.

[24] A. Vedaldi, A. Zisserman. "Delving deep into convolutional nets" arXiv preprint arXiv:1405.3531, 2014.

[25] S. R. Swarna, A. Kumar, P. Dixit, T. V. M. Sairam, "Parkinson's Disease Prediction using Adaptive Quantum Computing," 2021 Third International Conference on Intelligent Communication Technologies and Virtual Mobile Networks (ICICV), pp. 1396–1401, doi: 10.1109/ICICV50876.2021.9388628.

[26] G. Sasubilli, A. Kumar, "Machine Learning and Big Data Implementation on Health Care data," 4th International Conference on Intelligent Computing and Control Systems (ICICCS), 2020, pp. 859–864, doi: 10.1109/ICICCS48265.2020.9120906.

[27] S. M. Sasubilli, A. Kumar, V. Dutt, "Improving Health Care by Help of Internet of Things and Bigdata Analytics and Cloud Computing," 2020 International Conference on Advances in Computing and Communication Engineering (ICACCE), 2020, pp. 1–4, doi: 10.1109/ICACCE49060.2020.9155042.

Shubham Sharma, Pronika Chawla, Sanu Kumar, Mukul Nair

Chapter 2 Predictive maintenance of industrial machines using data collected through IoT sensors and analyzed by machine learning algorithms

Abstract: Industrial Internet of Things (IIoT) is using Internet of Things (IoT) technologies to a vast extent in different industries across all verticals. Especially in manufacturing industries, there is a great boom in using IIOT because manufacturing industry has seen a great potential in it, and they are heavily investing on it. Manufacturing industry is one such industry where physical devices are used and are important for production; there are no other alternatives yet. These physical devices are a very crucial part of the manufacturing industry because if there is any machine failure at any time, then it is a huge loss for the company in terms of revenue. Not able to fulfil their commitment will also be a huge loss in terms of brand value. In this competitive market where there are a number of organizations, no organization would want to compromise their brand value and profit because both go hand in hand. Therefore, in this chapter, we will be exploring and finding suitable ways where AI and IIoT can be helpful for the organizations to overcome these challenges, and it will surely become a sustainable growth path for organizations in all industries where physical devices' failure is cost high. This chapter mainly revolves around how sensors can be used to collect data from physical devices and how this data is moved over the internet to cloud, how deep learning model runs every time on that real-time collected data to predict the probability of machine failure, and how it raises alerts to concerned people if there is a high probability of machine failure. This system has only one challenge, which is latency, and this can be resolve using fog computing.

2.1 Introduction

It is observed from the experience of the past decades that the problem of learning and decision making is at the core level of argument in biological as well as in artificial aspects. So, scientists introduced Machine learning as a widely used concept in

Shubham Sharma, Consultant Technology-Virtusa, e-mail: shubhamsharma151998@gmail.com
Pronika Chawla, Assistant Professor-MRIIRS, e-mail: pronika.fet@gmail.com
Sanu Kumar, B. tech CSE-MRIIRS, e-mail: anukumar763390@gmail.com
Mukul Nair, B. tech CSE-MRIIRS, e-mail: m3mj2002@gmail.com

https://doi.org/10.1515/9783110791402-002

Artificial Intelligence. Machine learning is a branch of Artificial Intelligence (AI) and computer science that focuses on using data and algorithms to mimic the way humans learn and gradually improve its accuracy [5, 7]. Over the past few decades, technological advances in storage and computing power have enabled some innovative machine learning-based products, such as Netflix's recommendation engine and self-driving cars [1]. Machine learning is an important part of the growing field of data science. Using statistical methods, algorithms are trained to make classifications or predictions and to uncover key insights in data mining projects. These insights then drive decision-making within applications and businesses, ideally influencing key growth metrics. As big data continues to expand and grow, the market demand for data scientists will increase. They will need to help identify the most relevant business questions and the data to answer them. Machine learning algorithms are typically built using frameworks that accelerate solution development, such as TensorFlow and pyTorch.

In recent decades, the importance of machine tool analysis has increased and a large number of different models have been developed to optimize manufacturing processes. In this context, J. Lenz et al. propose a holistic approach to machine tool data analysis. They provide selected implementation examples according to the stated objectives of the analysis, including cross-domain analysis that focuses on the interface between domains. When it comes to the precise task of predicting tool life, extensive research has been done using ML to generate predictions as accurately as possible. Most of the previous work focused on tool wear prediction using artificial neural networks (ANN). Chungchoo and Saini developed an online fuzzy neural network algorithm that estimates the flank width and maximum crater wear depth.

2.1.1 Working of machine learning

i. **Decision process:** Generally, machine learning algorithms are used for prediction or classification. Based on some input data, which may or may not be labeled, your algorithm will make an estimate of a pattern in the data.
ii. **Error function:** The error function evaluates the model prediction. If there are known examples, the error function can perform a comparison and assess the accuracy of the model.
iii. **Model optimization process:** If the model fits the data points in the training set better, the weights are adjusted to reduce the discrepancy between the known example and the model estimate. The algorithm repeats this "evaluate and optimize" process and autonomously updates the weights until an accuracy threshold is reached [2].

2.1.2 Machine learning methods

A mathematical expression known as a machine learning algorithm, sometimes known as a model, describes data in the context of a problem, frequently a commercial problem. Some ways to tackle these problems with machine learning are:

- **Supervised learning**, also known as supervised machine learning, is defined by using labeled datasets to train algorithms to classify data or accurately predict outcomes. As input data is fed into the model, the model adjusts its weights until it is appropriately fitted. This occurs as part of the cross-validation process to ensure that the model avoids overfitting or underfitting. Supervised learning helps organizations solve various real-world problems at-scale, such as classifying spam into a separate folder from your inbox. Some methods used in supervised learning include neural networks, naive bays, linear regression, logistic regression, random forest, and support vector machine (SVM).
- **Unsupervised learning**, also known as unsupervised machine learning, uses machine learning algorithms to analyze and cluster unlabeled data sets. These algorithms reveal hidden patterns or groupings in data without the need for human intervention. This method's ability to uncover similarities and differences in information makes it ideal for exploratory data analysis, cross-selling strategies, customer segmentation, and image and pattern recognition. It is also used to reduce the number of elements in a model through a dimension reduction process. Principal component analysis (PCA) and singular value decomposition (SVD) are two common approaches. Other algorithms used in unsupervised learning include neural networks, k-means clustering, and probabilistic clustering methods.
- **Semisupervised learning** offers a happy medium between supervised and unsupervised learning. During training, it uses a smaller labeled dataset to guide classification and feature extraction from a larger, unlabeled dataset. Semisupervised learning can solve the problem of lack of labeled data for the supervised learning algorithm. It also helps if it is too expensive to label enough data [10].

2.1.3 Machine learning algorithm

Data scientists have created sophisticated data-crunching machines over the last five years by effortlessly implementing cutting-edge methodologies [11, 13]. The outcomes are astonishing. The most common machine learning algorithms developed in the past decades are:

Neural networks: Neural networks simulate the way the human brain works with a large number of interconnected processing nodes. Neural networks are good at pattern recognition and play an important role in applications including natural language translation, image recognition, speech recognition, and image generation.

Linear regression: This algorithm is used to predict numerical values based on a linear relationship between different values. For example, the technique can be used to predict real estate prices based on historical data for a given area.

Logistic regression: This supervised learning algorithm makes predictions for categorical response variables, such as yes/no answers to questions. It can be used for applications such as spam classification and quality control on the production line.

Clustering: Using unsupervised learning, clustering algorithms can identify patterns in data, to group them together. Computers can help data scientists by identifying differences between data items that humans have missed.

Decision trees: Decision trees can be used both for predicting numerical values (regression) and for classifying data into categories. Decision trees use a branching sequence of interconnected decisions that can be represented by a tree diagram. One of the advantages of decision trees is that they can be easily verified and audited, unlike the black box of a neural network.

Random forests: In a random forest, a machine learning algorithm predicts a value or category by combining results from several decision trees.

2.1.4 Concerns developed due through machine learning

As machine learning technology has developed, it has certainly made our lives easier. However, implementing machine learning in businesses has also raised a number of ethical concerns about AI technologies. Some of these include technological singularity. Although the topic is gaining a lot of public attention, many researchers are not interested in the idea that AI will surpass human intelligence in the near future. Technological singularity is also referred to as strong AI or superintelligence. Philosopher Nick Bostrum defines superintelligence as "any intellect that significantly exceeds the best human brain in virtually all areas, including scientific creativity, general wisdom, and social skills." Despite the fact that there is no threat of superintelligence in society, the idea of it raises some interesting questions when considering the use of autonomous systems such as self-driving cars. It is unrealistic to think that a driverless car would never have an accident, but who is responsible and liable under these circumstances? Should we still be developing autonomous vehicles, or will we limit this technology to semiautonomous vehicles that help people drive safely? The jury is still out on this one, but these are the types of ethical debates that arise with the development of new, innovative AI technology.

2.1.5 AI impact on jobs

While much of the public perception of AI revolves around job losses, this concern should probably be reframed. With every disruptive, new technology, we see market demand for specific job roles change. For example, if we look at the automotive industry, many manufacturers, such as GM, are moving to focus on producing electric vehicles to align with environmental initiatives. The energy industry is not disappearing, but the source of energy is shifting from a fuel economy to an electric one. In a similar way, artificial intelligence will shift the demand for jobs to other areas. There will need to be individuals to help manage AI systems. There will still need to be people to solve more complex problems in industries most likely to be affected by shifts in labor demand, such as customer service. The biggest challenge of artificial intelligence and its impact on the labor market will be helping people transition to new roles that are in demand.

2.1.5.1 Privacy

Privacy is often discussed in relation to data protection, data protection, and data security. These concerns have allowed policymakers to make more progress in recent years. For example, GDPR legislation was created in 2016 to protect the personal data of people in the European Union and the European Economic Area, giving individuals more control over their data. In the United States, states are developing policies, such as the California Consumer Privacy Act (CCPA), which was introduced in 2018 and requires businesses to inform consumers about the collection of their data. Legislation like this has forced companies to rethink how they store and use personal information (PII). As a result, investing in security has become an increasing priority for businesses as they strive to eliminate any vulnerabilities and opportunities for surveillance, hacking, and cyber attacks.

2.1.5.2 Bias and discrimination

Cases of bias and discrimination in a number of machine learning systems have raised many ethical questions regarding the use of artificial intelligence. How can we guard against bias and discrimination when the training data itself may be generated by biased human processes? While companies typically have good intentions for their automation efforts, Reuters highlights some unintended consequences of incorporating artificial intelligence into hiring practices. In an effort to automate and simplify the process, Amazon inadvertently discriminated against job applicants by gender for technical roles, and the company eventually had to cancel the project. The Harvard Business Review (link is external) has raised other pointed questions about the use of artificial intelligence in recruitment processes, such as what data you should be able to use when evaluating a candidate for a role. Bias and discrimination are not limited to the HR function either; they can be found in a range of applications from facial recognition software to social media algorithms.

2.1.5.3 Accountability

Since there is no significant legislation to regulate AI practices, there is no real enforcement mechanism to ensure that ethical AI is practiced. The current incentives for companies to be ethical are the negative impacts of an unethical AI system on the bottom line. To fill this gap, ethical frameworks have emerged as part of a collaboration between ethicists and researchers, to guide the construction and distribution of AI models in society. However, these only serve as a guide at this time. Some research shows that a combination of divided responsibility and a lack of foresight about potential consequences does not lead to the prevention of harm to society. In today's world, there is no Machine Learning without Artificial Intelligence. We may categorically state that machine learning is an area of or a kind of AI. Though one may consider all machine learning systems to be a form of artificial intelligence, not all AI is machine learning [14, 15].

2.1.5.4 Artificial intelligence

A number of definitions of artificial intelligence (AI) have emerged over the past few decades. "It is the science and engineering of creating intelligent machines, especially intelligent computer programs. It is related to the similar task of using computers to understand human intelligence, but artificial intelligence need not be limited to methods that are biologically observable." However, decades before this definition, the conversation about artificial intelligence began with Alan Turing's 1950 work "Computing Machinery and Intelligence." In this article, Turing, often referred to as the "father of computer science," asks the following question: "Can machines think?" From there, he offers a test, now known as the "Turing Test," where a human investigator would try to distinguish between a computer's response and human text. Although this test has undergone many reviews since its publication, it remains an important part of AI history.

2.1.5.5 Relation between artificial intelligence and machine learning

While there are differences between AI and ML, they both have a similar outcome – intelligent software that can handle more complex tasks. That's why these technologies have become so popular in recent years. They are useful, and individuals and companies alike are looking for computer programs that can take over human tasks while completing tasks faster. However, they do not do it in the same way. Although both ML and AI programs may be able to perform similar tasks, their design and approach differ. Each of these branches has created highly sophisticated programs that can handle complex tasks in seconds. However, they use different structures, coding, and approach to do so, which puts both AI and ML in a similar place in terms of their power. We can loosely say that Machine Learning is a subfield or type of AI. All machine learning programs can be considered part of artificial intelligence, but not all artificial intelligence is machine learning. They are inherently related, but that doesn't mean they are synonymous. AI is the parent of ML and both are sons of computer science. Simply put, everything in the field of machine learning falls under the umbrella of artificial intelligence. Machine learning can be used for AI applications and can improve artificial intelligence. However, this does not mean that it is always used. Simply put, the ability of machine learning to come up with answers on its own, without additional programming, is very convenient.

2.1.6 Cloud computing

Cloud computing is defined as computer software outsourcing technology that allows us to remotely access applications and data. It requires no software installation and no storage on your computer's hard drive. To use the services online, you must register.

2.1.6.1 Types of cloud computing

Cloud computing is mainly categorized into three types as follows:
- **Software as a Service (SaaS)**
- **Platform as a Service (PaaS)**
- **Infrastructure as a Service (IaaS)**

2.1.6.2 Cloud computing in machine learning

Although cloud computing and machine learning are new technologies, machine learning is relatively new. Both technologies play an important role in the growth of companies, but together they become stronger. Machine learning creates intelligent machines or software and, on the other hand, cloud computing provides storage and security to access these applications.

The main connection between machine learning and cloud computing is the demand for resources. Machine learning requires a lot of computing power, data storage, and many servers at the same time to work on the algorithm. Cloud computing then plays a significant role in providing new servers with predefined data and changing resources via the cloud (internet). With cloud computing, you can run any number of servers, work on the algorithm, and destroy the machine again when you're done. Cloud computing is primarily used for computing purposes. Machine learning needs a lot of computing power to create sample data, and not everyone has access to many powerful machines. Machine learning finds job scheduling and storage in cloud computing.

2.1.6.3 Advantages of machine learning with cloud computing

- The cloud works on the principle of "pay for what you need." The pay-as-you-go cloud model is good for companies that want to leverage ML capabilities for their business without spending a lot of money.
- Provides flexibility to work with machine learning features without having advanced data science skills.

2.2 Literature review

Predictive analytics is a branch of data analysis that predicts future outcomes of events based on past data and information. The results are calculated using a wide range of modern technologies, which include various mathematical processes, statistical modeling, machine learning, data mining, big data, and much more. Using predictive analytics, highly accurate predictions are made through several cycles of trial and error. This technique is used by businesses to gain deep insight into future events to improve decision making and facilitate sales maximization [12]. Gaining detailed knowledge of predictive analytics is possible only with a solid descriptive, diagnostic, and normative research background [3].

Predictive analytics works on a plan to use historical data to uncover real-time insights. It relies on repeating several steps in a cyclical order to increase the accuracy and viability of each predictive model. Gaining detailed knowledge of predictive analytics is only possible with a solid descriptive, diagnostic, and normative research background. The following are the steps involved in predictive analysis:

i. Understanding business

It is essential to understand the demand before providing a solution for its supply. The first step therefore involves gathering relative knowledge and information to determine a course of action. Next, you need to collect enough data to properly train the predictive model and identify predictive patterns.

ii. Data analysis and preparation

You need to analyze the data needed to train the model. This means eliminating all unwanted information or noise, and ensuring enough information for the flawless functioning of the model.

iii. Model preparation

This is the most important step. Here, you have to prepare the product according to the results of your research. Modeling is done using predictive analytics techniques such as machine learning, big data, data mining, statistical analysis, etc. At the end of training, the model learns from historical data and identifies trends accordingly.

iv. Evaluation of the model

By working with business analysts and conducting test runs, you can understand whether the model makes sense and meets the needs of the business. This step is necessary because complicated algorithms can lead to false predictions, and negatively impact the business.

v. Accuracy check

You can evaluate the accuracy by retraining the model using datasets. This is a continuous process that will gradually increase the effectiveness of the model, based on the feedback received.

vi. Deployment

After a while, when the model reaches a certain level of efficiency, it can be deployed for practical use in real-world situations to solve real-time problems.

2.2.1 Predictive analytics models

Predictive modeling is a statistical technique that uses machine learning and data mining to forecast future outcomes using past and existing data. It works by analyzing current and historical data, and projecting insights into models built to predict possible outcomes. Following are the predictive analysis models:

2.2.1.1 Classification model

The classification model is in some ways the simplest of the several types of predictive analytics models we will cover. It categorizes data based on what it learns from historical data.

Classification models best answer yes or no questions and provide broad analysis that is useful for guiding decision-making. These models can answer questions like:

- For the retailer: “Is this customer going to consume?”
- For lenders: “Will this loan be approved?” or “Is this applicant likely to default?”
- For online banking providers: “Is this a fraudulent transaction?”

The breadth of a classification model’s capabilities and the ease with which it can be retrained with new data means it can be used in many different industries.

2.2.1.2 Clustering model

A clustering model sorts data into separate, nested intelligent groups, based on similar attributes. If an e-commerce shoe company is looking to implement targeted marketing campaigns for its customers, it could go through hundreds of thousands of records and create a strategy that is tailored to each individual. But is it the most efficient use of time? Probably not. Using a clustering model, they can quickly segment customers into similar groups based on common characteristics and design strategies for each group at a larger scale. Other use cases for this predictive modeling technique might include grouping loan applicants into “smart buckets,” based on credit attributes; identifying high-crime areas in a city; and collating SaaS customer data into groups to identify global usage patterns.

2.2.1.3 Prediction model

One of the most widely used predictive analytics models, the predictive model, is concerned with predicting metric values – estimating a numerical value for new data based on insights from historical data.

This model can be used wherever historical numerical data is available. Scenarios include:

A SaaS company can estimate how many customers are likely to convert during a given week.

A call center can predict how many supports calls it will receive per hour.

A shoe store can calculate how much inventory it should keep on hand to meet demand during a certain sales period [4].

The forecast model also considers multiple input parameters. If a restaurant owner wants to predict the number of customers she is likely to receive in the coming week, the model takes into account factors that could affect this, such as: Is there an event nearby? What is the weather forecast? Is there any disease?

2.2.1.4 Outlier model

An outlier model is oriented toward anomalous data items within a dataset. It can identify anomalous numbers either by itself or in conjunction with other numbers and categories, such as:
- Noticing an increase in support calls that could indicate a product failure that could lead to a recall
- Finding anomalous data within transactions or claims to identify fraud
- Looking for unusual information in your NetOps logs and noticing signs of impending unplanned outages

The outlier model is particularly useful for predictive analytics in retail and finance. For example, when identifying fraudulent transactions, the model can assess not only the amount, but also the location, time, purchase history, and nature of the purchase (i.e., a $1,000 electronics purchase is not as likely to be fraudulent as a purchase of the same amount for books or common services).

2.2.1.5 A time series model

A time series model contains a sequence of captured data points using time as an input parameter. It uses data from the past year to create a numeric metric and uses that metric to forecast data for the next three to six weeks. Examples of uses for this model include the number of daily calls received over the past three months, revenue over the past 20 quarters, or the number of patients seen at a given hospital over the past six weeks. It is a powerful means of understanding the way a singular metric evolves over time, with a level of precision beyond simple averages. It also takes into account seasons or events that could affect the metric. If a salon owner wants to predict how many people are likely to visit his business, he might turn to a crude method of averaging the total number of visitors over the past 90 days. However, growth is not always static or linear, and a time series model can better model exponential growth and better align the model with the company's trend. It can also forecast multiple projects or multiple regions simultaneously, instead of just one at a time [6].

2.2.2 Techniques of predictive analytics

Predictive analytics uses various statistical techniques as well as data mining, data modeling, machine learning, and artificial intelligence to make predictions about the future, based on current and past data patterns. These predictions are made using machine learning models such as classification models, regression models, and neural networks.

- **Decision trees:** Simply put, this model places data into different segments, known as "branches," based on other variable parameters. Decision trees can be processed effortlessly and easily understood because the data is extracted according to the user's requirements in a short period of time.
- **Regression:** This statistical system facilitates the determination of patterns in data sets and creates pattern-based relationships between variables.
- **Neural networks:** This model was developed by mimicking the workings of the human nervous system and uses artificial intelligence to recognize patterns. It is highly effective when you have too much data to process but cannot formulate the relationship between inputs and outputs.

2.2.3 Predictive analytics and machine learning

One might be wondering if machine learning and predictive analytics use the same process to develop models with near-human accuracy. Although the original idea of these technologies is similar, there is a significant difference between them. Machine learning is aimed at making the system work completely independently and eliminating any dependence on human interaction. It strives to create an autonomously functioning ecosystem without the need for human intervention. However, predictive analytics is designed to be operated and modified with human experts in the loop according to the needs of the company. Without human intervention, predictive analytics is a stagnant technology and cannot thrive.

2.2.4 Applications of predictive analytics

Predictive analytics has found its way into various industries across various disciplines. From marketing and insurance companies to restaurant chains, every sector has embraced this emerging technology with open arms:

- **Investing**: Investors use this method to predict upcoming market movements to improve their portfolios and save themselves from potential losses.
- **Banking and finance:** Financial institutions, such as banks, use predictive analytics to ensure client satisfaction and security. Banking-specific models are used to identify cross-selling opportunities and detect fraud.
- **Energy and utilities:** Predictive analytics plays a vital role in the energy sector. Sophisticated models are used to predict the effects of weather changes. This helps in taking preventive measures to minimize catastrophic electricity losses due to power grid damage.
- **Manufacturing industry:** Production of any unit can be increased or decreased based on the results of product-specific work models. These models are built as a combination of predictive analytics and business analytics techniques.

- **Healthcare:** Healthcare organizations have shown keen enthusiasm over the past few years for adopting predictive analytics to improve their operations. According to surveys, healthcare organizations that use predictive analytics are likely to save (at least) a whopping 15% of their total budget over the next five years, if not more. Additionally, 93% of healthcare executives unanimously agreed that predictive analytics will play a key role in shaping their future. Areas in which healthcare organizations are implementing predictive analytics are:
  - Optimizing employee schedules
  - Procurement and supply management
  - Allocation of resources during saturated periods

Predictive analytics involves some manipulation of data from existing data sets to identify some new trends and patterns. These trends and patterns are then used to predict future results and trends. By performing predictive analytics, we can predict future trends and performance. Also defined as prognostic analysis (the word prognostic means prediction), predictive analytics uses data, statistical algorithms, and machine learning techniques to identify the likelihood of future outcomes, based on historical data.

2.2.5 Importance of predictive analysis

For years, organizations have recognized the importance of collecting data as a way to look back on industry trends, but business teams are only beginning to scratch the surface of what is possible when it comes to predictive analytics [8].

i. Define the problem statement:
This phase involves defining the project deliverables, scope of effort, objectives, and identifying the data sets to be used.

ii. Data collection:
Data collection involves gathering the necessary details required for analysis. It includes historical or historical data from an authorized source on which predictive analysis is to be performed.

iii. Data cleaning:
Data cleansing is the process by which we refine our data files. In the data cleaning process, we remove unnecessary and erroneous data. It includes removing redundant data and duplicate data from our datasets.

iv. Data analysis:
Data analysis includes data exploration. We examine the data and analyze it thoroughly to identify some patterns or new results from the data set. In this phase, we discover useful information, and finally identify any patterns or trends.

v. Building a predictive model:
In this phase of predictive analytics, we use various algorithms to build predictive models, based on observed patterns. It requires knowledge of python, R, statistics and MATLAB, and so on. We also test our hypothesis using standard statistical models.

vi. Verification:
It is a very important step in predictive analytics. In this step, we check the effectiveness of our model by performing various tests. Here, we provide sample input sets to check the validity of our model. At this stage, the model needs to be evaluated for its accuracy.

vii. Deployment:
In deployment, we let our model work in a real-world environment and it helps with everyday discussions and makes it accessible for use.

viii. Model monitoring:
This phase involves monitoring your models regularly to check performance and ensuring we are getting the right results. It tracks how model predictions perform compared to real data sets.

2.2.6 Predictive analysis and conditional monitoring

Predictive maintenance allows equipment users and manufacturers to assess the working condition of machines, diagnose faults, or estimate when the next equipment failure is likely to occur. When you can diagnose or predict failures, you can plan maintenance in advance, better manage inventory, reduce downtime, and increase operational efficiency [9].

Developing a predictive maintenance program requires a well-designed strategy for assessing the operational condition of machinery and early detection of incipient faults. This requires effective use of available sensor measurements and knowledge of the system. You need to consider many factors, including:

- Observed sources of failures and their relative frequency. Such sources can be the machine's main components (such as impeller vanes and flow valves in a pump), its actuators (such as an electric motor), or its various sensors (such as accelerometers and flow meters).
- Availability of process measurements through sensors. The number, type and placement of sensors, and their reliability and redundancy affect both algorithm development and cost.

- How the various sources of disturbances are reflected in the observed symptoms. Such cause-and-effect analysis may require extensive data processing from available sensors.
- Physical knowledge of system dynamics. This knowledge can come from mathematical modeling of the system and its errors, and from the knowledge of industry experts. Understanding system dynamics includes detailed knowledge of the relationships between various signals from machinery (such as input-output relationships between actuators and sensors), the operating range of the machine, and the nature of the measurement (for example, periodic, constant, or stochastic).
- The ultimate goal of maintenance, such as error recovery or the creation of a maintenance plan.

2.3 Exploratory data analysis and results

Predictive maintenance can be a game changer and can affect various industries and people across the globe. It would help us to know about the upcoming challenges. Predictive maintenance is helpful in every field, whether it is Life Sciences, Aerospace, Space missions, Travel Industries, Hospitality industries or Manufacturing industries. By using predictive maintenance, we can receive a breathing room or time to fix issues that can impact our critical process in the coming days. Predictive maintenance can be affordable because of low hardware cost, and due to availability and reachability of the cloud services.

There are a total 10,000 instances of data, out of which 9,661 instances belong to Not Failure of a machine and 339 instances belongs to Failure of Machine. Fig 2.1 represents data instances belonging to different classes.

Failure can be because of various reasons (Power Failure, Total Wear Failure, Overstrain Failure, No Failure, Heat Dissipation Failure, etc.) and Figure 2.2 represents the count of different types of failures.

Air temperature (K), process temperature (K), rotational speed [rpm], torque [Nm], and tool wear [min] are factors that can impact the machine, or we can say they are responsible for the failure of the machine. Figure 2.3 shows the factors that are responsible for predicting the probability of machine failure in the future.

Figure 2.4 shows the possibility of failure with respect to air temperature (K), process temperature (K), rotational speed [rpm], and torque [Nm].

Figure 2.5 represents the possibility of failure (%) with respect to tool wear [min].

Figure 2.6 shows the scatter plot between nf and rotational speed, where nf = tool wear [min] * torque [Nm]).

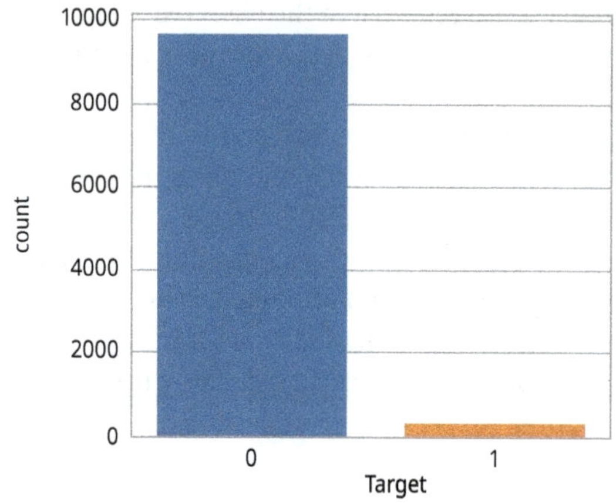

Target	Failure Type	count
0	No Failure	9643
	Random Failures	18
1	Heat Dissipation Failure	112
	No Failure	9
	Overstrain Failure	78
	Power Failure	95
	Tool wear Failure	45

Figure 2.1: Representation of data instances belonging to different classes.

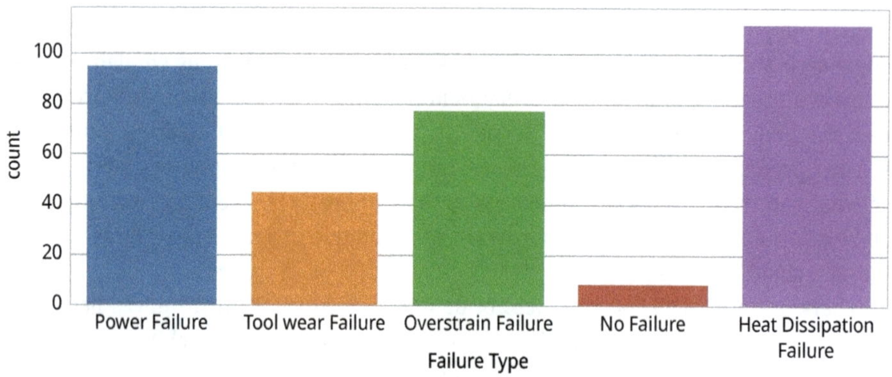

Figure 2.2: Count of different types of failures.

Air temperature [k]	Process temperature [k]	Rotational speed [rpm]	Torque [Nm]	Tool wear [min]	Target	Failure Type
298.1	308.6	1551	42.8	0	0	No Failure
298.2	308.7	1408	46.3	3	0	No Failure
298.1	308.5	1498	49.4	5	0	No Failure

Figure 2.3: Factors impacting machine failure.

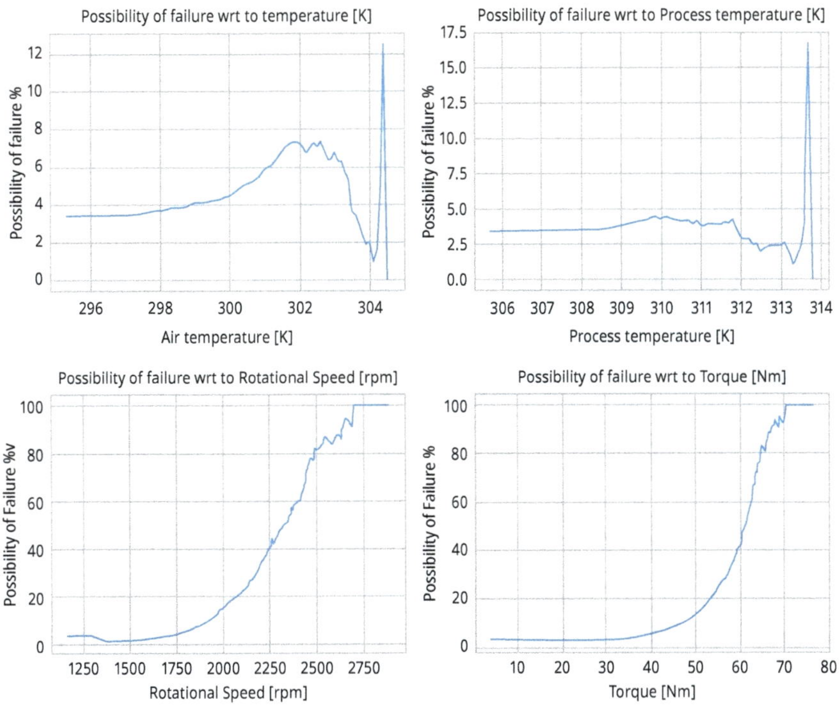

Figure 2.4: Possibility of failure (%) with respect to air temperature, process temperature, rotational speed, and torque.

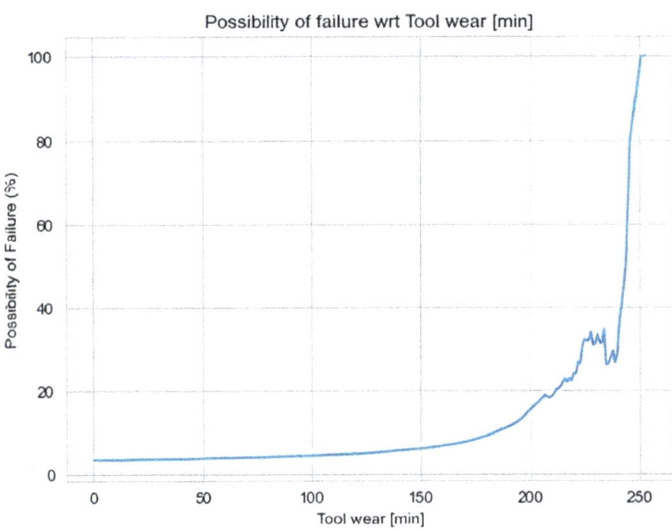

Figure 2.5: Possibility of failure (%) with respect to tool wear [min].

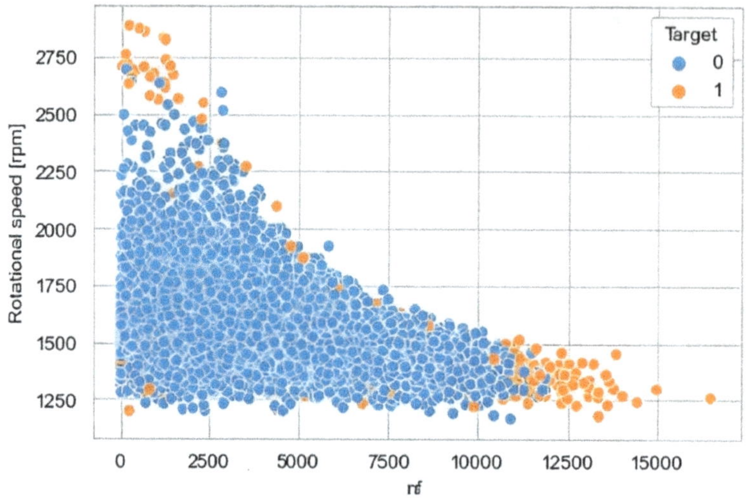

Figure 2.6: Scatter plot between nf and rotational speed.

2.4 Machine learning

Logistic regression

Logistic regression is an AI order calculation that is utilized to foresee the likelihood of an unmitigated ward variable. In strategic relapse, the reliant variable is a paired variable that contains information coded as 1 (indeed, achievement, and so on) or 0 (no, disappointment, and so on.). As such, the strategic relapse model predicts P(Y = 1) as an element of X. Figure 2.7 represents the confusion matrix of logistic regression.

Accuracy of Logistic Regression: 97.33333333333334%

Support Vector Machine (SVM)

Support Vector Machine or SVM is one of the most famous regulated learning calculations, which is utilized for arrangement as well as relapse issues. In any case, principally, it is utilized for arrangement issues in AI. The objective of the SVM calculation is to make the best line or choice limit that can isolate *n*-layered space into classes, so we can, without much of a stretch, put the new data of interest in the right classification later on. This best choice limit is known as a hyperplane. SVM picks the outrageous focuses/vectors that assist in making the hyperplane. These outrageous cases are called as help vectors, and the subsequent calculation is named as Help Vector Machine.

Accuracy of SVM Algorithm: 98.6666666666667

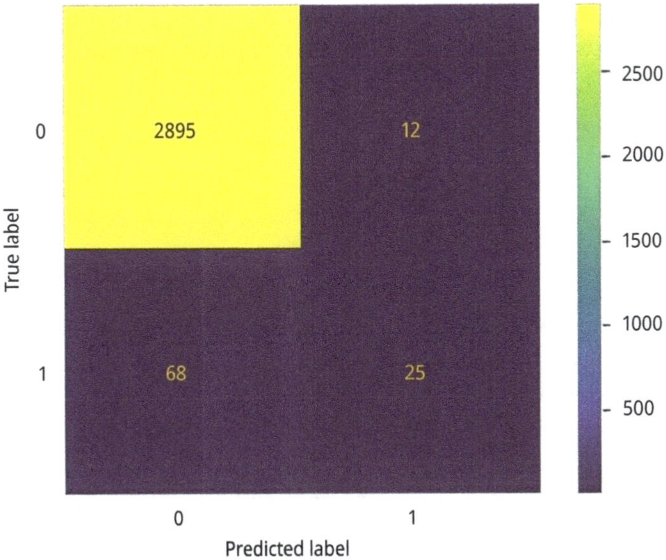

Figure 2.7: Confusion matrix of logistic regression.

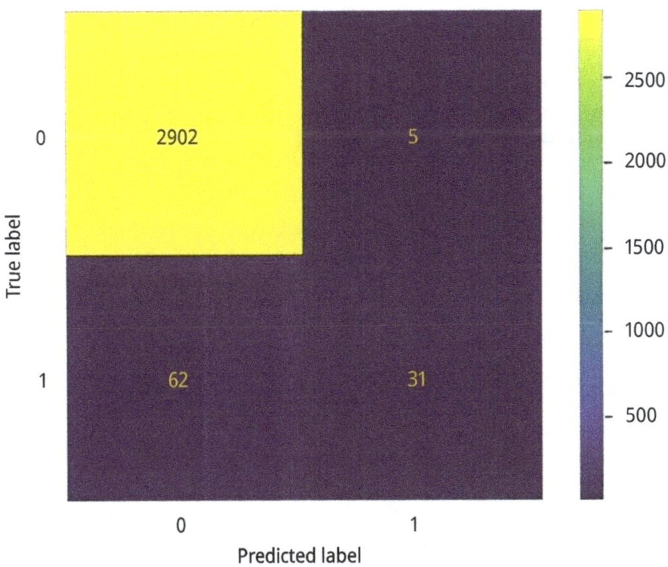

Figure 2.8: Confusion matrix KNN.

KNearestNeighbour (KNN)

KNN algorithm falls under the Managed Learning classification and is utilized for arrangement (most regularly) and relapse. It is a flexible calculation likewise utilized for crediting missing qualities and resampling datasets. As the name (K Nearest

Neighbor) proposes, it considers K Nearest Neighbors (Relevant pieces of information) to anticipate the class or consistent incentive for the new Datapoint. Figure 2.8 represents the confusion matrix of KNN.

Accuracy of KNN Algorithm: 97.76666666666667

2.5 Conclusion

Accuracy is characterized as the level of right expectations for the test information. It tends to be determined effectively by separating the quantity of right forecasts by the quantity of complete expectations.

We have summarized the accuracy of all machine learning algorithm used. Accuracy of logistic regression algorithm is 97.33%, that of KNN is 97.66%, and of SVM is 98.66%. Table 2.1 shows the machine learning algorithm used for prediction and the corresponding accuracy.

Table 2.1: ML algorithms and their accuracy.

Algorithm	Accuracy
Logistic regression	97.33 (%)
KNN	97.76 (%)
SVM	98.66 (%)

References

[1] S. Sharma, N. Chamoli, Machine Learning Approach for Network Intrusion Detection Systems, in: An Interdisciplinary Approach to Modern Network Security, CRC Press, 2022, pp. 35–49.

[2] S. Sharma, N. Chamoli, Real-time patient health monitoring, Comput. Intell. Predic. Anal. Med. Sci. Pragmatic Approach 6, 2021, 193.

[3] T. M. Mitchell, Artificial neural networks, Mach. Learn. 45(81), 1997, 127.

[4] S. Cho, S. Asfour, A. Onar, N. Kaundinya, Tool breakage detection using support vector machine learning in a milling process, Int. J. Mach. Tools Manuf. 45(3), 2005, 241–249.

[5] M. Bhatia, S. Sharma, M. Hooda, N. C. Debnath, A Machine Learning Approach toward Meal Classification and Assessment of Nutrients Value Based on Weather Conditions, in: Big Data Analytics and Intelligence, A Perspective for Health Care. Emerald Publishing Limited, 2020.

[6] S. Ahmed, S. Sharma, S. L. Yadav, Information Retrieval in the Hidden Web, in: New Opportunities for Sentiment Analysis and Information Processing, IGI Global, 2021, pp. 50–71.

[7] D. Wu, et al. A comparative study on machine learning algorithms for smart manufacturing: Tool wear prediction using random forests, J. Manuf. Sci. Eng. 139(7), 2017, 071018.

[8] M. Bhatia, S. Sharma, S. Bhatia, M. Alojail, Fog computing mitigate limitations of cloud computing, Int. J. Recent Technol. Eng. 2020, SCOPUS.

[9] R. Alswailem, S. Bhatia, Detection and Mathematical Modeling of Anxiety Disorder Based on Socio-economic Factors Using Machine Learning Techniques, in: Human-Centric Computing and Information Sciences, (SCI, IF: 6.41).
[10] A. Heena, N. Biradar, N. M. Maroof, et al. Machine learning based biomedical image processing for echocardiographic images. Multimed Tools Appl (2022). https://doi.org/10.1007/s11042-022-13516-5 (SCI, IF: 3.41)
[11] S. Laddada, et al., "A data-driven prognostic approach based on wavelet transform and extreme learning machine," 5th International Conference on Electrical Engineering, Boumerdes, 2017, pp. 1–4.
[12] W. Li, et al., Time varying and condition adaptive hidden Markov model for tool wear state estimation and remaining useful life prediction in micro-milling, Mech. Syst. Signal Process. 131(11), 2019, 689–702.
[13] S. M. Sasubilli, A. Kumar, V. Dutt, "Machine Learning Implementation on Medical Domain to Identify Disease Insights using TMS," 2020 International Conference on Advances in Computing and Communication Engineering (ICACCE), 2020, pp. 1–4, doi: 10.1109/ICACCE49060.2020.9154960.
[14] S. M. Sasubilli, A. Kumar, V. Dutt, "Improving Health Care by Help of Internet of Things and Bigdata Analytics and Cloud Computing," 2020 International Conference on Advances in Computing and Communication Engineering (ICACCE), 2020, pp. 1–4, doi: 10.1109/ICACCE49060.2020.9155042.
[15] P. K. Kotturu, A. Kumar, "Comparative Study on Machine Learning models for Early Diagnose of Alzheimer's Disease: Multi Correlation Method," 2020 5th International Conference on Communication and Electronics Systems (ICCES), 2020, pp. 778–783, doi: 10.1109/ICCES48766.2020.9137872.

Ananth J. P., G. Raghuraman, Premnath S P,
A. Francis Alexander Raghu

Chapter 3
A deep survey on quantum computing technologies

Abstract: There has been significant advancements in Quantum Computing research related to infrastructure and algorithmic complexity in recent years. Since the 1980s, the scientific community has been interested in the possibility of using quantum phenomena to tackle computer issues. Quantum computing systems get their power from the no-cloning theorem, quantum superposition, and quantum entanglement. Based on the fact that these events lack a classical equivalent, it is not possible to reach comparable conclusions using the framework of traditional computing. Already demonstrated are the experimental breakthroughs that can be made with quantum computing technology, and a number of investigations are currently being carried out. In this chapter, we analyze the most recent technological breakthroughs in the field of quantum computing and investigate some of the outstanding issues.

3.1 Introduction

The technology of quantum computing allows problem solution that is both more efficient and fundamentally different from what is conceivable with classical computations. It also offers completely distinct answers to computational issues. Positive experimental findings suggest that commercially viable quantum computers could be on the horizon in the next several years [1, 2]. The prime factorization method invented by Shor [6] is considered to be one of the most important and well-known instances of an algorithm that is based on the quantum computer-capability. Considering the regular computing environment, the computational problem would take billions of years to solve, but the quantum computer might theoretically resolve it in limited hours. This technique was responsible for the 1994 “great boom” in the field of quantum calculations. It also cleared

Ananth J. P., Sri Krishna College of Engineering and Technology, Kuniamuthur, Coimbatore, Tamil Nadu, India, e-mail: jude.ananth@gmail.com
G. Raghuraman, Sri Sivasubramaniya Nadar College of Engineering, Chennai, Tamil Nadu, India, e-mail: raghuramang@ssn.edu.in
Premnath S P, Sri Krishna College of Engineering and Technology, Kuniamuthur, Coimbatore, Tamil Nadu, India, e-mail: premnathsp@yahoo.com
A. Francis Alexander Raghu, Sri Krishna College of Engineering and Technology, Kuniamuthur, Coimbatore, Tamil Nadu, India, e-mail: francismtech@gmail.com

https://doi.org/10.1515/9783110791402-003

the technique towards the growth of technology linked to quantum computing as well as the assessment of quantum computers [7]. The gap between quantum computing and classical methods is exemplified by the fact that the Rivest-Shamir-Adleman (RSA) algorithm [6] can be broken. Quantum computers consist of a variety of unique parts. CPUs, memories, gates, registers, buses, and storage devices are all examples of functional components, and while they share certain similarities in function, their physical structures are quite different. Quantum operations in a quantum computing system are carried out on quantum registers. The quantum states that make up a quantum circuit are entangled, while the quantum states that make up a quantum register form quantum superposition. The confluence of these occurrences produces a system with fundamentally differing features from those of a typical computer. However, due to constraints of quantum hardware like the no-cloning theorem, new methods of designing circuits must be implemented. This is so because a quantum state cannot be in more than one quantum gate at once [5, 10].

Unitary operations on quantum systems are carried out by quantum computers using reversible quantum gates. Though quantum computers are now functional, only a small number of devices exist outside of the lab [3, 4]. However, in recent decades, a number of interesting new domains and outcomes have arisen, both of which can have the prospective to greatly augment the improvements. In a discrete atmosphere, smaller quantum computers can be connected through a quantum bus to create large-scale quantum computers. This permits the computers to communicate with one another. In the coming years, developments in technology will make it possible to shrink physically massive quantum computers down into devices of a more manageable size. In terms of both size and performance, the scenario bears a remarkable resemblance to the progression of classical computers. This chapter presents an overview of the current exploration articles and outcomes obtained through the use of quantum computing technology. In addition, each section discusses some of the subject's outstanding concerns. At the conclusion of each section, the Related Work section provides a summary of the references that are relevant to the current scenarios.

3.2 Quantum computations

The foundation of quantum computers is quantum information. Quantum computers use the many quantum effects made possible by quantum mechanics to store and process information in the form of quantum states. Effects like the no-cloning theorem, quantum superposition, interference, de-coherence, and entanglement are examples [7, 9]. There are several different ways in which quantum systems might materialize into the world (atom energy levels, polarization, and spin). A qubit system is a two-dimensional (2D) quantum system; hence, the collection of n quantum states (i.e. the quantum register) in superposition can concurrently signify dn different values that is classical. Each of the quantum system's state depends on the states of the other quantum systems, which is why entangled systems are the most intuitive way to describe the quantum states used in quantum circuit calculations. Important to quantum computing is the concept of reversible computation. Theoretically, the outcome of a reversible computation can be used to rebuild the initial state from scratch [11]. The classical system can also create reversible circuits [12]; a given reversible gate has the same number of inputs and outputs and each input is uniquely mapped to one of the outputs. A reversible gate must have an equal number of inputs and outputs, among other requirements. Since a quantum computing system must likewise comply with these conditions, the quantum circuit in which the quantum states input develop, is in a reversible manner via unitary operations. In actuality, a sequence of quantum gates are employed to achieve this reversibility. In quantum computing, "ancillary states" are ephemeral quantum systems that are discarded once the required result is obtained. In the end, a conventional numerical value is extracted from the quantum register.

Quantum algorithms embody the core ideas of quantum computational complexity. There have been many suggested quantum algorithms, and they all seem to agree that when compared to classical algorithms, they will achieve considerable speedups (exponential, polynomial, or super polynomial), thanks to the use of quantum mechanics' effects. In addition, many feel that quantum algorithms can provide answers to problems that would otherwise be intractable with more conventional means. The prime factorization issue is illustrative of this point. According to its core ideas, the DiVincenzo criteria [13] lay out the minimum standards for real-world realizations of quantum computers. A functioning quantum computer would feature quantum registers that can be extended, the capacity to initialize the registers that have quantum nature to an identified state, a universal gate set to execute any quantum algorithm, and sufficient coherence length and fidelity to perform extended operations. In addition, the outcomes of these quantum calculations must be observable on the quantum computer itself. These are required for the implementation of any form of effective quantum computing. Figure 3.1 depicts the hypothetical progression of quantum computing technology.

DiVincenzo requirements, reinforced by particular physical layer features, offer the physical underpinnings for quantum computing systems. The physical layer places constraints on quantum registers, gates, circuits, and memory that must be overcome in order to build quantum computers. Numerous approaches have been taken to the study of the complexity of quantum computing. Linear optics computational complexity was studied by authors, whose work is acknowledged in [14], lending credence to the claim that classical computers cannot mimic the capabilities of quantum computers.

Using contemporary photonics and optical technology, they also investigated the possibility of mass production. The authors conclude that their method is effective in problem searching and sample issues. In his research, Aaronson studied the "learnability" of quantum states [15]. These studies show that the number of observations necessary to define a quantum state is finite (or at most exponentially growing). Despite the

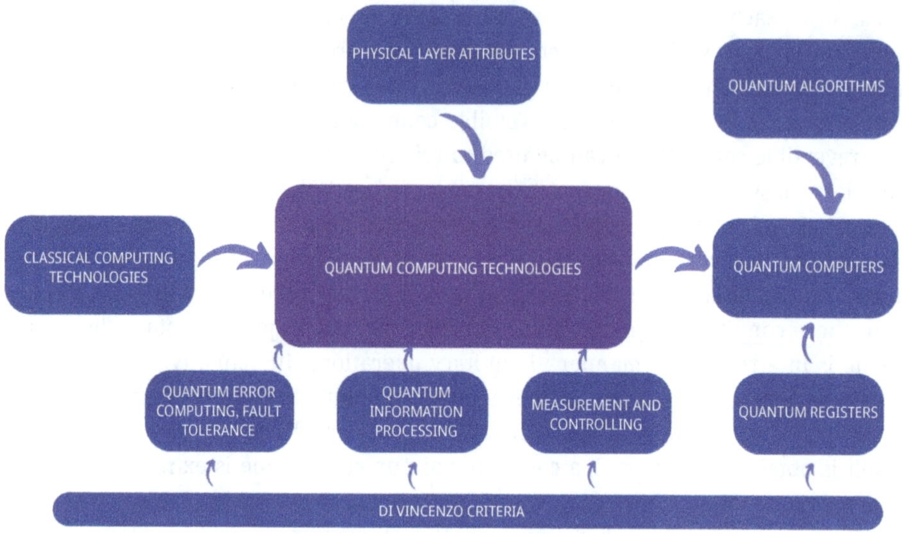

Figure 3.1: Development of quantum computing compared with classical computing.

complexity of quantum tomography, the author has found that innovative quantum one-way communication protocols can be simulated within the structure itself. It was also looked into how tough it is to use sound classical advice to confirm a flawed quantum piece of advice.

Browse the cited papers in [7, 9] to learn more about quantum computing and data, including foundational aspects of quantum communication networks and other related issues. Reference [16] cites Deutsch's key work on the quantum computer and also the quantum theory. Please refer to [17] for Feynman's thoughts on whether or not classical computers can effectively simulate quantum computations. See reference [18] for further information on the "no-cloning theorem." In [19], you may find a good review of quantum algorithms. In order to learn more about quantum algorithms and how they might be used to solve algebraic problems, you can read [20]. Read the essay [21] for a survey of current progress in quantum algorithms. Entanglement's contribution to the speedup of quantum computing is investigated in [22]. Quantum spin systems are investigated in [23] using a variety of theoretical frameworks, including projected entangled pair states, matrix product states, and other approaches. In the research, we looked into each of these in depth. The difficulties of quantum computing outside of any time framework is analyzed in [24]. In [25], the concept of global blind quantum computation is explored.

3.3 Quantum computing's relatively basic components

Using this classification scheme, we will discuss the most recent research on qubits and qubit circuits, the fundamental components of quantum computers. Recent discoveries include quantum control and measurement equipment, quantum memory, and error correcting systems.

3.3.1 The quantum gates

The Toffoli and Fredkin gates [7, 9, 11] are crucial because of the quantum circuit that may be deconstructed into a set of either Toffoli or Fredking gates. This can simulate CNOT (Controlled-NOT) or NOT gates. A group of Fredking gates can also substitute for any quantum circuit. In other words, both gates are applicable to the implementation of universal quantum computations. Due to the reversibility of quantum gates, only the useful outputs are kept while the auxiliary states are eliminated. In the physical layer, the quantum gates can be implemented using linear optic tools, ion traps, diamonds, superconductors, quantum dots, topological quantum computing elements, and donor-based systems. The utilization of topological quantum computing components is also possible.

Implementations of effective quantum error-correction codes are hampered by the absence of a solution to the problem of dealing with errors caused by quantum gates [26, 27]. The authors of [28] examined the possible quantum benefits of shallow circuits. The authors of the study found that comparing quantum and classical computing, the classical counterparts are more powerful, considering the quantum circuits. The authors also found that this problem has a solution that is right 100% of the time if it is solved using a constant-depth quantum circuit. In order to show a gate model structure that is practically implementable, this quantum circuit must only consist of qubit gates (one and two) that are locally acting on a two-dimensional grid. Despite the presence of some physically based limits, the authors of the study cited as [29] explored the potential of Instantaneous Quantum Polynomial-time (IQP) circuits. Considering that it would be difficult to construct a conventional IOP circuit, the results are all the more noteworthy. Researchers found that a specific depth square lattice of n qubits can support a certain family of IQP circuits. One of the results of their study was this discovery. They also found that, given polynomial amounts of time, a classical algorithm can reproduce sampling from the distribution created, when a constant noise of arbitrary small magnitude is supplied to each qubit.

Researched in [30] by Toffoli, Controlled-Not-Not, and Fredkin, these high-fidelity three-qubit gates have several applications, including quantum information processing and error correction in quantum systems. As a basis for the suggested model, essentials of machine learning were applied. The processes can be applied with a system consisting of three neighboring superconducting artificial atoms, as discovered by the researchers. The authors determined that their proposed approach was 99.9% reliable. This research is extremely useful since it provides a minimum level of accuracy for quantum computing that is error-tolerant.

3.3.2 Memory in quantum computing

Quantum memories for use in quantum computing are typically formulated with n stationary quantum states. Many different strategies have been proposed in the academic literature for actualizing quantum memory. Intriguingly, data can be stored in a torus or array of quantum states using a technique called topological quantum memory [31]. These quantum systems have been entangled in a wide range of configurations in an effort to build a coherent and well thought out quantum system. Even though the results seem promising, expanding the memory lifetimes remains a problem that needs to be tackled in terms of the practical implementations of quantum memories. The publication [33] demonstrated a quantum bit memory that could store information for more than one second at normal temperature. Using atomic memory and photonic interconnects, the study [34] describes the architecture for a massively parallel quantum computer. Quantum random access memory is discussed in [19, 32]. The authors' definition of a qRAM states that in order to access any quantum superposition including N memory cells, n qubits are needed. The authors proposed a structure with a design that would reduce the need for a memory call by a factor of 10 or more. The research results, as reported by the authors, allow for the development of a more stable qRAM technique, and result in a reduction in addressing power consumption that is exponential in nature. The project was successfully completed, thanks to a quantum optics implementation.

In [35], researchers examine the storage capacity in an exponential form of associative memories, accessed using quantum annealing. The authors of this study provided evidence that using quantum annealing toward the recall tasks increases the storage capacity of associative memory models by a factor of exponential. Some theoretical models suggest that these types of models can store memories at a sublinear rate. Researchers in [36] look into the possibility of using recurrent neural networks to achieve optimal dynamical decoupling in quantum memory. These authors used the tried and true methods of machine learning. In order to optimize the dynamic coupling sequences, the idea of recurrent neural networks is used. The simple approach called “decoupling” can be utilized to lessen the frequency with which the quantum memory fails.

Resistors with a memory known as Quantum memristors are used. The history of the charges related to crossing is found based on the resistance, and is described in detail in reference [37]. A feedback system is used in the decoherence mechanism to control, and that makes use of continuous measurements. The authors also established the reality of memory effects in quantum systems and the suitability of superconducting circuits for exploiting these effects in the actual world. The basic component is the proposed quantum memristor. This is used as a basic component in neuromorphic quantum computing. In [38], the analysis is done by potential strategies for incorporating quantum memristors into existing designs, based on superconducting circuits. As a result of cancelling out supercurrents, the authors presented a quantum device with memristive behavior. To explain this phenomena, we turn to tunneling induced by quasiparticles. The authors have succeeded in their mission because they have provided a Quantum model for quantifying the memory retention, and determined that hysteretic behavior is possible using the current accessible measuring methodologies in superconducting quantum circuits. Qubit-based meminductors and memcapacitors are discussed in reference [39]. Researchers have shown that both capacitive and inductive devices have amazing capabilities for use in quantum computers. This research shows that qubit-based meminductors and memcapacitors exhibit unusual hysteresis curves in response to selected inputs. The most important finding from this study is that conventional quantum devices can be made to include memcapacitive and meminductive components.

3.3.3 Central processing unit in quantum perspective

Quantum central processing units (CPUs) use a communication mechanism known as a quantum bus to exchange information with the other quantum computer components. The computational building blocks of quantum central processing units (CPUs) are quantum adders, and an examination of these adders is the key to comprehending how these CPUs work. Numerous varieties of reversible quantum adders have been designed, allowing for the realization of quantum computations in a wide range of architectural models. Additional forms of adders are used. Quantum adders, like their conventional counterparts, can be constructed for use in quantum computing. They are reversible in structure and can be implemented in parallel. All quantum adders have different circuit depths, different algorithms latencies, and performances, which are reversible and use ancilla quantum states [40, 41]. depicts how to implement their collaboration. In order to facilitate the quantum circuits parallelization, two key network models are used [11]. In the initial concept, communication is carried out among the collections of quantum states. In divergence, the second model allows for just local contacts, meaning that interactions can take place only between people who are physically adjacent to one another in a linear configuration. Using these network models, researchers were able to characterize the performance of several configurations of quantum adders and found that they are on par with one another.

Quantum communications plays a vital role. Communication among the quantum functional elements and the computer's quantum central processing unit (CPU) can be implemented with the help of the numerous experimental quantum error correcting methods. Shor's prime factorization method examined five qubus interconnect topologies. The circuit that is part of the factorization algorithm is depicted in Figure 3.2. The purpose of this quantum algorithm is to make the solutions in quantum register A more trustworthy.

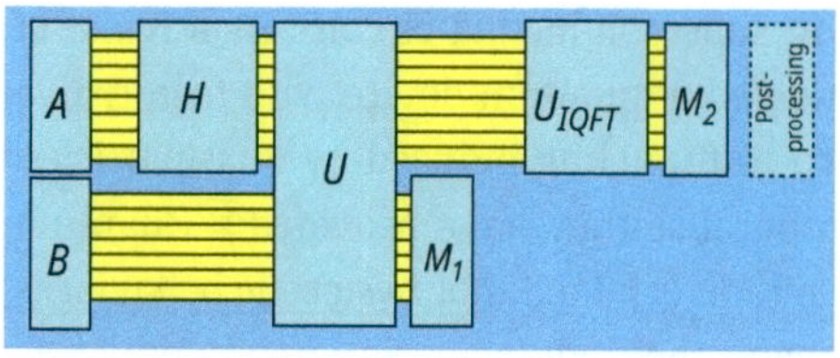

Figure 3.2: The circuit that is part of the factorization algorithm.

Shor's prime factorization algorithm using a quantum circuit is demonstrated. Dashed boxes signify register A (t qubits) and register B (n qubits) of the quantum registers that make up the quantum circuit (n qubits). First, the H Hadamard operation is carried out on the qubits in register A to get them ready for superposition. Next, the U unitary operation is applied; this is crucial when determining periods. The M1 measurement operator is then applied to the register B qubits. The inverse quantum Fourier transformation (IQFT), also known as UIQFT, is then performed on the quantum register A. After everything has been exhausted, the M2 operator does its measurements on the A qubits.

The authors of the study [11] discovered a methodology in contrast to the BCDP (Beckman-Chari-Devabhaktuni-Preskill) method. This method employs modular exponentiation. According to the findings provided in [11], the best performance can be obtained using a quantum carry-ripple adder. Quantum adders make the process of factoring faster, with reference to the efficient operation of these quantum adders. In addition, the authors explored the quantum adder properties [41], specifically the utilization of an approximate quantum adder in a protected state transfer mechanism. Through the use of depolarizing quantum channels, this study defines a decoherence-protected protocol for the transmission of quantum states in photonics. By using spontaneous parametric down-converters, a rudimentary quantum adder was built, allowing the protocol to be implemented. Scientists found that for distances below a threshold value, the procedure has a higher chance of success than more refined quantum teleportation approaches.

Using genetic algorithms, the authors of [43] investigate the approximate quantum adders. Considering the theoretical properties, the approximate quantum adders are described and outlined an optimization technique based on evolutionary algorithms. Additionally, it was analyzed, and the authors have discussed the applications of approximate quantum adders. In the end, the authors state that their research makes it possible to improve the efficacy and consistency of many preexisting treatment protocols. Using IBM's Quantum Experience, the authors constructed a working model of an approximate quantum adder. This study demonstrates that approximation quantum adders can contribute to the development of cutting-edge QI/CT techniques. The authors of [44] developed a protocol for the secure transmission of a photon's quantum state from one cavity to another via an endless, one-dimensional waveguide. This initial cavity serves to preserve the quantum states of the photons. Researchers discovered that this exchange is feasible, with the transmitted data remaining unaffected by random noise. The authors also examined using atoms or atomic ensembles as the quantum memory in a cavity QED setup. The authors conclude that the models and conclusions they have provided can be applied to many other areas, including photonic quantum information processing.

Quantum auto encoders were investigated in paper [42], which explored the combination of quantum adders and evolutionary algorithms. According to the abstract, the authors of this work propose a realistic link between approximate quantum auto encoders and quantum adders. Quantum auto encoders built on programmable quantum platforms can be optimized for specific input data. The study's authors draw a variety of inferences about how these findings apply to the actual world. In [45], the authors attempt to solve the issue of designing a reversible floating point adder for quantum computing that does not require excessive physical space. The authors claim that reversible logic is necessary for both quantum computing experiments and low-power computing studies. The suggested approach has applications in both binary quantum computation and the optimization of expensive hardware architectures. In addition to describing strategies for minimizing circuit costs, the authors also examined methods for constructing fault-tolerant circuits.

The article [46] delves into the rival and complementary technological advances surrounding the use of quantum computers in experiments and in the real world. As this article shows, classical computers will continue to play a significant role in the years to come, but their performance will be limited by the laws of thermodynamics.

3.3.4 Measurement and controlling

Control methods are essential for the successful manipulation of quantum states, readouts, error correction techniques, and fault-tolerant quantum computing. Major advancements have been made in many fields since the ion trap-based implementations, such as superconducting quantum circuits, linear optics, topological quantum computing, quantum dots, donor-based quantum implementations, and any on-based quantum computing. Preparing quantum systems for computation and implementing measurement that extract significant information from quantum states are two examples of key unsolved problems in these areas. In their study [47], the authors explored the potential of superposing two pure quantum states, utilizing prior knowledge in a practical setting. Since it is challenging to construct a superposition of any two unknown pure states, the discoveries are important in the field of experimental quantum information processing. The reason being that doing so necessitates prior knowledge of the states being input. Even then, it can only be done probabilistically. The authors used a three-qubit nuclear magnetic resonance device to probabilistically superpose two pure states. They also determined the average degree of faithfulness between the prepared states and utilized that data to construct an input state family with which to evaluate the technique. Due to the high level of authenticity achieved, the authors conclude that the proposed work may be prolonged to incorporate additional complex circumstances and sophisticated quantum circuits.

The problem of superposing quantum states with unknown values is also investigated by the authors of [48]. In particular, the authors of this study developed a mechanism for superposing numerous pure states of a d-dimensional quantum system that can be tested empirically. However, the scientists didn't stop at the theoretical level but also gave a practical implementation using a two-qubit nuclear magnetic resonance (NMR) quantum computer. The authors of this paper [77] present a concise description of the reasoning behind their conclusion that classical computers struggle to model boson sampling problems. Finally, settling on a taxonomy for each subtype, the authors go on to explore the most recent experimental realization linked to the quantum advantage of boson sampling.

Researchers in the field of photonic quantum information processing examined the application of differential equations in their study [49]. They presented a model for the realization of photonic quantum circuits, in which the dynamics of the circuits is regulated by a set of differential equations. The authors conclude that the intentional model allows for the establishment of both feedforward and quantum feedback without the need for any intermediate measurements to be included in the quantum circuit. In [50], a machine learning based model is analyzed to see if it can be used to foresee and fix qubit decoherence in real time. The authors used various time-tested methodologies from the disciplines of control theory and machine learning to foresee a qubit's future state. The research team came to the conclusion that the experimental models under consideration significantly enhanced qubit phase stability, as compared to the conventional measurement-based feedback strategies. The provided models, the authors find, may be used to qubit systems with no additional changes needed on the hardware-layer for projective measurement-based measurements.

Reference to the aforementioned study can be found in [51], which also discusses the use of quantum artificial intelligence (QAI) to the study of superconducting circuits. Using superconducting circuit implementations, the investigation goes into key protocols of quantum reinforcement learning. These circuits are considered to have a feedback-loop control system throughout the book. The research also details further findings related to superconducting circuits. The authors of [52] explored the problem of finding and correcting the persistent biases that plague quantum annealers. In this research, we investigate a few of the problems that have been observed recently while using quantum annealers. To resolve discrepancies between the expected and the actual values of programmable parameters, the authors developed a method to detect and fix them. The authors have put their method to test with D-wave quantum devices in extensive laboratory experiments. Research reveals that the proposed model may help the practical quantum gadget operate better. In [53], the authors employed the median number of copies to examine the possibility of quantum state discrimination. The authors believe that the probability of error and the available resources are the two most significant variables to consider when trying to discriminate between nonorthogonal quantum states that utilize many copies.

In order to accommodate a defined maximum acceptable error probability, the authors designed a new state-discrimination task, which requires significantly fewer resources, on average. They judged that past strategies utilized to do this new assignment were not optimal and so they devised and tried out some fresh alternatives. The authors of [54] deliberate the potential of unknown quantum states being formed by superposition. The research presented a probabilistic approach to creating superposition of two a priori unknowable states. If you use this technique, the quantum states will always be superimposed on the pure reference state. In their research, the authors discovered that the protocol is applicable to any Hilbert space. This opens up a broad variety of uses for the method.

An examination of superposition of nonobservable photonic quantum states is conducted in reference [55]. The results of the study are presented in this reference. The average fidelity was found to be 0.99, suggesting that putting this technology into practice could have a major impact on experimental quantum computing. The protocol could be used, for instance, in the context of coherent superposition-based information compression. In [56], the authors elaborate on an earlier research by explaining how two unknown states can emerge through superposition.

3.3.5 Quantum error correction

Now that we have developed quantum error-correction methods, we can meet the stringent standards for quantum computing that necessitate high fidelity and tolerable coherence times. When compared to traditional computing systems, quantum computing is more susceptible to quite distinct kinds of errors. Numerous physical technologies were developed for the various quantum hardware implementations in order to provide reliable, fault-tolerant quantum computing. Quantum error-correction systems frequently leverage both the quantum states of the input data and the syndrome quantum states to localize the erroneous data. In experimental quantum computations, the logical and physical representations of quantum states can be distinguished. This is because several distinct quantum states are used to encode the logic of a quantum system. Developed quantum error-correction codes (topological error-correction design, etc.) can be easily modified to suit the needs of specific physical implementations.

The authors of [57] examined the surface code error correction problem on a faulty lattice as part of their research. These error-correcting techniques were inspired by the realization that quantum computer implementations will be prone to loss due to quantum states that are physically flawed. Similarly, the study's authors arrived at the same result. This implies that adaptive error correction algorithms will play a crucial role in the development and the eventual implementation of quantum computing systems. The authors of this study did simulations of single-fault lattices and statically placed lattices. The authors of the study determined that a static loss in the lattice's peripheral would have less of an effect than one in the structure's center. In addition, the scientists investigated a variety of performance metrics for quantum processors. Specifically, they discovered a substantial correlation between postcorrection residual error rates and the depth of the lattice's deepest stabilizer circuit.

The challenges of implementing a deformation-based surface code are examined in [58], including state injection, lattice surgery, and dense packing. Considering the challenges, the authors created a surface deformer based on a super stabilizer that searches for the shortest error chains. These findings suggest that there is a possibility to place logical qubits in close proximity during actual quantum computing operations. The authors also investigated the procedure of converting from the defect-based surface code, and established a placement design for the deformation-based surface code. The results are relevant to large-scale quantum computation experiments. Erroneous learning's quantum sample complexity was studied by the authors of [59]. The researchers concluded that a powerful quantum learning algorithm exists, and that it can solve learning problems with polynomial sample and time complexity. Statistical evidence confirms the significance of the authors' work in cryptography.

The authors of [60] looked into the challenges of implementing surface-code quantum computers on silicon. The research established the viability of surface-code quantum computing using solid state spins, which could resolve the issue between nanoscale separations for qubit-qubit interactions. We hoped that by finding this answer, we could settle the argument over the nanoscale gaps. The suggested method employs probe spins aimed toward mechanical separation, and contains an error-correction technique to account for the inevitable misalignment of the spins. In addition to providing much higher qubit densities than other systems by many orders of magnitude, the authors' research indicates that the specified architecture can solve a number of problems plaguing solid-state quantum computing. The importance of these Majorana zero modes [61] cannot be emphasized, as they could form the backbone of fault-tolerant quantum computers. Coulomb-blockade spectroscopy was employed as a methodology by the authors of the study. This method creates a superconducting Coulomb island, also called a "Majorana island [78]," in a section of nanowire. The authors conclude that the proposed model aids in the investigation of more advanced facets of topological security and the elucidation of the transition from minor to topological infinite systems.

According to the study by the authors, the results show that hardware-efficient qubit encoding is preferable to more traditional methods of quantum error correction. Quantum supremacy is the idea that "quantum devices without error correction can complete a well-defined computational task beyond the capability of currently available classical computers". The authors tackled the challenge of randomly sampling from the distributions of results produced by (pseudo)random quantum circuits. They concluded that a costly numerical simulation of the circuit is required for traditional sampling of the distribution. In other words, the more qubits your system has, the more money you'll have to spend on it. Up to 42-qubit quantum circuits were studied by the authors. The authors conclude that quantum devices containing about fifty superconducting qubits are adequate for establishing quantum supremacy in the near future. For their study on the characterization of decohering quantum systems, the authors of the paper [62] turned to a machine learning technique. They looked at how decoherence affected adaptive characterization approaches, which include adjusting measurement parameters in response to data. The authors found that by tweaking the measurements, they could generate estimates, whose error rate went down as the number of measurements increased. The authors have constructed a framework for decohering quantum systems that takes into consideration the presence of defects in the readout and the presence of a substantial uncertainty in the initial value.

In [63], the quantum auto encoders as a technique was discussed by the authors for efficient quantum data compression. Auto encoders have the theoretical responsibility of reducing the number of dimensions by mapping the original input to a new location, from which the original input may be recovered. The authors developed a model of a quantum auto encoder to perform the aforementioned functions using quantum data. The authors conclude that conventional optimization techniques can be used to train the parameters of the quantum auto encoder.

3.4 Quantum computing on a massive scale

Advanced quantum algorithms can only be implemented on a quantum computer of supercomputer size or larger. Larger physical distances and, thus, larger building blocks in the physical layer are required by the structural characteristics of quantum devices due to the spatial position of quantum states. This means that rather than relying on a single, enormous quantum computer (macro-architecture), a set of smaller quantum computers (micro-architecture) will interface with each other in a set network topology to carry out distributed computations. Design options for large-scale quantum structures vary, depending on the model used to implement quantum error correction at the physical layer.

3.4.1 A quantum computer based on the gate model

Most quantum computations can be performed on a gate model quantum computer using its numerous layers of quantum gates. The quantum systems sent into each quantum gate undergo an atomic operation. Hardware limitations, such as the no-cloning theorem (which states that a given quantum system cannot participate in more than one quantum gate at once), necessitate applying the gates in many rounds. The feasible circuit depth is a function of the fidelity of the quantum gates, and current realistic implementations of gate model quantum computers are based on qubit systems with a certain qubit-to-qubit coupling topology. In [79], the authors suggested a quantum method, they dubbed, the "Quantum Approximate Optimization Algorithm" to effectively manage combinatorial optimization issues. According to the authors, the quantum algorithm can only work with positive integers, and relies on approximations to find solutions. They argue that a simple increase in this positive integer is all that is needed to improve the approximation of the quantum approach. Within this approach, the authors demonstrated that the depth of quantum circuits develops linearly. Researchers tested the quantum algorithm on standard graph types and found it to be competitive, with or even superior, to classical methods, in some cases. A well-defined combinatorial problem was used to implement the Quantum Approximate Optimization Algorithm in [65]. The model under discussion receives input from a large number of systems, each of which is a set of linear equations containing exactly three boolean variables, the solutions of which are either zero or one, depending on whether or not the sum of the variables mod 2 is zero or one.

The authors proved that their quantum algorithm not only excels above classical approaches but also provides other benefits. Quantum Approximate Optimization Algorithm's authors delved into the significance of quantum supremacy in [66]. The authors assert that their approach is well-suited for use with gate-model quantum computers. The input to the algorithm is an optimization problem in combinatorics, and the output is a text. The authors claim that many more clauses than not are satisfied by this output string. The authors verified the existence of well-characterized performance bounds for a subset of scenarios while using their proposed quantum approach. Furthermore, the authors showed that the output distribution of the quantum algorithm cannot be adequately reproduced by a classical device.

In order to verify this statement, the Quantum Adiabatic Algorithm was applied. The authors conclude that the proposed architecture can be utilized to demonstrate quantum supremacy in the near future, and is compatible with existing quantum computers. In [5], the authors go into the theoretical underpinnings of quantum algorithms in fixed qubit designs. The study is performed using quantum computers built on the gate concept. The authors have come up with a method to program gate-model quantum computers without resorting to error correction or compilation. It is one of the authors' central hypotheses that the total number of logical qubits will match the total number of physical qubits. By superimposing a succession of parameterized unitaries on the qubit layout, the proposed methods generate quantum states as a function of parameters. The authors additionally outlined and analyzed the usefulness of parameter optimization techniques. These results are especially applicable to gate-model qubit quantum computer architectures and have far-reaching implications for the future of quantum computing.

Large-scale quantum computing [10, 11] appears to require distributed topologies due to the spatial distance between the quantum states. Smaller quantum computers talk to each other over a quantum bus, which can be realized via optical fibers, wireless quantum channels, etc., allowing for the distributed implementation of quantum algorithms and error-correction procedures in a distributed topology. For quantum applications, quantum error correction, and quantum control and measurement to function on a distributed scale, robust protocols are required. In practice, a specialized SAN is used to connect multiple smaller quantum computers (SAN). With the right protocols in place, quantum computers in a SAN architecture are free to use any available quantum hardware and any flavor of quantum coding (the model also implements classical protocols). For more extensive areas, there are potential answers, such as quantum wide area networks (Q-WAN) and quantum metropolitan area networks (Q-MAN). Here, in [10], the ideas and procedures of local and distributed quantum computing are examined. Many of the experimental quantum error correction models and the construction methods that are crucial to scalable quantum computing are explored.

In [64], the construction of a million-qubit quantum computer is outlined. Given the need for safe hardware technology and the high degree of integration seen in quantum computers, the authors argue that a middle ground must be found. The authors found that equilibrium could be evaluated using appropriate modelling tools. To see how efficiently quantum logic operations might operate on practical quantum hardware with limited computational resources, they ran simulations. Also, a 1,024-bit number was factored using the Shor algorithm on a baseline budget of 1.5 million qubits, demonstrating the accuracy of the results.

3.5 Algorithms for use in quantum computing

Considering one of the most significant quantum algorithms is Shor's prime factorization algorithm [8], which also serves as a benchmark for characterizing the performance of quantum computers [11], several physical methods have been used to construct the prime factorization algorithm, each having its own limitations in terms of scalability. Distributed quantum computations rely heavily on noncomputational techniques such as quantum Fourier transform, quantum teleportation, quantum key distribution, quantum error-correction methods, and quantum communication protocol. The Upcoming experimental quantum calculations, such as the creation of the quantum Internet, will rely heavily on their successful real-world applications. The topic of quantum machine learning includes certain quantum algorithms that have been put into practical use. There is also the distinct field of quantum programming languages, the goal of which is to create functional programming languages, considering the quantum computers. The extensive quantum computing will be discussed further below.

3.5.1 Extensive quantum computing

Classical algorithmic tools to manage massive quantum computations are also an important area of study. The distributed quantum computing mechanism and procedures include quantum error correction in the nodes, protocols for communication between quantum nodes, information transmission between the nodes and the quantum bus, decoding operations, optimization algorithms, and many others. Classical data is expected to be incorporated into the system as a supplement to quantum algorithms and quantum communication protocols. As a result, rather than being exclusively quantum systems, these algorithms and techniques will be combined "quantum-classical" systems, augmenting classical information processing. Please refer to the section titled "Related Work" for further reading suggestions.

The methodologies for molecular energy quantum computing are proposed in [76]. To simulate molecule energies, the model employs the variational quantum eigensolver (VQE) technique. The VQE algorithm approximates the ground state energies of quantum systems by using the computational power of quantum computers to efficiently find values using a classical optimization procedure. The authors conclude that the proposed methodologies allow for better wavefunction optimization and a reduction in the quantum circuit depth that is required to implement the algorithm. The authors of [75] investigated the potential for optimizing hybrid quantum-classical algorithms in practice. Since the state preparation and measurement processes need to be performed several times, the authors examined the necessary number of repetitions that are required for a precise estimation. A subset of hybrid quantum-classical algorithms forms the basis of their analysis. This category is favored because of its usefulness in solving quantum chemistry and combinatorial challenges. Quasi-Newton optimization techniques for use in hybrid algorithms were also examined.

You can find a review of existing quantum algorithms in [67]. The analysis is not concerned with the technical description of the algorithms, but rather with their potential applications. Some potential future uses of quantum algorithms in experimental quantum computing are also discussed. The methodology developed by the authors in [68] simulates the biological actions of people who are subject to natural selection. Scientists observed that the designed evolution of the quantum living units is indicative of the essential properties of life. The authors conclude that these findings have potential application in the development of quantum technologies for the creation of artificial life and embodied evolution. Nonsparse low-rank matrices can be decomposed into singular values using the procedure outlined in [69]. In order to perform matrix exponentiation on a quantum computer, the authors created a quantum algorithm. The authors conclude that the new approaches allow them to locate singular values and the associated singular vectors tenfold faster than is possible using classical algorithms.

3.5.2 Toward a quantum approach to machine learning

The study of these protocols is very important for experimental quantum information processing because of the developing fields of quantum artificial intelligence and quantum machine learning [2, 82]. In [51], the authors investigated how to use superconducting quantum circuits to execute several fundamental protocols of quantum reinforcement learning. The superconducting quantum circuits offer a workable method for bringing quantum computing and quantum information processing into the real world. The authors of this article outline potential setups for proof-of-principle tests with the use of existing superconducting circuit technologies.

In [70] you will find a model that can be used to show how the benefits of quantum computing may be used to machine learning. The writers demonstrated that a five-qubit superconducting processor can solve a learning parity with noise, which is an Oracle-based problem, and implement the solution. The authors provided empirical evidence for the well-established conclusion that the query count for a given oracle is drastically different for classical and quantum algorithms. Based on their analysis, the writers determined that as the mistake rates and the size of the challenge increased, so did the feasible gap. The authors arrive at the same conclusion on the complexity of fault-tolerant designs required for experimental universal quantum computing. Quantum recommendation systems were the subject of research in [71]. The suggested approach, the authors found, efficiently samples from a preference matrix approximation, resulting in good suggestions. Their technique does not necessitate a full matrix reconstruction. The authors conclude that their scheme is the first example of a quantum machine learning algorithm in practice, and that it is the first method for recommendation systems to run in time polylogarithmic in the dimensions of the matrix.

A deep learning framework, considered to be quantum-classical for industrial datasets on near-future devices, was investigated in [72]. The writers created a framework that is quantum-classical, which may be able to handle real-world machine learning datasets that are high-dimensional with continuous variables. The developers of the suggested method used deep learning to create a binary representation of data with few dimensions. The writers conclude that the suggested technique is amenable to training on unsupervised generative models using relatively tiny quantum processors. The authors also suggested doing experimental verification of the concept using a real-world dataset, and provided examples using a quantum annealer to drive home the point. In [73], a method is described using Boltzmann machines (a type of stochastic neural network used in machine learning) for learning thermodynamics. Thermodynamic observables for thermodynamically balanced physical systems can be modeled with the help of the Boltzmann machine, which was introduced in the paper. The writers' tend to train the Boltzmann machine-based unsupervised learning with spin-configured data sets; it is the paper's key contribution.

After training the Boltzmann machine, the authors found that it could be used to produce spin states. Here, the scientists also demonstrated that the machine accurately reproduces physical system observables and that the number of neurons needed for reliable findings grows as the system approaches criticality. The authors of [74] investigated the wide scope of the analysis to include some machine learning uses. In order to solve the issue of employing a quantum annealer for Boltzmann sampling successfully, the authors devised a model. Their method relies on a basic algorithm for estimating the effective temperature of a system. The paper also specified additional algorithmic solutions for the quantum domain and analyzed the effects of effective temperatures on the training of particular Boltzmann machines implemented on quantum hardware [80, 81].

3.6 The final thoughts

In this chapter, we have surveyed some of the most recent writings and lines of inquiry in the developing topic of quantum computing. Here, we have described the parts that make up a quantum computer, talked about what needs to be in place before large-scale quantum computers can be built, and provided the latest discoveries from studies into putting computers, algorithms, and the quantum devices into practice. We also dealt with a few of the open questions that have been plaguing the industry recently. The research results are hopeful, and the tone of the entire chapter is constructive. Quantum information processing, which makes use of the information's underlying qualities, opens up novel possibilities for research in the computer, networking, and communication sciences. When it comes to future computations and communications, the technologies behind quantum computing continue to hold a lot of promise.

References

[1] R. Barends, et al., Logic gates at the surface code threshold: Superconducting qubits poised for fault-tolerant quantum computing, Nature 508, 2014, 500–503.

[2] J. Biamonte, et al., Quantum machine learning, Nature 549, 2017, 195–202.

[3] S. Debnath, et al., Demonstration of a small programmable quantum computer with atomic qubits, Nature 536, 2016, 63–66.

[4] L. DiCarlo, et al., Demonstration of two-qubit algorithms with a superconducting quantum processor, Nature 460, 2009, 240–244.

[5] E. Farhi, J. Goldstone, S. Gutmann, H. Neven, Quantum algorithms for fixed qubit architectures, 2017, arXiv:1703.06199v1.

[6] R. Rivest, A. Shamir, L. Adleman, A method for obtaining digital signatures and public-key cryptosystems, Commun. ACM 21(2), 1978, 120–126.

[7] S. Imre, L. Gyongyosi, Advanced Quantum Communications – An Engineering Approach, Wiley-IEEE Press, 2013.
[8] P. W. Shor, Polynomial-time algorithms for prime factorization and discrete logarithms on a quantum computer, SIAM J. Comput. 26(5), 1997, 1484–1509.
[9] S. Imre, F. Balazs, Quantum Computing and Communications – An Engineering Approach, John Wiley and Sons Ltd, 2005.
[10] R. Van Meter, S. J. Devitt, Local and distributed quantum computation, IEEE Comput. 49(9), 2016, 31–42, arXiv:1605.06951v1.
[11] R. Van Meter, Architecture of a Quantum Multicomputer Optimized for Shor' states Factoring Algorithm, (Ph.D. thesis), Keio University, 2006, arXiv: quant-ph/0607065.
[12] C. H. Bennett, Logical reversibility of computation, IBM J. Res. Dev. 17, 1973, 525–532.
[13] D. P. DiVincenzo, The physical implementation of quantum computation, Fortschr. Phys. 48(9–11), 2000, 771–783.
[14] S. Aaronson, A. Arkhipov, "The computational complexity of linear optics," Proceedings of the Forty-third Annual ACM Symposium on Theory of Computing, STOC '11, ACM, 2011, pp. 333–342, San Jose, California, USA.
[15] S. Aaronson, The learnability of quantum states, Proc. R. Soc. Lond. Ser. A Math. Phys. Eng. Sci. 463 (2088), 2007, 3089–3114.
[16] D. Deutsch, Quantum theory, the Church-Turing principle and the universal quantum computer, Proc. R. Soc. Lond. Ser. A. 400, 1985, 97–117.
[17] R. Feynman, Simulating physics with computers, Internat. J. Theoret. Phys. 21, 1982, 467–488.
[18] W. K. Wootters, W. H. Zurek, A single quantum cannot be cloned, Nature 299(5886), 1982, 802–803.
[19] M. Mosca, Quantum algorithms, 2008, Arxiv preprint arXiv:0808.0369.
[20] A. M. Childs, W. van Dam, Quantum algorithms for algebraic problems, Rev. Modern Phys. 82, 2010, 1–52.
[21] D. Bacon, W. van Dam, Recent progress in quantum algorithms, Commun. ACM 53(2), 2010, 84–93.
[22] R. Jozsa, N. Linden, On the role of entanglement in quantum-computational speed-up, Proc. R. Soc. Lond. Ser. A Math. Phys. Eng. Sci. 459 (2036), (2003) 2011–2032.
[23] F. Verstraete, V. Murg, J. I. Cirac, Matrix product states, projected entangled pair states and variational renormalization group methods for quantum spin systems, Adv. Phys. 57(2), 2008, 143–224.
[24] D. Shepherd, M. J. Bremner, Temporally unstructured quantum computation, Proc. R. Soc. Lond. Ser. A Math. Phys. Eng. Sci. 465(2105), 2009, 1413–1439.
[25] A. Broadbent, J. Fitzsimons, E. Kashefi, "Universal blind quantum computation," 2009 50th Annual IEEE Symposium on Foundations of Computer Science, 2009, pp. 517–526.
[26] J. Combes, et al., In-situ characterization of quantum devices with error correction, 2014, arXiv:1405.5656.
[27] A. D. Corcoles, et al., Demonstration of a quantum error detection code using a square lattice of four superconducting qubits, Nat. Commun. 6, 2015, 6979.
[28] S. Bravyi, D. Gosset, R. Koenig, Quantum advantage with shallow circuits, 2017, arXiv:1704.00690.
[29] M. J. Bremner, A. Montanaro, D. J. Shepherd, Achieving quantum supremacy with sparse and noisy commuting quantum computations, Quantum 1(1), 2017, 8.
[30] E. Zahedinejad, J. Ghosh, B. C. Sanders, Designing high-fidelity single-shot three-qubit gates: A machine-learning approach, Phys. Rev. Appl. 6, 2016, 054005.
[31] E. Dennis, A. Kitaev, A. Landahl, J. Preskill, Topological quantum memory, J. Math. Phys. 43, 2002, 4452–4505.
[32] V. Giovannetti, S. Lloyd, L. Maccone, Quantum random access memory, Phys. Rev. Lett. 100(16), 2008, 160501.

[33] P. C. Maurer, et al., Room-temperature quantum bit memory exceeding one second, Science 336 (6086) (2012), 1283–1286.

[34] C. Monroe, R. Raussendorf, A. Ruthven, K. R. Brown, P. Maunz, L. M. Duan, J. Kim, Large-scale modular quantum-computer architecture with atomic memory and photonic interconnects, Phys. Rev. A. 89, 2014, 022317.

[35] S. Santra, O. Shehab, R. Balu, Exponential capacity of associative memories under quantum annealing recall, 2016, arXiv:1602.

[36] M. August, X. Ni, Using recurrent neural networks to optimize dynamical decoupling for quantum memory, Phys. Rev. A. 95, 2017.

[37] P. Pfeiffer, I. L. Egusquiza, M. Di Ventra, M. Sanz, E. Solano, Quantum memristors, Sci. Rep. 6, 2016, 29507.

[38] J. Salmilehto, F. Deppe, M. Di Ventra, M. Sanz, E. Solano, Quantum memristors with superconducting circuits, Sci. Rep. 7, 2017, 42044.

[39] S. N. Shevchenko, Y. V. Pershin, F. Nori, Qubit-based memcapacitors and meminductors, Phys. Rev. Appl. 6, 2016, 014006.

[40] H. Buch, S. Mahapatra, R. Rahman, A. Morello, M. Y. Simmons, Spin readout and addressability of phosphorus-donor clusters in silicon, Nat. Commun. 4, 2013, 06.

[41] G. Gatti, D. Barberena, M. Sanz, E. Solano, Protected state transfer via an approximate quantum adder, Sci. Rep. 7, 2017, 6964.

[42] L. Lamata, U. Alvarez-Rodriguez, J. D. Martin-Guerrero, M. Sanz, E. Solano, Quantum autoencoders via quantum adders with genetic algorithms, 2017, arXiv:1709.07409v1.

[43] R. Li, U. Alvarez-Rodriguez, L. Lamata, E. Solano, Approximate quantum adders with genetic algorithms: An IBM quantum experience, Quantum Meas. Quantum Metrol. 4, 2017, 1.

[44] B. Vermersch, P. O. Guimond, H. Pichler, P. Zoller, Quantum state transfer via noisy photonic and phononic waveguides, 2016, arXiv:1611.10240.

[45] T. D. Nguyen, R. Van Meter, A space-efficient design for reversible floating point adder in quantum computing, 2013, arXiv:1306.3760.

[46] R. Van Meter, Quantum computing's classical problem, classical computing's quantum problem, Found. Phys. 44(8), 2014, 819–828.

[47] K. Li, et al., Experimentally superposing two pure states with partial prior knowledge, Phys. Rev. A. 95, 2017, 022334.

[48] S. Dogra, G. Thomas, S. Ghosh, D. Suter, Superposing pure quantum states with partial prior information, 2017, arXiv:1702.02418.

[49] U. Alvarez-Rodriguez, et al., Advanced-retarded differential equations in quantum photonic systems, Sci. Rep. 7, 2017, 42933.

[50] S. Mavadia, V. Frey, J. Sastrawan, S. Dona, M. J. Biercuk, Prediction and real-time compensation of qubit decoherence via machine learning, Nat. Commun. 8, 2017, 14106.

[51] L. Lamata, Basic protocols in quantum reinforcement learning with superconducting circuits, Sci. Rep. (ISSN: 2045-2322) 7(1), 2017, 1609.

[52] A. Perdomo-Ortiz, B. O'Gorman, J. Fluegemann, R. Biswas, V. N. Smelyanskiy, Determination and correction of persistent biases in quantum annealers, Sci. Rep. 6, 2016, 18628.

[53] S. Slussarenko, M. M. Weston, J. G. Li, N. Campbell, H. M. Wiseman, G. J. Pryde, Quantum state discrimination using the minimum average number of copies, Phys. Rev. Lett. 118, 2017, 030502.

[54] M. Oszmaniec, A. Grudka, M. Horodecki, A. Wojcik, Creating a superposition of unknown quantum states, Phys. Rev. Lett. 116, 2016, 110403.

[55] X. M. Hu, et al., Experimental creation of superposition of unknown photonic quantum states, Phys. Rev. A. 94, 2016, 033844.

[56] S. Sami, I. Chakrabarty, A note on superposition of two unknown states using deutsch CTC model, Modern Phys. Lett. A. 31, 2016, 1650170.

[57] S. Nagayama, A. G. Fowler, D. Horsman, S. J. Devitt, R. Van Meter, Surface code error correction on a defective lattice, New J. Phys. 2017.
[58] S. Nagayama, T. Satoh, R. Van Meter, State injection lattice surgery and dense packing of the deformation-based surface code, Phys. Rev. A. 95, 2017, 012321.
[59] A. B. Grilo, I. Kerenidis, Learning with errors is easy with quantum samples, 2017, arXiv:1702.08255.
[60] J. O'Gorman, N. H. Nickerson, P. Ross, J. J. L. Morton, S. C. Benjamin, A silicon-based surface code quantum computer, NPJ Quantum Inf. 2, 2016, 16014.
[61] S. Boixo, et al., Characterizing quantum supremacy in near-term devices, 2016, arXiv:1608.00263.
[62] M. P. V. Stenberg, O. Kohn, F. K. Wilhelm, Characterization of decohering quantum systems: Machine learning approach, Phys. Rev. A. 93, 2016, 012122.
[63] J. Romero, J. Olson, A. Aspuru-Guzik, Quantum autoencoders for efficient compression of quantum data, 2016, arXiv:1612.02806.
[64] E. Farhi, J. Goldstone, S. Gutmann, A quantum approximate optimization algorithm, 2014, arXiv:1411.4028.
[65] E. Farhi, J. Goldstone, S. Gutmann, A quantum approximate optimization algorithm applied to a bounded occurrence constraint problem, 2014, arXiv: 1412.6062.
[66] E. Farhi, A. W. Harrow, Quantum supremacy through the quantum approximate optimization algorithm, 2016, arxiv:1602.07674.
[67] A. Montanaro, Quantum algorithms: An overview, NPJ Quantum Inf. 2, 2016, 15023.
[68] U. Alvarez-Rodriguez, M. Sanz, L. Lamata, E. Solano, Artificial life in quantum technologies, Sci. Rep. 6, 2016, 20956.
[69] P. Rebentrost, A. Steffens, S. Lloyd, Quantum singular value decomposition of non-sparse low-rank matrices, 2016, arXiv:1607.05404.
[70] D. Riste, M. P. da Silva, C. A. Ryan, A. W. Cross, A. D. Corcoles, J. A. Smolin, J. M. Gambetta, J. M. Chow, B. R. Johnson, Demonstration of quantum advantage in machine learning, NPJ Quantum Inf. 3(1), 2017, 16.
[71] A. Kerenidis, A. Prakash, Quantum recommendation systems, in: Innovations in Theoretical Computer Science, ITCS'17, 2017.
[72] M. Benedetti, J. Realpe-Gomez, A. Perdomo-Ortiz, Quantum-assisted Helmholtz machines: A quantum–classical deep learning framework for industrial datasets in near-term devices, 2017, arXiv:1708.09784.
[73] G. Torlai, R. G. Melko, Learning thermodynamics with Boltzmann machines, Phys. Rev. B 94, 2016, 165134. arXiv:1606.02718.
[74] M. Benedetti, et al., Estimation of effective temperatures in quantum annealers for sampling applications: A case study with possible applications in deep learning, Phys. Rev. A 94(2), 2016, 022308.
[75] G. G. Guerreschi, M. Smelyanskiy, Practical optimization for hybrid quantum–classical algorithms, 2017, arXiv:1701.01450.
[76] J. Romero, et al., Strategies for quantum computing molecular energies using the unitary coupled cluster ansatz, 2017, arXiv:1701.02691.
[77] A. P. Lund, M. J. Bremner, T. C. Ralph, Quantum sampling problems, boson sampling and quantum supremacy, NPJ Quantum Inf. 3(1), 2017, 15.
[78] S. M. Albrecht, A. P. Higginbotham, M. Madsen, F. Kuemmeth, T. S. Jespersen, J. Nygard, P. Krogstrup, C. M. Marcus, Exponential protection of zero modes in Majorana islands, Nature 531, 2016, 206–209.
[79] M. Ahsan, R. Van Meter, J. Kim, Designing a million-qubit quantum computer using resource performance simulator, 2015, arXiv:1512.00796.

[80] G. Sasubilli, A. Kumar, "Machine Learning and Big Data Implementation on Health Care data," 4th International Conference on Intelligent Computing and Control Systems (ICICCS), 2020, pp. 859–864, doi: 10.1109/ICICCS48265.2020.9120906.
[81] S. M. Sasubilli, A. Kumar, V. Dutt, "Improving Health Care by Help of Internet of Things and Bigdata Analytics and Cloud Computing," 2020 International Conference on Advances in Computing and Communication Engineering (ICACCE), 2020, pp. 1–4, doi: 10.1109/ICACCE49060.2020.9155042.
[82] P. K. Kotturu, A. Kumar, "Comparative Study on Machine Learning models for Early Diagnose of Alzheimer's Disease: Multi Correlation Method," 2020 5th International Conference on Communication and Electronics Systems (ICCES), 2020, pp. 778–783, doi: 10.1109/ICCES48766.2020.9137872.

Sourabh Sharma, Poonam Chaudhary

Chapter 4
Machine learning and deep learning

Abstract: From the evolution of the human race, the stories that have engulfed the curiosity of all the civilizations consist of nonliving things acting like human beings or, in simple terms, equating humans in terms of wisdom and adapting human behavior. Stories were plotted around objects like talking trees, dolls, and statues that could talk and behave like human beings, who actually accompanied or even guided the living entity to success, leading to a happy ending.

This is how the early concept of Artificial Intelligence (AI) came into imagination; it was later extended by Alan Turing and, most recently, the Hollywood movie character Iron Man introduced to the kids and youngsters through the AI bot Zarvis, who accompanies the lead character to gain superpowers through technology.

AI has now become an integral part of the world, knowingly or unknowingly, it has made huge advancements especially through the evolution of machine learning (ML) and deep learning (DL) algorithms and rich computation resources. The music that you listen to on YouTube is being used to train an ML model, which further predicts music for you; social media pages suggest different people and pages according to your likes and dislikes; self-driving cars, your shopping apps, and many other platforms are using AI to predict consumer behavior, plan efficiently, and thus, increase their revenue.

AI is a fast moving technology with impacts and implications on both individuals and society. With a gentle introduction to history and some important algorithms used in AI, from ML to DL, we will be covering the major aspects of AI. The chapter takes a balanced look at AI, its impact on the existing world, as well as its growing potential in future.

4.1 Introduction

Artificial means "man-made" and *Intelligence* means "thinking power," so we can say Artificial Intelligence (AI) means "a man-made thinking power." It denotes intelligence demonstrated by machines, as natural intelligence is to humans or animals. AI exists when a machine can perform human tasks like learning, solving problems, reasoning,

Sourabh Sharma, Department of CSE, The NorthCap University, Haryana, India,
e-mail: sourabh.714sharma@gmail.com
Poonam Chaudhary, Department of CSE, The NorthCap University, Haryana, India,
e-mail: poonamchaudhary@ncuindia.edu

https://doi.org/10.1515/9783110791402-004

and many more. In the emerging era of technologies, AI is one of the fastest-growing fields, transforming how we work, live, and play. It is all set to create a new revolution in the world by making intelligent machines. AI is all around; you can see it in almost each and every domain. It has become one of the most fascinating fields in Computer science that gets machines to work as a human. With AI, you do not need to preprogram a machine to do work; instead, you create a machine with programmed algorithms that can work with its own intelligence and that is the beauty of it.

4.1.1 History of AI

AI is not a newly coined term but is much older than you could have imagined. The given timeline demonstrates the evolution of AI with time. The history of AI has had cycles of success, misplaced optimism, and resulting cutbacks in enthusiasm and funding [1]. There have also been cycles of introducing new creative approaches and systematically refining new ones as shown in Figure 4.1.

Artificial Intelligence's Inception:
- 1943: The first **artificial neurons** model was proposed by Warren McCulloch and Walter Pits in which each neuron is characterized as being "on" or "off," like a switch, in response to stimulation by a sufficient number of neighboring neurons.
- 1950: Alan Turing proposed the **Turing Test** to check the machine's ability to exhibit intelligent behavior comparable to human intelligence and articulated a complete vision of AI in his article "Computing Machinery and Intelligence."
- 1951: First neural network computer named **SNARC** was built by Marvin Minsky and Dean Edmonds.

The Birth of the Artificial Intelligence:
- 1955: The first artificial intelligence program named **Logic Theorist** was created by Allen Newell and Herbert A. Simon that proved 38 out of 52 mathematical theorems.
- 1956: **Artificial Intelligence** was first adopted at the Dartmouth conference by American computer scientist John McCarthy and was made a separate field.

The Growth of Artificial Intelligence:
- 1959: **Geometry Theorem Prover,** made by Herbert Gelenter was able to prove many mathematical theorems.
- 1966: The first chatbot named **ELIZA** was created by Joseph Weizenbaum.
- 1972: The first intelligent humanoid robot named **WABOT-1** was built in Japan.

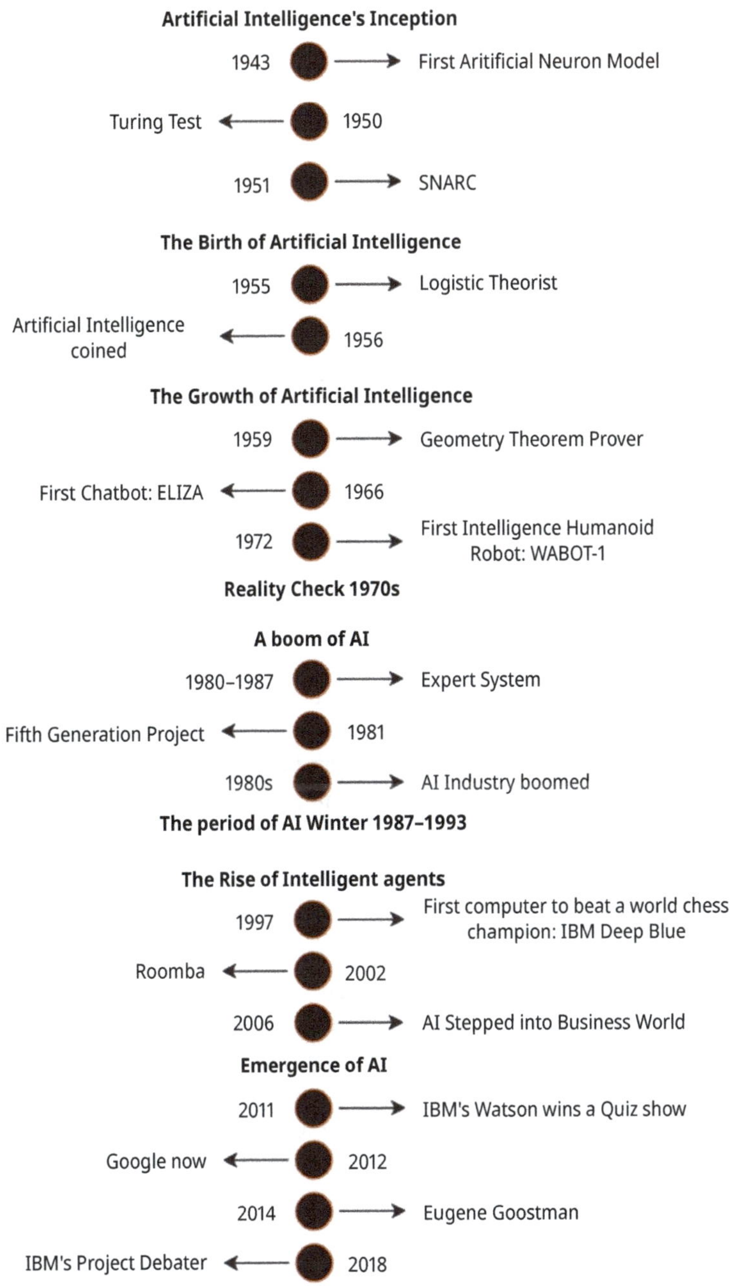

Figure 4.1: Timeline of AI evolution.

Reality Check:

- 1970s: During this time computer scientists dealt with a huge shortage of funding for AI research from the government, as the machines were not giving desirable and expected results and the money spent was huge.

A boom of AI:

- 1980–1987: **Expert System** brought back AI. These systems were programmed to imitate the decision-making ability of a human expert.
- 1981: **Fifth Generation** project was announced, a 10-year plan to build intelligent computers running for a prolonged period.
- The AI industry overall boomed from a few millions dollars to billions of dollars.

The period of AI winter:

- 1987–1993: Funding and investment again stopped for the AI research because of high cost and not getting efficient results, and many companies suffered as they failed to deliver on extravagant promises.

The rise of Intelligent agents:

- 1997: **IBM Deep Blue** becomes the first computer to beat the world chess champion Gary Kasparov.
- 2002: AI made its entry to homes in the form of **Roomba**, a vacuum cleaner.
- 2006: AI stepped into the business world when companies like Facebook, Netflix, and Twitter started using it.

Emergence of AI:

- 2011: IBM's **Watson** won a quiz show named Jeopardy, where it solved the complex questions as well as riddles. It proved that natural language could be understood and tricky questions can be solved easily by a machine.
- 2012: **Google Now**, an android app was launched by Google to provide information to the user as a prediction.
- 2014: Chatbot **Eugene Goostman** won a competition of the Turing test.
- 2018: IBM's **Project Debater** debated with two master debaters and performed unbelievably well.

AI has advanced more rapidly in the past decade, with the coming of DL and Big data and Data science. The subfields of AI have become more integrated and AI has found common ground with other disciplines. Companies are working with AI to create amazing devices.

4.1.2 Advantages and disadvantages of AI

The advantages of AI [2] are countless, but to name a few:

1. Less error, high accuracy: Humans are known to be statues of mistakes, but this can be avoided in machines if they are properly programmed. As the decisions are made on the previously gathered information applying a certain set of algorithms, errors are reduced and chances of reaching higher accuracy increase.
2. Useful in risky areas: This is the major advantage the AI-developed robots are providing. We are able to reach Mars, even the deepest parts of oceans are discovered, and robots are helping humans reach where they are not capable of or are very risky zones.
3. Daily applications: Siri, Cortana, and OK google are used very frequently and carry out many of the command tasks, be it making a call, sending a mail, or many more.
4. Digital assistants: You must have seen chatbots, the best example you see around on many websites. It can interact with humans and resolve their problems without letting us know who you are talking to – a machine or a person. It saves the need for a human resource for such tasks.
5. Easy repetitive tasks: You can easily automate the daily repetitive and boring tasks for humans and keep them free for more creative tasks, as machines are still not capable of using their creativity.
6. Availability 24 × 7: Humans, on average, can work for 6 h excluding the breaks. They even need proper sleep and refreshment time to get their brains to function in a healthy way, but by using AI, you can make machines work 24 × 7 tirelessly, without any breaks. This has been a plus point in helpline centers that address people's queries, 24 × 7.

With pros come the cons, the same goes with AI. Despite making our lives easy, it is even making us get second thoughts about its use due to the following reasons:

1. Emotionless: No doubt machines can work very efficiently, but they cannot be a full replacement for humans. They cannot work when it comes to team management tasks, brainstorming sessions, decision-making, and many more.
2. High cost: To be in touch with the AI updates, you have to get the hardware and software updated from time to time for meeting the requirements; this increases the maintenance and repair costs too.
3. Making humans dependent: AI is making humans so much dependent on them by automating even the simpler tasks, thereby making them lazy and deteriorating their health.
4. Lack of mental ability: Machines can only perform the work they are programmed to do; any invalid input can lead to its failure or give us undesired outputs.

5. Unemployment: Every organization is looking to automate its work, so that its labor cost is reduced and AI robotics are helping it automate the majority of the work that does not involve much of the mind.

Every new technology will bring both advantages and disadvantages with itself, but humans have to take care of them and make sure to exploit only its positive side to create a better place to live in.

4.1.3 Symbolic AI and computational AI

Symbolic AI, also called Classical AI or good old-fashioned AI, is the branch of AI that approaches to train your machines just the way our human brain learns [3, 12–13]. It represents problems using symbols and then uses the logic to search for the solutions. This practice had shown impressive results in the early decades of AI research, but has fallen wayside, today, with the uprising of neural networks that involves self learning rules established from the correlation between the inputs and outputs.

4.1.3.1 Working

Symbolic AI takes strings of characters that represent real world entities and then these strings are stored manually or incrementally in any appropriate data structure [14, 15]. They are made available to the interfacing human beings or machines as and when requested and used to make intelligent decisions and conclusions, based on the memorized facts and rules put together by propositional logic or first-order predicate calculus techniques. This does not require any training or massive amounts of data or guesswork. The best part is that it can explain its decisions by showing which parts were evaluated as true or false, unlike the black box of machine learning.

4.1.3.2 No longer popular

Symbolic AI is useful for settings where the rules are simple and straightforward, and users can easily obtain input and transform them into symbols but fail to deal with the messy data that exists in the real world. Thus, this technology has had minimal success, and neural networks that imitate human intelligence have mostly taken over the field [4].

Though several efforts are being made to combine neural networks and symbolic AI, many leading scientists still believe that symbolic AI will continue to remain a very important part of AI. A hybrid AI system Neuro-Symbolic Concept Learner is developed by MIT_IBM Watson AI Lab that uses both rule-based programs and neural networks to solve visual question-answering problems. This model can learn new tasks with fewer data, is explainable, and does not struggle with analyzing the content of images.

Computational Intelligence (CI), also called soft computing, is the ability of a computer to learn a specific task from data or experimental observations. It provides the solution to solve complex real-world problems that cannot be solved using traditional or mathematical modeling with the methods that are close to the human's way of reasoning, i.e., it uses incomplete and inexact knowledge to produce control actions in an adaptive way [5]. CI makes use of a combination of five main complementary techniques: The *fuzzy logic* that makes the computer understand the natural language;, *artificial neural networks* that enables the system to learn experiential data by operating like the biological one; *Evolutionary computing* that is based on the process of natural selection, learning theory, and probabilistic methods, which helps in dealing with uncertainty and imprecision; *swarm intelligence* and *artificial immune systems*, which are biologically inspired algorithms that can be seen as a part of evolutionary computation, data mining, natural language processing, image processing, and AI. CI and AI scout for similar goals to reach general intelligence, i.e., the intelligence of a machine that could perform any intellectual task like human beings. But there is a clear distinction between them. CI is based on fuzzy logic that is much closer to the way the human brain works, by aggregating data to partial truths – one of the main exclusive aspects of CI.

Its main application can be seen in engineering, computer science, data analysis, and biomedicine.

4.2 Algorithms in AI

The advancements of computing infrastructure allowed the easy use of AI algorithms.

Though AI algorithms were available mathematically [6], their computation cost was something that hindered the development of AI computation.

4.2.1 AI has found a lot of application in various fields

1. Social Media

- Identifying fraud accounts and malicious activities:
  Social Media platforms have provided human society a platform to express their thoughts and eliminate the barriers of communication [7]. They have also acted as wheels of revolution for democratic ideas and voices. However, fake accounts and activities with malicious intent have proved to be serious challenges; for instance, social media platforms have become a major source of false information that result in violence and riots. AI-enabled tools are being used to detect and eradicate fraud accounts and prevent malicious activities.

- Recommendation
  AI-based recommendation systems have added a shot in the arm of business intelligence forming smart environments that govern the whole e-commerce system [8]. Amazon and Flipkart have developed their AI-based recommendation systems to better address the needs of customers and serve them better.
- Translation
  Social media platforms such as Meta and Twitter have users from across the globe. This, in turn, requires in the use of different languages; so, AI-based algorithms are being used by social media platforms to translate the posts, messages, and comments.
- Suicide prevention
  Social Media has become a platform where people actively share day-to-day activities and feelings. This involves instances when people commit suicides either declaring it on their social media platforms or going live. Social Media giants such as Meta are actively working on AI models to identify such instances and prevent them by informing authorities and stakeholders on time.

2. Chat bots

Chat bots have made the web world really interactive. They not only allow the websites to answer user specific queries on time and without the aid of human resources, saving time, effort and money, but also enhance user experience on the website. AI-based smart chatbots with IoT have made life easy and they are becoming helpful for people with disabilities.

3. Finance

AI-based algorithms and tools are being used by organizations to make better decisions by assessing more factors to identify and calculate financial conditions of borrowers to make responsible lending. AI-based models are also being used to identify banking frauds such as fraud credit /debit card transactions and identifying and reducing risks in business.

4. Gaming

Gaming industry is also using AI models to better engage the user and provide him with better approximations of vectors in 3D space. AI-based behavior in games generally includes responsive and intelligent activities exhibited by characters that are not players.

This results in better gaming experience and more engagements from the player side.

5. Automobiles

Automobile sector is one prominent sector that is leveraging AI in multiple stages, from manufacturing to the final product. AI-based models are being used to generate better designs and run simulations. They are also used to enhance user experience; smart AI assistants in cars help the driver connect with the car and use the features in an efficient way . Smart AI systems in the cars learn about the driver – his destinations points, favorable car temperatures, and playlists and provide results accordingly. For instance, they suggest routes with lesser traffic and taxes, etc., setting favorable temperature and playing music that are more aligned to the driver's preferred tastes Automobile buying and selling platforms are also leveraging AI to build smart recommendation systems to better serve the customer.

6. Agriculture

AI has found numerous applications in the field of agriculture. There are mobile-based applications that leverage AI algorithms to suggest best practices for farming, based on geographical locations and season. They also provide weather forecasts. AI-based tools are being used for crop and soil monitoring and crop yield predictions, along with input quantity and prices.

7. Weather

Weather forecasting is crucial for every sector of the economy. It is also crucial for mitigating the risks caused by natural calamities such as cyclones and floods. Even we need it for our day-to-day life for scheduling our business trips, holidays, etc.

Traditional forecasting models are very heavy – there is tedious workload from collecting the data, integrating with data from other sources, and then simulating it. This requires higher storage and computation resources, such as supercomputers. Still, it takes a lot of time to process it, along with IoT of power and proportionately cost also tends to increase.

Researchers are now working on DL-based models to better learn the patterns, which is something that AI models are highly specialized in. Results have shown that AI-based models have shown significant improvement in terms of accuracy and power efficiency.

8. Robotics

Robots have been in existence for more than a century, but with the introduction of AI, it has provided robots an ability to think on their own. It has started interacting, performing tasks that would otherwise require human intervention. This has been made possible through AI technologies, such as natural language processing, computer vision, advanced motion control, etc.

9. Navigations

Navigation was never as efficient as it is now in this modern world. Cab and logistics companies are actively utilizing AI for lane detection, road and traffic analysis, and optimum route selection.

Researchers are also working on developing AI models that provide optimum routes with the most efficient time, keeping in mind the safety indicators.

4.3 Machine learning

ML is a subset of AI, where an algorithm learns from preexisting/historical data to create a program that predicts the output of new/ unseen data [9]. The majority of ML algorithms may be categorized into three: supervised learning, unsupervised learning, and reinforcement learning. Figure 4.2 represents the algorithms and techniques for machine learning.

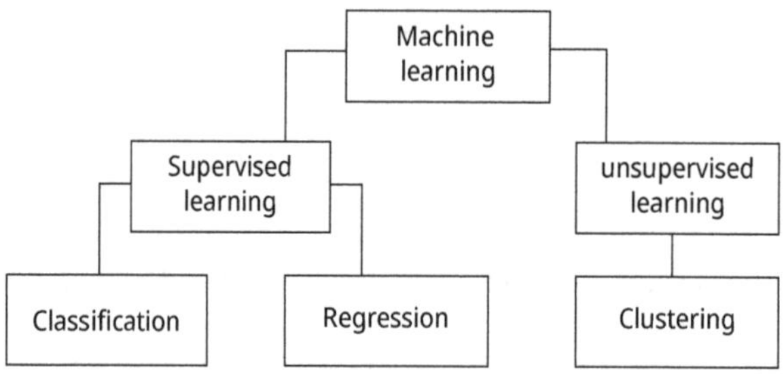

Figure 4.2: Algorithms and techniques for machine learning.

4.3.1 Supervised machine learning

In supervised learning (SL), a variety of techniques are employed, including neural networks, support vector machines, decision trees, and Bayesian networks [2–4]. A key reason to consider the SL a precision model is that the model predicts the value of one dataset based on the values of the other datasets. Based on pattern recognition, predictive models prove to be phenomenal in dealing with uncertainty, which is a big problem for off-road traffic. Algorithms are generally applied to produce models using input data and known results of that data, i.e., the outputs, and to train that model to make practical predictions regarding the response to new data [1]. An SL model is shown in Figure 4.3 in its exact working state.

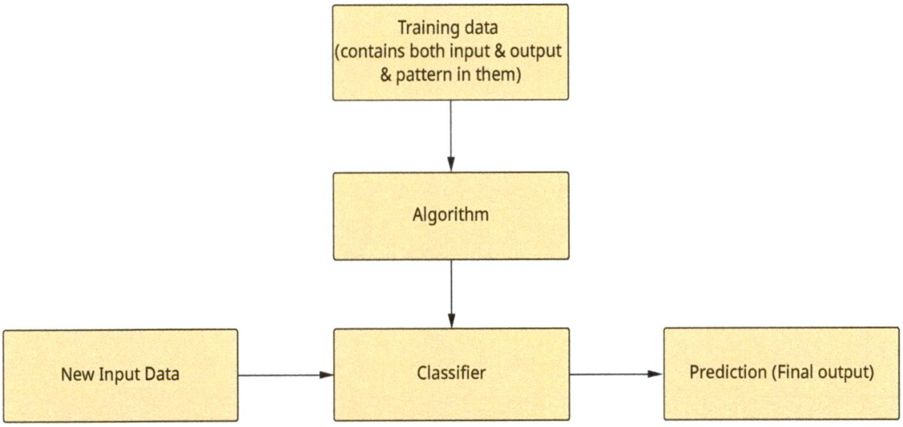

Figure 4.3: Model of supervised learning.

4.3.2 Using unsupervised machine learning

The most essential quality of Unsupervised Learning [6] is that it will master and learn linked to the data on its own, i.e., learning from the environment. Three main tasks involve using unsupervised learning models: clustering, association, and reduction of dimensionality. Clustering, which is predominantly employed in the electrical signal's framework, and density-based k-means algorithms are among the most commonly used. Clustering, which employs a metric or attribute to show similarity and fabricates dimensionality reduction to observe and evaluate high-dimensional data, is one of the most widely used approaches. This can also be referred to as feature selection [16].

4.3.3 Reinforcement learning

Through frequent contact with a dynamic environment that either rewards or penalizes the agent's action, Reinforcement Learning (RL) [11] is an ML paradigm, where an agent learns the best course of action for a particular task. However, the recent success of DL algorithms has rekindled interest in the topic of RL among scientists, who are now effectively using this to solve extremely difficult problems that were previously thought to be intractable. The rise of AI, which may soon surpass human intelligence, has garnered attention globally as a result of events like artificial agents like AlphaGo defeating world champion Lee Sedol or IBM Watson winning the game of Jeopardy.

Building such intelligent systems that can learn from their experiences over time requires the use of the RL paradigm. Robotics, healthcare [17, 18], recommender systems, data centers, smart grids, financial markets, and transportation are all areas where reinforcement algorithms are being used more and more, these days.

4.3.4 Deep learning

Neural Network is a subset of ML that aims to mimic the behavior that a human brain shows during learning. It uses a network of connected nodes called Neurons that perform some computations on received information and pass on their outputs to other neurons as inputs. Artificial Neural Network is the oldest type of Neural Network, where one or more fully connected hidden layers are present between the input and the output [8]. These layers are responsible for learning from input data, using back propagation techniques.

4.4 Humans and AI

AI models have changed the very dynamics of the virtual world– on the one hand it has proved really handy and it has acted as a catalyst in the field of industry automation; on the other hand, it has sent turbulence across the sectors that it might reduce employment opportunities replacing humans. Furthermore, AI models are not fully perfect; there are always chances of errors from highly trained models, because machines lack the emotional component and ethical wisdom that humans possess [9].

To clear these misconceptions, we need to understand that AI models are just an automation asset to utilize human resources more efficiently, so that we can use them on more sophisticated tasks rather than underutilizing them on repetitive tasks. This would enable human society to solve more complex problems that our society and world are coping with and which require more human attention and wisdom.

Secondly, we cannot allow AI models to work independently, as they lack emotional and ethical wisdom that only humans possess. Thus, we need smart systems that have human inputs at certain points; and specially, there should be specific checkpoints where functioning of models should be cross-checked by a team of humans to ensure sustainability of AI models to ensure they do not cause any harm.

For instance, a team of sustainers ensures the ethical dilemma of the data that is being fed to the machine and another team of sustainers anticipates safety and harm during the functioning. We will read more about this in ethics and AI. Concluding the discussion, AI is not just a standalone making decisions themed from some fiction movie, instead, it is a collaborative concept where humans and machines assist each other to save resources and time and solve complex social, health, security, and economic problems more efficiently.

4.5 Ethics and AI

AI has led a revolution in the world of computing. It has enabled computer tasks that could have been only imagined or written in comics [10]. The emergence of AI technology has resulted in saving a lot of time and computation power. Combining all these factors along with higher different evaluation metrics achieved by the SOTA (state-of-the-art) models have made the computer systems smarter and reduced intervention of humans in those tasks.

These achievements are really astonishing, but an ethical code of conduct is the most essential pillar of any technology. At this very point, ethics of AI comes into the picture. Over a period of time, multiple stakeholders from various backgrounds have raised legal and social issues pertaining to AI that might result in fatal undesirable outcomes. AI models, like other statistics models, are prone to biases – they might be discriminating in nature, they can be manipulative to change our behavior, and might compromise our security.

AI ethics can be described as systemizing, defending, and recommending concepts of right and wrong conduct to foster responsible use of AI.

AI ethics can be classified into three subfields:

- Meta Ethics: It is related to the study of meaning of ethical concepts, existence of ontology, and epistemology.
- Normative Ethics: This field deals with the use of conceptual knowledge in practice for determining a moral philosophy.
- Applied Ethics: This field concerns the action of an ideal person acting in situations of ethical dilemma.

These three subfields empower an AI practitioner with the wisdom to design better and more efficient models, which are in compliance with the ethical and moral values of society.

4.6 Conclusion

ML and DL algorithms have evolved over the years from conceptual algorithms enclosed on pieces of paper to real-world applications helping the world achieve remarkable results. With the introduction of quantum computing hardware, which relies on laws of physics from quantum theory, models trained on these hardware are supposed to outperform the models trained on existing computing resources. However, along with the growth of technology, wise and judicious use for the betterment of society is an equally important facet in the world of emerging technologies.

References

[1] L. Fanti, D. Guarascio, M. Moggi, From Heron of Alexandria to Amazon's Alexa: A stylized history of AI and its impact on business models, organization and work. J. Ind. Bus. Econ., 49(3), 2022, 409–440.

[2] L. Diop, S. Samadianfard, Annual rainfall forecasting using hybrid artificial intelligence model: Integration of multilayer perceptron with whale optimization algorithm, Water Resour. Manag. https://doi.org/10.1007/s11269-019-02473-8,2020.

[3] H. Wang, H. Zhang, K. Wang, C. Zhang, C. Yin, X. Kang, Off-road path planning based on improved Ant Colony Algorithm, Wireless Pers. Commun., 2018, doi: 10.1007/s11277-017-5229-5.

[4] W. Aidan, A brief introduction to supervised learning, Towards Data Sci. 2019.

[5] P. Kaur, S. Harnal, R. Tiwari, S. Upadhyay, S. Bhatia, A. Mashat, A. M. Alabdali, Recognition of leaf disease using hybrid convolutional neural network by applying feature reduction, Sensors. 22(2), 2022, 575.

[6] J. E. Van Engelen, H. H. Hoos, A survey on semi-supervised learning, Mach. Learn. 109(2), 2019, 373–440.

[7] M. Abouyoussef, S. Bhatia, P. Chaudhary, S. Sharma, M. Ismail, Blockchain-enabled online diagnostic platform of suspected patients of COVID-19 like pandemics, IEEE Internet Things Mag. 4 (4), 2021, 94–99.

[8] P. Chaudhary, V. Kalra, S. Sharma, "A Hybrid Machine Learning Approach for Customer Segmentation Uszing RFM Analysis," International Conference on Artificial Intelligence and Sustainable Engineering, 2022, pp. 87–100. Springer, Singapore.

[9] N. Yildirim, A. Kass, T. Tung, C. Upton, D. Costello, R. Giusti, S. Lacin, S. Lovic, J. M. O'Neill, R. O. R. Meehan, E. Ó Loideáin, "How Experienced Designers of Enterprise Applications Engage AI as a Design Material," CHI Conference on Human Factors in Computing Systems, 2022, April, pp. 1–13.

[10] L. Zhu, X. Xu, Q. Lu, G. Governatori, J. Whittle, AI and Ethics – Operationalizing Responsible AI, in: Humanity Driven AI, Springer, Cham, 2022, pp. 15–33.

[11] S. Safdar, M. Rizwan, T. R. Gadekallu, A. R. Javed, M. K. I. Rahmani, K. Jawad, S. Bhatia, Bio-imaging-based machine learning algorithm for breast cancer detection, Diagnostics. 12(5), 2022, 1134.

[12] V. Van Roy, D. Vertesy, G. Damioli, AI and robotics innovation. Handbook of labor, human resources and population economics, 2020, pp. 1–35.

[13] I. H. Sarker, Ai-based modeling: Techniques, applications and research issues towards automation, intelligent and smart systems, SN Comput. Sci. 3(2), 2022, 1–20.

[14] S. Bhatia., A. K. Dubey, R. Chhikara, P. Chaudhary, A. Kumar, Intelligent Healthcare, Springer International Publishing, New York, US, 2021.

[15] G. Alam, I. Ihsanullah, M. Naushad, M. Sillanpää, Applications of artificial intelligence in water treatment for optimization and automation of adsorption processes: Recent advances and prospects, Chem. Eng. J. 427, 2022, 130011.

[16] S. R. Swarna, A. Kumar, P. Dixit, T. V. M. Sairam, "Parkinson's Disease Prediction using Adaptive Quantum Computing," 2021 Third International Conference on Intelligent Communication Technologies and Virtual Mobile Networks (ICICV), 2021, pp. 1396–1401, doi: 10.1109/ICICV50876.2021.9388628.

[17] G. Sasubilli, A. Kumar, "Machine Learning and Big Data Implementation on Health Care data," 4th International Conference on Intelligent Computing and Control Systems (ICICCS), 2020, pp. 859–864, doi: 10.1109/ICICCS48265.2020.9120906.

[18] S. M. Sasubilli, A. Kumar, V. Dutt, "Improving Health Care by Help of Internet of Things and Bigdata Analytics and Cloud Computing," 2020 International Conference on Advances in Computing and Communication Engineering (ICACCE), 2020, pp. 1–4, doi: 10.1109/ICACCE49060.2020.9155042.

Stewart Kirubakaran S, R. Senthil Kumar, Anusha Bamini A. M., Chitra R.

Chapter 5 From evolution to revolution: the contemporary development of quantum computing

Abstract: It is likely that you were familiar with quantum computing before beginning to read this book chapter. The fundamental building blocks of quantum computing are derived from the mind-boggling scientific field of quantum physics. Our society, particularly the fields of security, finance, and research, will be greatly impacted by the enormous potential effects of quantum computing. As a result, you may find articles about quantum computing in both specialized scientific journals and widely read lifestyle publications. But what does quantum computing mean to modern software developers? The prospective effects of quantum computing on developer's lives were discussed in this book. In order to focus on the aspects of quantum computing that are pertinent to developers, we quickly discuss its principles and implications. The fundamental concepts are first introduced. Then, you will learn how to build a straightforward Java program that makes use of quantum computing. We also introduce the strange library, which enables you to continue using quantum principles while writing in Java (or other high-level languages). The qubit, a key component of quantum computing, is also introduced. Nothing is stopping you from beginning to explore quantum computation in your projects right now, thanks to quantum simulators and early prototypes. The likelihood that your software will be prepared by the time the hardware is available will rise as a result.

Stewart Kirubakaran S, Department of Computer Science and Engineering, Karunya Institute of Technology and Sciences, Karunya Nagar, Coimbatore, Tamil Nadu, India, e-mail: stewartkirubakaran@gmail.com
R. Senthil Kumar, Department of Computer Science and Engineering, Saveetha School of Engineering, Saveetha Institute of Medical and Technical Sciences, Tamil Nadu, India, e-mail: rsenthilmecse@gmail.com
Anusha Bamini A. M, Department of Computer Science and Engineering, Karunya Institute of Technology and Sciences, Karunya Nagar, Coimbatore, Tamil Nadu, India, e-mail: anushabamini@gmail.com
Chitra R., Department of Computer Science and Engineering, Karunya Institute of Technology and Sciences, Karunya Nagar, Coimbatore, Tamil Nadu, India, e-mail: chitrajegan5@gmail.com

https://doi.org/10.1515/9783110791402-005

5.1 Introduction

5.1.1 A brief history

In his 1960 “There’s Plenty of Room at the Bottom” lecture, Caltech Professor (and MIT alumni) Richard Feynman initially proposed the idea of a quantum computer by proposing the use of quantum effects for computation. But the real exploration of the hypothesis did not start until the late 1970s. The first article on quantum computation was released in May 1980 by Paul Benioff, who was, at the time, a researcher at the Centre de Physique Th’eorique, CNRS. The publication of “Computable and Incomputable,” a book by the Russian mathematician and Steklov Mathematical Institute professor, Yuri Manin, in the same year served as inspiration for the creation of quantum computers.

The Deutsch-Josza method was put forth in 1992 by David Deutsch and Richard Josza. It is one of the first examples of an algorithm that might be resolved quickly in a quantum processer, compared with classical (or nonquantum) systems, even if the algorithm itself was not very useful, and could be considered a toy problem. The concept of the quantum advantage holds that quantum computers might have a higher computational complexity than classical computers. More research in the area of quantum algorithms was conducted in the years that followed, and one of the most significant findings in the field was made in 1994. Significant advancements in quantum computation were made, both theoretically and experimentally, in the years that followed the publication of Shor’s algorithm. The quantum Grover’s approach, which uses a method now known as amplitude amplification, was created by Lov Grover of Bell Labs in 1996, and offers a quadratic speedup for search in unstructured situations.

The US government’s first public request for research proposals in the area of quantum information was made that same year, demonstrating the growing interest in quantum processing. In addition, David P. DiVincenzo, who was then with IBM Research, outlined the five essential conditions for building a physical quantum computer. Among these were the difficulties of separating noisy quantum systems from their surroundings, and precisely regulating unitary transformations of the system. This was swiftly followed by improvements in quantum annealing, a physical implementation of the Grover’s algorithm, and the creation of a 3-qubit NMR quantum computer. Researchers have thus proved the physical feasibility of quantum computing systems by the turn of the twentieth century.

It is feasible to mimic a noise-free or “fully error corrected” quantum computer through executing a QEC algorithm on the device, despite the fact that the operations of a physical quantum computer’s qubits are noise-sensitive. It seems improbable that a difficult quantum program, like the one that used Shor’s algorithm, could ever run perfectly on a quantum computer without QEC. But QEC experiences noteworthy outlays in terms of the quantity of corporeal qubits, essential to simulate a supplementary healthy and reliable qubit, known as a “logical qubit”. It also shows how many simple qubit actions need to be carried out on physical qubits in order to replicate a quantum operation on this logical qubit.

5.1.2 Background

Paul Benioff's suggestion of a quantum mechanical model of the Turing machine signaled the beginning of quantum computing in 1980. A quantum computer might be able to mimic things that a classical computer is not actually capable of, claim Richard Feynman and Yuri Manin. IBM has long been a leader in the realm of quantum computing [1]. A significant revolution is to increase the level of control over individual quantum systems, such as charged molecules, allowing for even more potent uses of quantum information. The technique uses a quantum logic strategy that was developed for an experimental NIST atomic clock [2]. A new field of science and technology is discovered as a result of evolution in one. Figure 5.1 represents a single qubit computational base as shown below:

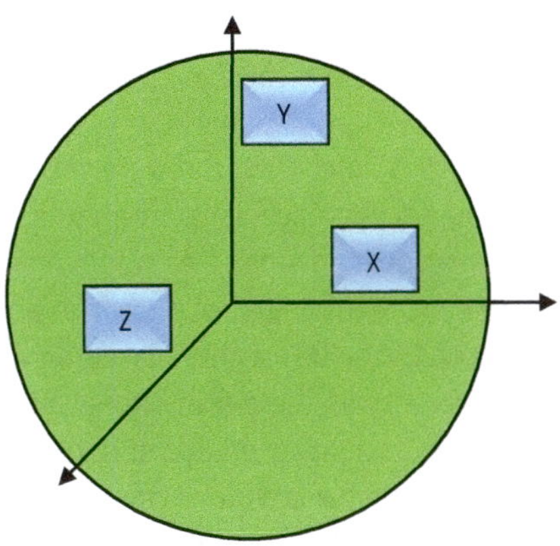

Figure 5.1: A single qubit computational basis state.

The ability to perform the same computation in numerous ways using various physical systems is one of the defining characteristics of computing [3]. As a simple example, think about adding the numbers two and three. You've most likely already calculated the answer in your head using mental arithmetic, which is five. You may alternatively shuffle the beads on an abacus or use the calculator on your smartphone, for instance. Despite being three very distinct physical objects – brains, smartphones, and abacuses – all three can carry out the same calculation.

Despite the fact that digital silicon technology today dominates for more potent computation, when human brains were still used for the majority of computation, and analog devices were used to solve differential equations, silicon transistor-based computers were considered a ground-breaking invention. Now, computers built with this dominant silicon technology can calculate at a fast rate. Up until about 2009, silicon chips could be made quicker by shrinking the transistors. Because the electrical signals must travel further while retaining the same speed, smaller transistors enable faster operation. If not, the heat from the calculation's waste heat would cause the chips to melt [4].

Functional computing research and development has revolutionized science, technology, and the nation in less than a century. During the end of the twentieth century, the first practical computer was unable to do mathematical operations on its own. Theoretical principles must be implemented physically well in practical devices. Today's computers are capable of swiftly and accurately resolving problems if the given input is applicable and the set of directives is favorable. The "Universal Turing Machine," a real all-purpose computer, with a storable and a programmable model, was created by Alan Turing during World War II [5], and that is when everything started. Von Neumann revised it, and it is today the most crucial architecture for practically every computer. The performance and capabilities of computers and their physical components grew over time. The military department that started the computer sector eventually lost out to it in size. We now have the newest technological gadgets, thanks to humankind's increasing empathy and control over nature and corporeal systems [6].

5.1.3 Computing methods

Modern computers are less expensive, faster, more efficient, and even more powerful than the big, expensive, and power-hungry early computers. Advancements in architecture, hardware, and the software that utilizes them have made it possible. Electronic computer circuits are getting smaller and smaller every day. Signals that are electric or electronic are switched and amplified using tiny semiconductor transistors placed on a silicon piece they had previously been made. To create the circuit, these transistors were assembled onto a solitary Si superficial. At the same time, each layer of silicon had the shape of the circuits present in an IC, written on it. By making the circuit as small as an atom, the concept of the tiniest computer is born. However, as electrons inside of atoms have the capacity to exist concurrently in two separate positions, this will prevent these circuits from acting as switches. The different types of quantum computing were represented in Figure 5.2. To accomplish this, they can vanish from one side of a barrier and reappear on the other. In quantum physics, a teleportation phenomenon called "Quantum

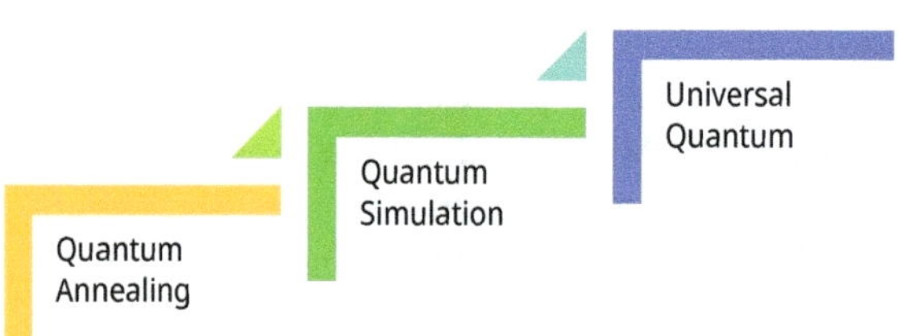

Figure 5.2: Types of quantum computing.

Tunnelling" is what causes this. It designates that the scope of the journeys in a conventional computer reaches its maximum beyond 5–7 nanometers.

The random and unpredictable nature of the physical world is addressed by a revolutionary type of computer known as quantum computing, which is based on quantum physics. Just as quantum mechanics offers a more complete representation of physical science than traditional mechanics, quantum computing also offers a more complete representation of computing and has a larger potential to address problems that classical computing is unable to. They employ their individual quantum bits, commonly referred to as "Qubits," to stock and process the information, in contrast to ordinary standard computers, which are based on traditional computing and use binary bits 0 and 1 separately [7]. Computers that use this type of computing are known as quantum computers. Such small computers are unable to utilize the transistor, logic gate, or integrated circuit circuitry. They make use of bits made of subatomic particles like atoms, electrons, photons, and ions as well as data on their spins and states. They can also be layered on top of one another to make new combinations. As a result, they are more potent since they can efficiently utilize memory when operating in parallel. Only quantum computing can ignore the Church-Turing theorem, allowing for operation in orders of magnitude quicker than traditional computing [8].

5.2 Why we need quantum computers?

One of the main drivers for growing the field of quantum computation has been the major prospect of creating a quantum computer that is capable of carrying out Shor's algorithm for huge numbers. However, it is critical to recognize that quantum computers will probably only significantly speed up a small subset of problems if one wants to acquire a more comprehensive perspective on them. Researchers are striving to create algorithms that demonstrate quantum speedups as well as identify the issues that are most suited. Quantum computers will be of great assistance with problems involving optimization, which are crucial to everything, from financial trading to defense as represented in Figure 5.3.

To execute quantum algorithms, quantum computers take advantage of the quantum-evident characteristics of matter and light. Existing software cannot be made faster using quantum computers, and they cannot take the place of conventional computers because they lack any special powers. No quantum computer has yet been demonstrated to operate more quickly than quantum computer simulators on conventional processors. Some scientists believe we will never understand it, and they might be right [9].

Everyone is using high-speed computing, I guess. The answer given here is incorrect. Yes, quantum computers are designed to be quick (thanks to qubits), but one would wonder why we cannot utilize traditional computers to attain the same speed. Furthermore, conventional computers can be readily engineered to outperform quantum computers in

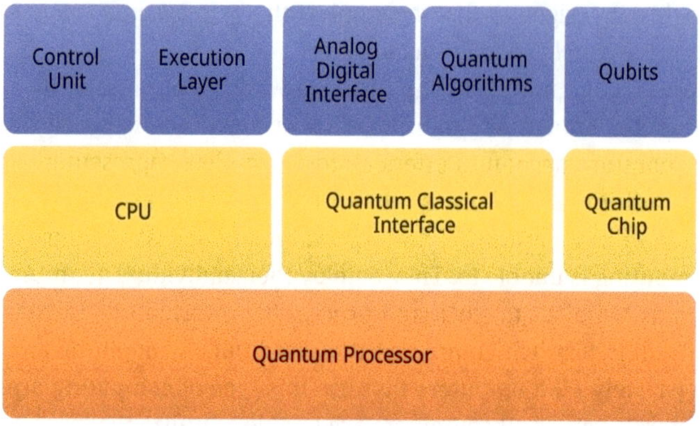

Figure 5.3: An overview of quantum construction with diverse throttles.

terms of speed. In fact, it is possible to make conventional vacuum tube-based computers from the 1960s outperform quantum computers in terms of speed. Speed bottlenecks are not caused by specific hardware configurations (whether it is based on bit or qubit). The actual bottlenecks are size, cost, energy requirements, heat control, etc. represented in Figure 5.4.

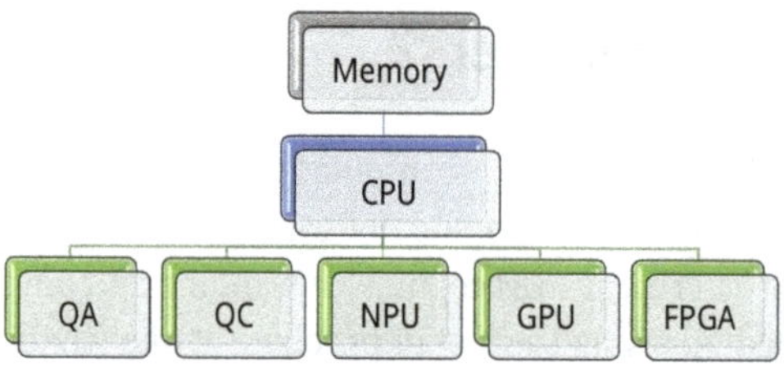

Figure 5.4: Different accelerators of quantum architecture.

Supercomputers have amazing power, but many issues remain unsolved because the issues are beyond the capacity of supercomputers. We urgently need to create new tools to examine or process the data that is being transmitted in massive chunks from the universe to human DNA due to the growing amount of data. Due to this flaw, quantum computers were created, allowing humans to process data extraordinarily quickly. Quantum computers are the ideal synthesis of physics, engineering, and computer science. At 1,000 qubit D-WAVE, the processor can consider 21,000 possibilities simultaneously which is one of the most cutting-edge technologies in quantum computer.

5.3 Quantum versus classical computing

5.3.1 Bits and qubits

This section goes into great detail about qubits, gates, and circuits. Qubits, the building blocks of quantum computers, are equal to conventional or digital computer bits, with the additional property of being in a superposition of states. A quantum register with two qubits can simultaneously store four numbers in superposition, in contrast to a classical register with two bits, which can only hold two, while a register with 300 qubits can hold more numbers than there are atoms in the universe. As a result, at the time of computation, an unlimited amount of data is stored, but we are unable to retrieve it. There was an issue with reading out an output that is in a superposition state and had so many different values [10].

5.3.2 The ket (| >)

The symbol **(**| >**)** used in quantum mechanics to represent a system's state is called a ket. A ket's symbols directly denotes or represents precisely known system information in an abstract way. The ket (| >) is a fragment of Dirac's notation. The ket is basically a vector notation. Every qubit's state is an element path in C2. So

$$\begin{bmatrix} \alpha \\ \beta \end{bmatrix}$$

It is a path, and the same can also be rewritten as,

$$\alpha|0>\beta|1$$

With

$$|0> = \begin{bmatrix} 1 \\ 0 \end{bmatrix}$$

Despite the fact that quantum theory has been around for more than a century, the current quantum revolution is founded on the more recent understanding that uncertainty, a crucial aspect of quantum particles, may be an extremely useful tool. It is impossible to determine with absolute certainty every property of a quantum particle, such as an electron or a photon (a particle of light), at any given time. For instance, the GPS in your automobile can provide precise location, speed, and direction information all at once to help you reach your destination. However, a quantum GPS would not be able to simultaneously and precisely display all of those characteristics of an electron; this is not due to poor design, but rather to a prohibition imposed by the principles of quantum physics [11].

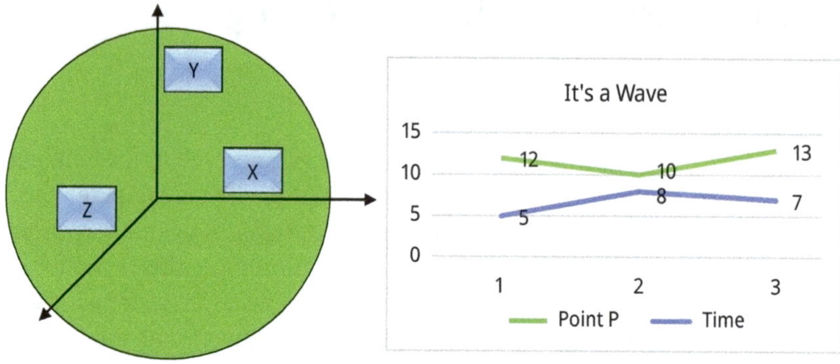

Figure 5.5: Behavior of a qubit compared with an upsurge.

It is not as simple as it sounds to make 50 or more qubit function though. A qubit is quite prone to mistake due to interference, a fairly fundamental phenomenon. I compared qubit and their superposition's to a spinning coin earlier. Imagine that two qubits are suddenly bouncing about on a table, like coins. A wave-like pattern can be detected by looking at the location of a point on one of the coins at each point in time (see Figure 5.5). Similar to how sound and water waves travel through space and time, these waves have a specific amplitude and wavelength.

Sometimes two waves can interfere destructively, leaving you with no measurements. You can conceive of it in terms of the two coins: picture them both spinning at the same speed and, say, in a clockwise direction, on a table top. Imagine further that both coins are being focused on at the same time, but that one is perpetually one-half turn behind the other (see the "destructive interference" part in the picture below). Since we frequently cannot distinguish between the two measurements in quantum computing, we are essentially monitoring two waves at once. But in this case, nothing can ever be measured because the coin locations are equal.

Destructive interference (DEI) of two waves is a phenomenon that occurs as a direct result. It is a problem that has not been fully resolved yet. DE coherence grounds errors, which we definitely do not need that while developing software that could have a substantial influence on people's breaths.

The last basic principle of quantum computing is predicament. Even though there is no analog in the world of classical computing, when two quantum particles get tangled, it continuously returns to the similar state when they are restrained. It appears as though the two coins that you were holding were mysteriously combined. As long as you completed the measurements concurrently, the outcome would be the same, whether you could stop one coin from spinning in Tokyo and the other in London, which would be either heads or tails.

This event has implications for quantum encryption since it suggests that we may soon be able to entangle two quantum particles on different computers. Since any interruption is immediately apparent, the connection between these two particles is

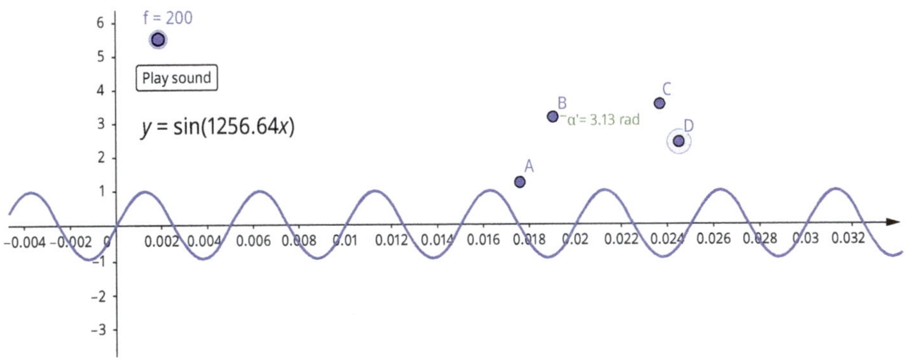

Figure 5.6: Quantum computing helps in solving destructive interference.

inherently secure. Quantum entanglement, which ultimately may lead to a quantum internet, may allow people to instantly send data between computers as represented in Figure 5.6.

At first, such uncertainty is unsettling. 0s and 1s in conventional computers are related to switches and electronic circuits turning on and off. From a computational perspective, it would not make much sense to not know for sure whether they are on or off. In reality, doing so would result in calculation errors. However, the ground-breaking notion underpinning quantum information processing is that quantum uncertainty, or the fuzzy state existing between 0 and 1, is actually a feature rather than a flaw. It offers new controls for stronger means of data processing and communication.

5.4 Quantum computing: principles & applications

The influence of quantum computing on industry appears to be substantial. Numerous quantum algorithms have been created since the 1980s. The integer factoring algorithm of Shor and Glover's database search algorithm are the two most important quantum algorithms. They are capable of defeating encryption schemes (such AES and ECC), which were frequently cast-off on the Internet, and are known to perform far better than the algorithms for normal computers. Administrations have been increasing grants for major computing research and development in order to advance computer technology and safeguard their national security. The various vital foundations for quantum computing were represented in Figure 5.7.

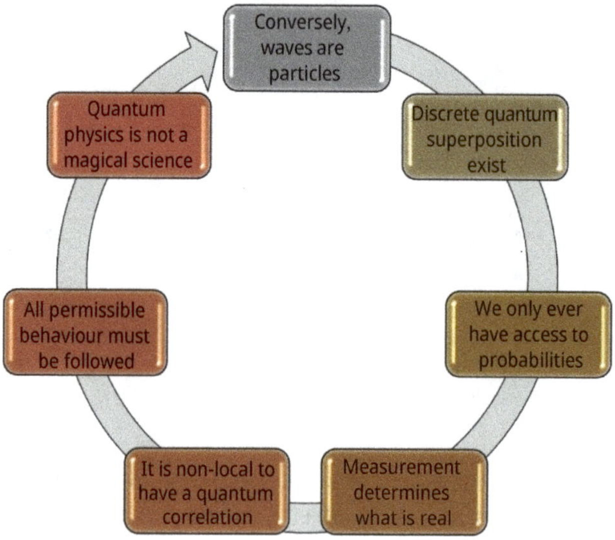

Figure 5.7: Vital foundations for quantum computing.

5.4.1 Essential elements

1. **Conversely, waves are particles:** Everything in the cosmos possesses both particle-like and wave-like characteristics, according to quantum theory. It is not that everything is actually formed of particles that occasionally give the impression of being waves, or that everything is made of particles that sometimes give the impression of being waves. Every object in the cosmos is a brand-new type of object known as a "quantum particle," which combines aspects of particles and waves but is neither.

Because they are distinct and (in theory) countable, quantum particles act like other particles in their behavior. When searching for an atom or a photon of light, you will only be able to find it at that specific location. Matter and energy are discrete units, in that, they exhibit diffraction and interference effects; quantum particles also behave like waves. A small slit will cause a beam of electrons or a beam of photons to disperse on the other side. When the beam is directed at two closely spaced slits, it will result in a pattern of alternating bright and dark spots on the opposite side of the slits as if water waves were travelling simultaneously through both slits and interfering on the opposite side. Even though each individual particle is only ever observed as a particle at a single location, this is nonetheless the case.

2. **Discrete quantum superposition exist**: Everything in quantum physics exists in discrete amounts, which is referred to as the "quantum." Only integer numbers of photons can exist in a beam of light: 1, 2, 3, 137, but never 1.5 or 22.7. Only specific discrete energy values for an electron in an atom are possible: −13.6 electron volts or, in the case

of hydrogen, −3.4 electron volts; −7.5 eV is never possible. You can only ever detect a quantum system in one of these unique permitted states, regardless of what you do.

3. **We only ever have access to probabilities**: The only thing that physicists can predict when using quantum mechanics to anticipate an experiment's outcomes is the likelihood of discovering each potential result. However, we can never say with certainty that a given electron will absolutely end up at site A or unquestionably end up at location B in an experiment where an electron would finally wind up in one of the two locations. Instead, we can say that there is a 17% probability of finding it at location A and an 83% probability of finding it at location B. We can never know exactly how the experiment will come out, no matter how carefully we prepare each particle in the same way.

4. **Measurement determines what is real**: Up until the moment of measurement, the precise state of a quantum particle is unknown; in fact, it can be thought of as being distributed among all possible outcomes. After a measurement, the state of the particle is unquestionably established, and any ensuing measurements on that particle will yield the same outcome.

It is worth mentioning again that this is unequivocally proved by experiment, despite the fact that it seems impossible to believe, as it is the issue that gave rise to various problems. This indeterminacy is demonstrated by the double-slit experiment, which shows that an electron resides in a superposition of both potential routes up until it is finally measured at a single location on the other side of the slits. A result of the superposition of several states is the interference pattern that is seen when many electrons are recorded one after another. Another illustration of a consequence of quantum measurement is the Quantum Zeno Effect, which states that a quantum system can be prevented from changing its state by taking repeated measurements of it.

The system is in a superposition of two conceivable states in which one is more likely than the other to occur between measurements. The system enters a single, distinct state after every measurement, and the evolution must restart. Although the results of measurement can be understood in a variety of ways – as the physical "collapse" of a wave function, as the division of the universe into numerous parallel universes, etc. – the outcome is same in each case. Prior to being measured, a quantum particle can and will exist in several states; following the measurement, it exists in just one state.

5. **It is nonlocal to have a quantum correlation**: One of quantum physics' most intriguing and important phenomena is entanglement. When two quantum particles interact properly, their states will be dependent on one another, no matter how far apart they are. Unquestionably and entirely, the outcome of the measurement at Princeton will have an impact on the outcome of the dimension in the capital of France, and vice versa, even if you measure both particles simultaneously while firming one particle in Princeton and distributing the other to the capital of France. Any local theory in which the particles have distinct states is unable to account for the correlation between their

states. These states are uncertain until one is measured, at which point, the states of both are unquestionably ascertained, regardless of how far away they are.

These nonlocal quantum correlations exist. Regardless of how far apart two quantum particles are, when they interact properly, their states will be dependent on one another. One particle can be held at Princeton while the other is sent to Paris, and both particles can be measured simultaneously. The results of the measurements in Princeton and Paris are, without a doubt, inextricably linked. Any local theory in which the particles have distinct states is unable to account for the correlation between their states. Prior to one being measured, both of these states are uncertain, but after one is measured, both of these states are unquestionably determined, regardless of the speed of the light.

6. **All permissible behavior must be followed**: A quantum particle travelling from point A to point B will simultaneously travel every single route between points A and B. This covers scenarios where electron-positron pairs suddenly arise and then vanish again, among other exceedingly implausible phenomena. Even the most insanely implausible processes can contribute to the entire theory of quantum electrodynamics (QED). It is important to stress that this is not some theoretical jargon that has no practical application. An electron's interaction with a magnetic field can be accurately predicted by QED to 14 decimal places. It is one of the most well-studied hypotheses in the annals of science, strange as the notion may appear.

7. **Quantum physics is not a magical science**: Yes, this was on the other list as well, but it bears repeating because of how crucial it is. Without a doubt, quantum physics is very unusual, but that does not mean that all common-sense principles are suspended. Nothing can change sooner than the speed of light; then, the fundamental laws of physics such as energy conservation and entropy growth remain in place. A perpetual motion machine, telepathy, or clairvoyance cannot be produced by using quantum effects.

I must nevertheless strongly disagree with any assertion that reality is determined by measurement. The fact that many individuals hold this opinion does not necessarily imply that it is accurate, just as many people hold the opinion that science and religion can coexist. As with Bell's theorem for no locality, one would need to demonstrate some form of "measurement determines reality" thesis in order to be rigorous rather than rely on the pronouncements of the Copenhagen school, which we now know were a little overzealous – things like the Kochen-Specker theorem, the "free will" theorem, and the latest testing of Leggett's inequalities get near, but not without assumptions that are easily contested.

5.4.2 Principles of quantum computing

Qubit

In order to address an issue, a physical system's states are modified using computation. In quantum computing, the medium for storing and transmitting digital information is a minuscule entity (such as an electron, photon, or ion). One-bit information can be prearranged using the two immaterial states of a small entity (i.e., zero or one). One type of quantum bit is a quantum two-state system (or qubit). A quantum computer may be used to solve a problem by setting qubits in starting states and then changing the states so that the expected result appears on the qubits.

$$\psi 1 = \alpha\beta|1,0\rangle + \beta\alpha|0,1\rangle,\ \{\alpha\beta,\ \beta\alpha \in \mathbb{C}_0\} \tag{5.1}$$

where α and β were composite numbers ($\mathbb{C}_0$), called likelihood bounty, and it gratifies

$$|a|^2 + |b|^2 + = 1 \tag{5.2}$$

$|a|^2$ – possibility of receiving the state $|0,1\rangle$ as the outcome of the dimension on the qubit $|\psi\rangle$

$|b|^2$ – possibility of attainment$|1,0\rangle$.

"$|\rangle$" – column vector, a commonly used notation in quantum mechanics for describing states.

The normal base $|0\rangle$ and $|1\rangle$ can stand on broadside as

$$|0\rangle = [0,1]^T,\ |1\rangle = [1,0]^T \tag{5.3}$$

In this way, a column vector with 2^n elements can represent an n-qubit state.

Superposition state

Quantum computing requires superposition states, which are crucial. For the sake of simplicity, a photon is used in this section as an example of a qubit. One bit of information is represented by the polarization, which is the symmetrical positioning of the photon. The old-style bit 0 is represented by a horizontally polarized photon, and the bit 1 by a vertically polarized photon. Each photon was directed at the emitter, which is shown in Figure 5.8, and is passed via a Filter-V that only allows vertically polarized photons to pass through.

We make the assumption that only one photon can enter the PBS at a time period in order to make the illustration much easier. According to measurements made at the detector DV, PBS deflects horizontally polarized photons while it communicates with steeply polarized photons. Finally, in Figure 5.8, all of the active photons are driven to be restrained at DV.

Quantum circuit and measurement

Computing, as previously described, is the act of changing the states of a physical system. The probability that the solution will be present in the computation's output state is increased by manipulating the quantum state that is used as the input by a quantum computer. For instance, Shor's technique uses quantum state manipulation

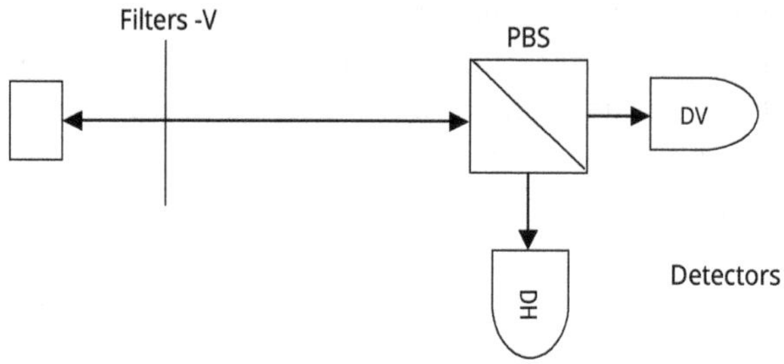

Figure 5.8: All the photons were detected at DV.

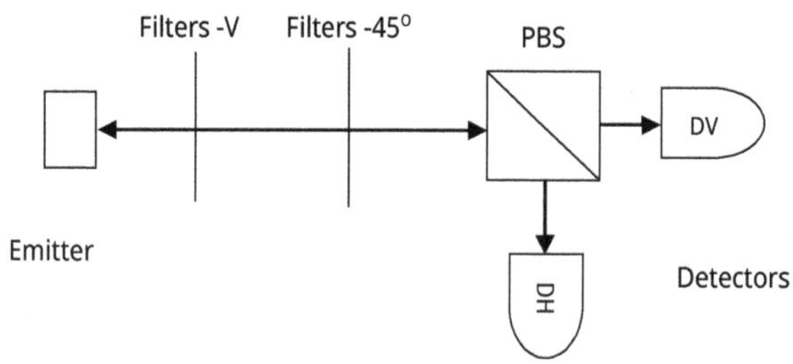

Figure 5.9: One half of the photons were originated at DV and the rest at DH.

to determine the prime factors of a large integer. A crucial contrast between a traditional circuit and a quantum circuit is the presence of in-between positions in the circuits. We are able to measure the intermediate state in a classical electronic circuit because binary zero and one are represented by voltages on a circuit node [12].

Because of this, an error (like 2.5 V) can be found by monitoring the voltage in the circuit. However, it is most likely that the quantum circuit's intermediate states are superposed. The quantum state is besmirched and turns into one of the two immaterial base states (i.e., |left.0rightrangle (or) |left.1rightrangle) at the time of the measurement, if the superposition state is measured, according to the probability $|a|^2$ and $|b|^2$.

Quantum gate and reversibility

The basic actions that a quantum circuit does to change an input state are referred to as quantum gates. Each gate action can be quantitatively represented using a matrix. Figure 5.9 shows how the Filter-45° changes from one of the orthogonal basis states (i.e., $\frac{1}{3\sqrt{2}}|0\rangle + \frac{1}{3\sqrt{2}}|1\rangle$.

In addition to the above example, another example is the cNOT gate (represented in Figure 5.10) that performs like the traditional XOR gate, revealed in Table 5.1.

This contradicts the entire run toward the second qubit $|\psi_{in2}\rangle$. Whereas the first qubit $|\psi_{in1}\rangle$ is $|1, 0\rangle$, $|\psi_{in1}\rangle$ is unchanged (i.e., $|\psi_{in1}\rangle = |\psi_{out1}\rangle$). The same can be inscribed like:

Table 5.1: Quantum XOR gates with 2 basic IO (in and out) operations – A Truth Table.

ψ_{in1}	ψ_{in2}	ψ_{out1}	ψ_{out2}
0	0	0	0
0	1	0	1
1	0	1	1
1	1	1	0

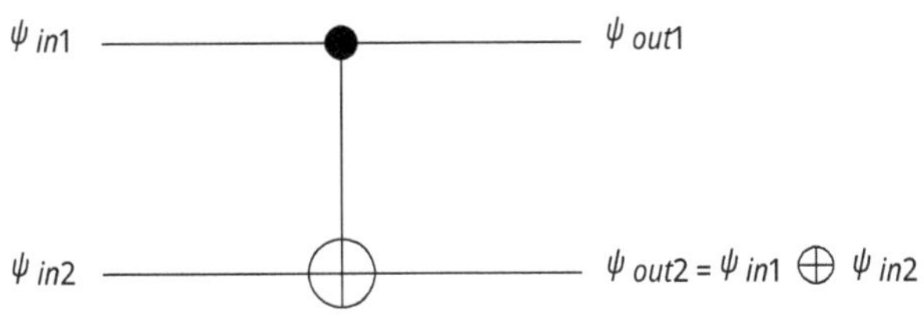

Figure 5.10: cNOT gate representation that outperforms a traditional XOR gate.

$$|\psi_{out1}\rangle|\psi_{out2}\rangle = U_{\mathrm{cNOT}}|\psi_{in1}\rangle|\psi_{in2}\rangle = \begin{bmatrix} 1 & 0 & 0 & 0 \\ 0 & 1 & 0 & 0 \\ 0 & 0 & 0 & 1 \\ 0 & 0 & 1 & 0 \end{bmatrix} \begin{bmatrix} \psi_{1,0} & \psi_{2,0} \\ \psi_{1,0} & \psi_{2,1} \\ \psi_{1,1} & \psi_{2,0} \\ \psi_{1,1} & \psi_{2,1} \end{bmatrix} = \begin{bmatrix} \psi_{1,0} & \psi_{2,0} \\ \psi_{1,0} & \psi_{2,1} \\ \psi_{1,1} & \psi_{2,1} \\ \psi_{1,1} & \psi_{2,0} \end{bmatrix} \tag{5.4}$$

Any logic gate (for example, AND, OR, NOT) can be made using a gate called NAND since the NAND gates were universal. Table 5.2 represents NAND gate's Truth Table

Table 5.2: Traditional NAND using two inputs – A Truth Table.

$\mathbf{IN_1}$	$\mathbf{IN_2}$	**Out**
0	0	1
0	1	1
1	0	1
1	1	0

Given that the inputs are two qubits and the NAND gate is constructed using the aforementioned truth table $|0\rangle|1,0\rangle = [1,0,0,1]^T \otimes [0,1,0,0]^T = [0,1,0,0,1]^T$, the equation can be represented as:

$$\text{NAND:} \begin{bmatrix} 1 & 1 & 1 & 1 \\ 0 & 1 & 0 & 0 \end{bmatrix} \tag{5.5}$$

$$U_{\text{NAND}}|0\rangle|1\rangle = \begin{bmatrix} 1 & 1 & 1 & 1 \\ 0 & 1 & 0 & 0 \end{bmatrix} [0,1,0,0,1]^T = [1,0,0,1]^T = |1,0\rangle \tag{5.6}$$

This NAND gate is not realizable with quantum gates because the two bits of information before the gate are reduced to one bit after the gate (10). Quantum physics prevents a system (such as a quantum circuit) from losing information unless its quantum states are carefully watched. As a result, quantum gates must be both, have an equal number of inputs as outputs, and be reversible without sacrificing any information. The NOT gate is the only one of the classical gates that is a one-way function, meaning that some of the input data is lost upon leaving. This requirement is another significant deviation from conventional computing.

Table 5.3: Quantum NAND gate Truth Table.

ψ_{in1}	ψ_{in2}	ψ_{in3}	ψ_{out1}	ψ_{out2}	ψ_{out3}
0	0	1	0	0	1
0	1	1	0	1	1
1	0	1	1	0	1
1	1	1	1	1	0

Quantum NAND gate is made of cNOT gate, represented by Figure 5.11 and Table 5.4. A studied truth table for Quantum NAND gate is given in Table 5.3

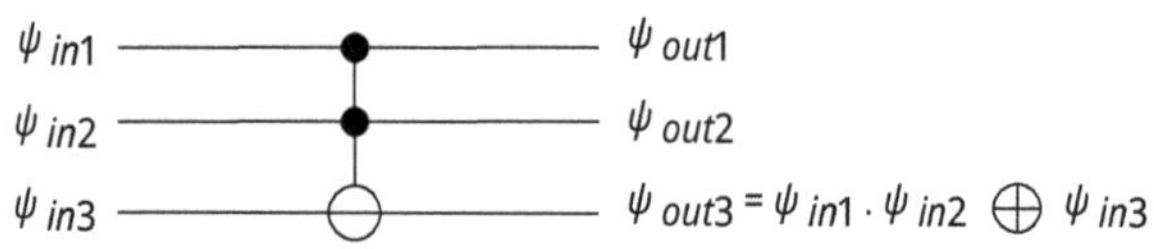

Figure 5.11: Controlled cNOT gate.

Table 5.4: Controlled cNOT – A Truth Table.

ψ_{in1}	ψ_{in2}	ψ_{in3}	ψ_{out1}	ψ_{out2}	ψ_{out3}
0	0	0	0	0	0
0	0	1	0	0	1
0	1	0	0	1	0
0	1	1	0	1	1
1	0	0	1	0	0
1	0	1	1	0	1
1	1	0	1	1	1
1	1	1	1	1	0

5.4.3 Essence of quantum mechanics

Life and the universe are not assemblages of mechanisms. Their mechanisms are evolutionary pathways. The classical physics end-target/states are not specified or determined by the mechanisms. They are pathways for the evolution of classical physics states. In order to move from one classical physics condition to another WITHIN the space from the cosmic singularity to the most enlarged cosmos and back, quantum mechanics is a method that is likely, plausible, and actually exists. The architecture diagram for quantum computing is represented in Figure 5.12.

5.4.4 Architecture

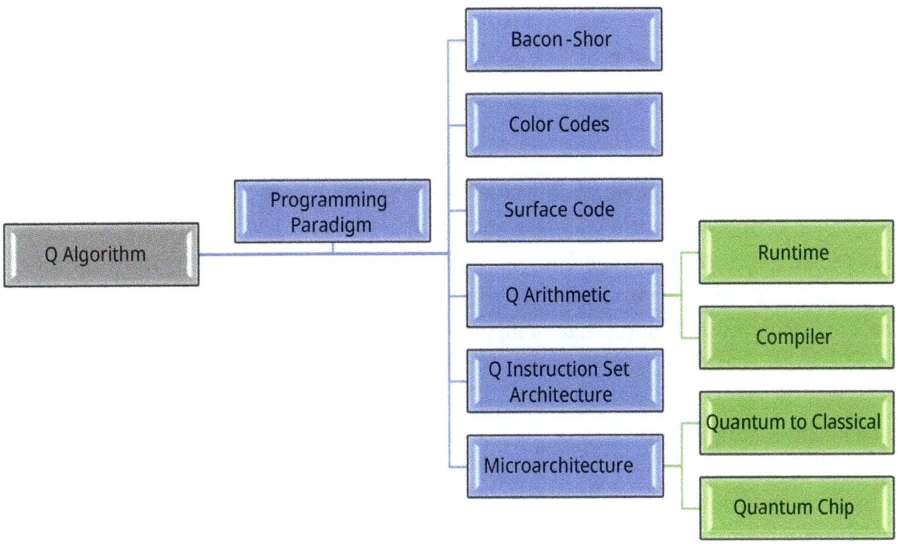

Figure 5.12: Quantum architecture.

5.4.5 Applications of quantum computing

5.4.5.1 Quantum calculations

One of the major advantages of quantum computers is its capacity for massively parallel computation, as was already mentioned. By using a quantum superposition state, $2n$ inputs can be concurrently stored in n qubits. Since universal quantum gates allow us to construct any quantum circuit, the n qubits can be used as the input for a quantum circuit that performs any computation [13]. For instance, two qubits can concurrently store the conventional numbers 0, 1, 2, and 3. The status can be used to describe this. For example, a circuit could be developed to determine f(*x*) = *x* + 5. Putting the qubits in the superposition states in the circuit seems to make it possible to finish four computations in one go.

Despite having four possible output values (5, 6, 7, and 8), the actual output is a superposition of all four. The result of the measurement on the output qubits is one of the four possible outputs. In a word, the superposition of 2*n* outputs for 2*n* inputs is what makes up the output qubits of a quantum circuit that uses classical logic.

Among the 2*n* possible output states, the measurement result has a probability of 1/2*n*. Therefore, to enable measurements to produce results with probabilities greater than 1/2*n*, a quantum circuit is required to change the probability amplitudes of the qubits. The two well-known quantum algorithms that are introduced in this section perform considerably better than classical algorithms [14].

5.4.5.2 Grover's procedure

This database search algorithm is used to find a specific item in an unordered list. It can be used to speed up brute force key searches for symmetric key encryption systems, for example, AES. Using this method, it is possible to search unsorted arrays of size N in O(N) operations, as opposed to the O(N) operations required by more conventional techniques. The following idea serves as the foundation for Grover's algorithm. When the number *xa* that satisfies f(*xa*) = 1 needs to be located in a sizable unsorted database, the qubits are made to be in a superposition state of all potential ID numbers *x* = 0,1,2, . . ., N 1.

5.4.5.3 Shor's algorithm

According to Peter Shor's 1994 demonstration, a quantum computer could factor a large number in polynomial time. Because the most popular public-key encryption system, RSA, can be broken by a quantum computer, security organizations were particularly interested in his strategy [15]. Both conventional and quantum components are present in Shor's algorithm. A brief explanation of how Shor's strategy combines quantum computation with classical computation comes after the introduction of the RSA algorithm.

5.4.6 Challenges

5.4.6.1 Quantum processers

Numerous methods to realize quantum commuting have been investigated for decades, as noted in "The European Quantum Technologies Roadmap," and certain methods have effectively proven quantum operations. With the help of trapped ions, superconductors, and optical devices, the c-not gate procedures have been made imaginable. A quantum computer based on superconducting has 53 qubits, which is the most qubits that have been tested in labs as of 2019. Commercial quantum computers, with a superconducting foundation, are available from IBM. In their research facilities, quantum computers are also used by Google, IonQ, and Rigetti [16].

If universal quantum gates were physically realized, theoretically, every sort of quantum circuit could be created. But implementing the hardware for gate-based quantum computation, with many qubits, presents numerous technological difficulties. Any substance chosen for a qubit, for instance, must be able to be accurately represented as a stable two-level system and have a longer decoherence period than the gate operations. The coupling of a qubit's environment is known as decoherence [17]. The gate operation's output state is likely to contain mistakes if the wave-particle duality time is insufficient. The environment around each qubit, including other neighboring qubits, must therefore be appropriately segregated from each qubit (like an ion or electron). Some faults can be fixed using a process called quantum error correction.

It is theoretically possible to create any kind of quantum circuit if universal quantum gates are realized as hardware. The hardware for gate-based quantum processing with a large number of qubits, however, faces a number of technical challenges. Any material chosen for a qubit, for instance, must be able to accurately represent itself as a stable two-level system and have a longer decoherence period than the gate operations and the connection between decoherence and the environment of the qubit. The output state of the gate operation is probably faulty if the decoherence time is not enough. Thus, each qubit (such as an ion or electron) requires sufficient isolation from its surroundings, including other qubits close by. Some errors can be fixed using quantum error correction [18].

5.4.6.2 Quantum key dispersal

QKD techniques have been used in optical fiber and free space. In a typical telecommunication optical fiber network, a QKD system with phase encoding was constructed. The first authentic bank transfer using a QKD method happened in 2004. The QKD system was experimentally realized across 260 km of conventional telecom fiber in 2012. The free-space QKD system was tested in 2017 over 53 kilometers in daylight. However, the community of information security practitioners has not broadly shared them.

The absence of a commercially viable quantum repeating firearm to increase the choice and follower out crossways a system is one of the major problems with a QKD for the practitioners. A photon's tendency to disrupt its surroundings is a result of its nature. Since the information carried through a quantum channel is made up of haphazard bits that can be rejected when they contain faults, it is not a serious issue for a short-range QKD technique due to the high SNR (i.e., a photon) that is required for long-distance communication (signal-to-noise ratio). The no-cloning theorem prevents duplication of an unknown photon broadcast by a quantum repeater, making its creation extremely difficult, if not impossible.

Although quantum repeaters have been suggested, the majority of QKD systems that are currently in commercial production were built and utilized with devoted links among different networks. QKD's use was still constrained to networks in urban areas. It was proven in 2018 that entangled photon pair distribution may be accomplished via a satellite. It made it possible to transfer pairs of photon amongst 2 places that were like 1,200 kilometers apart. Both entanglement-based QKD and QKD networks may employ the satellite system as a repeater station [19].

The security of QKD schemes is another problem. In general, the QKD techniques do not ensure the authenticity of the message's origin. Man-in-the-middle attacks against QKD methods are feasible if a listener is able to compromise both quantum and classical channels.

5.5 Threats to quantum computing

Superposition, entanglement, and indeterminacy are examples of "spooky" quantum phenomena that are used in quantum computing. It was first proposed by Richard Feynman as a viable method to model physical chemistry many years ago. Experimental development lagged behind but is now picking up speed. A functioning quantum computer called Sycamore created by Google reached a noteworthy concert landmark known as "quantum supremacy" in September 2019. Cryptography is crucial towards cyber security that has become a big issue for the governments, businesses, and civil society around the world. Quantum computing has significant ramifications for these fields. Today, a select few cryptographic methods, including RSA (Rivest-Shamir-Adleman), provide the foundation for almost all secure digital communication. The RSA standard is the one that is most frequently used to establish public key infrastructure (PKI), which links real people and organizations to the cryptographic keys they use for secure communications and digital signatures.

Before such a machine can be constructed, many engineering obstacles must be solved. It may take decades before a functional large-scale universal quantum computer is developed since quantum error correction is very difficult. However, the threat of quantum computing to cryptanalysis is sufficiently real that "quantum secure," "quantum-resistant,, or "postquantum" methods have already been developed.

Both classical and quantum computers struggle with these mathematical problems, and the only way to improve performance is if an appropriate quantum method can be developed. The US National Institute of Standards and Technology is currently reviewing various viable classical protocols toward setting up a technical standard-setting organizations.

5.6 Conclusion

The branch of physics known as quantum mechanics, which defines the behavior of incredibly small elements, serves as the basis for a new model in computation. The original purpose of the idea of quantum computing (QC) was to improve computer modeling of the behavior of incredibly small ("quantum") corporeal systems. The 1990s saw a rise in interest in the field due to the Shor's algorithm, which was used on a quantum computer. It could exponentially accelerate the vital class of cryptanalytics and possibly upend some of the cryptologic techniques used to protect stored data and communications between the government and the general public. In this paper, the fundamental ideas behind quantum calculations were introduced along with three famous quantum submissions. Quantum computation is a potentially game-changing expertise. Businesses may address a variety of optimization difficulties, including those involving data analytics, and database search has significantly advanced, thanks to quantum computing.

Actually, quantum computers are the only computing paradigm that are now known to exist that could offer rapid speedups over present computers. Even though these discoveries were very intriguing in the 1990s, no one knew how to build a computer using quantum systems; so they were only relevant theoretically. About 25 years later, progress has been made in the creation and manipulation of quantum bits of material, or "qubits," to the point that many investigate teams have established small quantum supercomputers that demonstrate the operation of such systems. This approach has led to significant remote segment investment, revitalizing the industry.

References

[1] https://www.quantum-inspire.com/kbase/what-is-a-qubit/.

[2] https://www.cioinsight.com/innovation/quantum-computing-companies/.

[3] https://www.nist.gov/physics/introduction-new-quantum-revolution/second-quantum-revolution.

[4] Generation of Multiple Key Based on Monitoring the User Behavior, in: Innovation Communication and Computational Technologies, part of Lecture notes in Networks and Systems, Springer, (LNNS, volume 89). https://doi.org/10.1007/978-981-15-0146-3_37.

[5] https://www.ncbi.nlm.nih.gov/books/NBK538701/#:~:text=Quantum%20computers%20have%20the%20potential,great%20progress%20is%20under%20way.

[6] "Secure User Authentication Using Honeywords," International Conference on Computer Networks, Big Data and IoT, part of the Lecture Notes on Data Engineering and Communication Technologies (LNDECT, volume 31). https://doi.org/10.1007/978-3-030-24643-3_105.

[7] Cyber Security Policies, Quantum Computing and Classical Politics: The Ambiguity of Advantage in Signals Intelligence, 1st edition, 2022, doi: 10.4324/9781003110224-7.

[8] Internet of things-based energy-efficient optimized heuristic framework to monitor sportsperson s health, Technol. Health Care, 29(6), 2021, IOS Press, 1st November. https://doi.org/10.3233/THC-213007.

[9] Simulating physics with computers, Int. J. Theor. Phys., 21, 1982, 467–488. https://doi.org/10.1007/BF02650179.

[10] "Testing the Connection Drop and Improving 4G Telecommunication Network And Accessing The Performance With 4G Network," 2022 International Conference on Business Analytics for Technology and Security (ICBATS), Dubai, United Arab Emirates, doi: 10.1109/ICBATS54253.2022.9759045.

[11] Grover's algorithm in a 4-qubit search space, J. Quantum Computing, 2021, Tech Science Press, doi: 10.32604/jqc.2021.018114.

[12] "An Improved AODV Algorithm for Efficient Building of 4G Networks to Enhance the Communication in ZigBee Networks," 2022 International Conference on Business Analytics for Technology and Security (ICBATS), Dubai, United Arab Emirates, doi: 10.1109/ICBATS54253.2022.9759002.

[13] Quantum cryptography – A theoretical overview, J. Quantum Computing, 2021, Tech Science Press, doi: 10.32604/jqc.2021.019864.

[14] Towards developing privacy-preserved data security approach (PP-DSA) in cloud computing environment, Comput. Sci. Eng., 2022, Tech Science Press, doi: 10.32604/csse.2023.026690.

[15] A blueprint for building a quantum computer, Rodney Van Meter and Dominic Horsman, Communications of the ACM, Volume 56, Issue 10, October 2013, pp 84–93. https://doi.org/10.1145/2494568.

[16] Enhancing Accuracy Performance in Novel Drug Delivery System for Composite Microbeads Using Hadoop Ecosystem over Conventional Drug Delivery Systems, in: Innovations in Computer Science and Engineering, part of Lecture notes in Networks and Systems, Springer, (LNNS, volume 385). https://doi.org/10.1007/978-981-16-8987-1_33.

[17] S. M. Sasubilli, A. Kumar, V. Dutt, "Machine Learning Implementation on Medical Domain to Identify Disease Insights using TMS," 2020 International Conference on Advances in Computing and Communication Engineering (ICACCE), 2020, pp. 1–4, doi: 10.1109/ICACCE49060.2020.9154960.

[18] S. M. Sasubilli, A. Kumar, V. Dutt, "Improving Health Care by Help of Internet of Things and Bigdata Analytics and Cloud Computing," 2020 International Conference on Advances in Computing and Communication Engineering (ICACCE), 2020, pp. 1–4, doi: 10.1109/ICACCE49060.2020.9155042.

[19] P. K. Kotturu, A. Kumar, "Comparative Study on Machine Learning models for Early Diagnose of Alzheimer's Disease: Multi Correlation Method," 2020 5th International Conference on Communication and Electronics Systems (ICCES), 2020, pp. 778–783, doi: 10.1109/ICCES48766.2020.9137872

Anusha Bamini A. M., Chitra R., Jeba Priya S.

Chapter 6 Real-time big data analytics

Abstract: Real-Time Big Data Analytics (RTBDA) is a tool or a software feature that has the capability of analyzing huge volumes of data gathered and stored by IT enterprises. RTBDA is a combination of real-time analytics and big data. It is a good initiative in today's business world. Big data that comes in from websites, logs, and enterprises can be processed as they arrive, and a decision can be taken in real time, based on the data. Enterprise Security Software and Enterprise Event Management software have huge sets of real-time data to be processed. Enterprises deploy their applications on cloud. Each application creates an event and log records. This real data can be processed efficiently to make better decisions. Examining big data is possible due to enhanced analytical capabilities, efficient access to various data sources, and inexpensive and improved computing power in the form of cloud computing. Mathematical and logical concepts are applied over the huge volume of data to make better real-time decisions. This big data analytics can also be business analytics, app analytics, web data analytics, and so on.

6.1 Introduction

Huge data sets gathered from a variety of sources, including scientific research, sensor systems, social network activity, telecommunications data, cameras and surveillance systems, regular business and financial transactions data, are referred to as "big data." Large amounts of data are gathered by businesses regarding their clients, suppliers, processes, and transactions [1, 3]. Trillions of bytes of real-time data are produced by sensors implanted in buildings, cars, smart energy meters, mobile phones, and other devices. Additionally, through social media platforms and applications, billions of people every day contribute to the growth in scale and accessibility of big data [2, 4]. A fantastic chance to comprehend and apply such vast amounts of data exists. However, the magnitude of such data sets prevents the development of new technologies [5, 6].

Big data allows companies and organizations to get more insight into their operations and business processes, which, in turn, helps them make better decisions and

Anusha Bamini A. M, Department of Computer Science and Engineering, Karunya Institute of Technology and Sciences, Coimbatore, Tamil Nadu, India, e-mail: anushabamini@gmail.com
Chitra R., Department of Computer Science and Engineering, Karunya Institute of Technology and Sciences, Coimbatore, Tamil Nadu, India, e-mail: chitrajegan5@gmail.com
Jeba Priya S., Department of Computer Science and Engineering, Karunya Institute of Technology and Sciences, Coimbatore, Tamil Nadu, India, e-mail: jebapriya@karunya.edu

https://doi.org/10.1515/9783110791402-006

perform better [11, 12]. Big data accessibility can be used to improve functionality across a range of application domains, including healthcare, banking, transportation, education, security, and governmental planning. To improve the effectiveness of their operations or services or to boost profitability, certain unique sorts of big data applications must make judgments in real time. Intelligent transportation, financial market interchange and monitoring, military operations decision-making, smart grids, and emergency response are a few examples of these uses. These applications must handle both historical and current huge data sets in order to execute quick analyses that support immediate decision-making. Big data applications can perform badly when data is collected and analyzed slowly or incorrectly, which delays decisions. Due to the enormous amount of data that is accessible for optimal or nearly optimal decision-making, such applications face a variety of technical hurdles in real-time processing.

Real-time big data analytics is a cutting-edge innovation that is revolutionizing how IT firms gather valuable business intelligence, identify cyber security threats, and evaluate the performance of critical web and cloud-based applications and services [9, 10]. Businesses can become aware of data and act on it as soon as it enters their system, thanks to real-time analytics. Answers to real-time app analytics questions are obtained quickly. They have rapid reaction times and the ability to quickly comprehend large amounts of info. For instance, real-time big data analytics analyses data from bank databases to assist in trading decisions.

Analytics can be done continuously or on demand. On-demand analytics alert users when they need results. Continuous remodeling can be set up to react automatically to specific circumstances as they arise [13, 14]. For instance, real-time web analytics could reenergize an administrator, if the page load presentation goes over the predetermined threshold. A software function or program called "real-time big data analytics" enables one to examine massive amounts of incoming data as they are being created or stored by the IT infrastructure.

6.2 Big data

In the digital era, huge amounts of big data were everywhere, including information from e-commerce sites, IT, medical, energy, finance, and governance. Big data has the responsibility of storing, processing, and analyzing data. People are switching from manual to automatic data calculation [7]. Big data processing requires the use of standard computing techniques, since both organized and unstructured data must be handled. Big data may originate from a variety of sources [8].

6.2.1 Dimensional view of big data

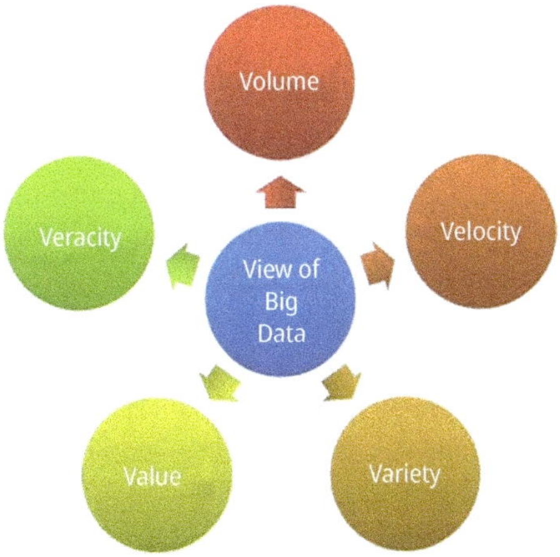

Figure 6.1: Dimensional view of big data.

Dimensional view of big data is shown in Figure 6.1. It may also be represented as 5Vs of big data.

Volume: Organizations gather data from different range of sources, like e-sites, Internet of Things devices, machines, social media, banking, governance, etc. Data from these various sources is called Volume. Holding all these data is an expensive task. But Data Lakes, Hadoop, and the Cloud have made storage more affordable [15].

Velocity: How fast the data is generated and moved is termed as velocity. This is an important concept of business, because business decisions are made faster, when there is faster data flow [16]. The data must be available at the correct time in business decisions. Data can flow from social media, websites, banks, organizations, medical industry, etc. This data must be analyzed faster in real time.

Variety: Variety denotes the type of data. Data from various sources may be in different formats. It may be a text file, image, audio, video, etc. There can be a variety of unstructured data, semi-structured data, and structured data.

Value: It refers to the value given to the big data. It specifies what an organization can do with the received data. Businesses must know social media trends and handling of peak data daily. It is difficult to value seasonal and unseasonal data loads.

Veracity: Data quality and accuracy are termed as veracity. Missing and inaccurate data is not good for maintaining big data. Trusted data is important for veracity. Relationships, hierarchies, and numerous data linkages need to be associated and correlated in business. If not, data quality may easily go out of control.

6.3 Components of big data ecosystem

The Big Data Ecosystem (BDE) is a complex network, which contains three basic components: data source, data storage, and data analysis and consumption [7]. The big data components are always arranged in the form of layers [8]. It is not a single-step process to convert the available data into the useful data for the specific application. The big data analytical tools play a vital role in the process of converting the raw data into data insights. In conventional big data analysis, the big data can be initially taken from the available sources, transformed into the form suitable for storage, and then the stored image analyzed and prepared in the new format, based on the application. This process takes a long time, say, months or even a year. The introduction of real-time data analytical tools changes the scenario and new technologies are adopted in the data resolution. Social networks such as Twitter, LinkedIn, Google, and Amazon are some examples of BDE. It is necessary to extract, transform, and load the data before the application [17, 18]. The components of big data eco system are described in Figure 6.2.

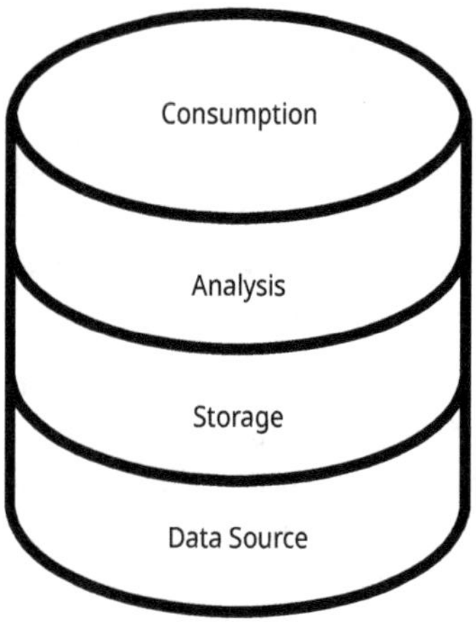

Figure 6.2: Components of big data system.

6.3.1 Data source

The data may be private or public data obtained from various sources like cloud database, e-commerce application, social networking, phone calls, and so on. The obtained data is homogeneous in nature. It can be obtained from either the relational data base or private streaming. The data can be consumed in two different ways, Batch collection and Streaming. In batch collection, the bulk amount of data is transformed once by initializing the data transfer. It is a demand-based data transfer. Whenever the data is sent to the BDE, it is triggered initially. In the Streaming process, the data stream is continuously transferred to BDE. This is most suitable for real-time data analytics and requires continuous monitoring to check the availability of data pool. In batch collection, the data is sent through packets, generally. The data sources of big data are further broadly classified into three categories:

6.3.1.1 Social data

These are raw information collected from various clients of the social media and include profile information and communication information. These data are mainly helpful for the marketers to promote their advertisements as well as ratings. This data is also useful for decision-making in business intelligence platforms. It is also helpful in marketing analytics to improve the marketing by analyzing the sentiments of the customer.

6.3.1.2 Machine data

This is the data obtained from industrial equipment, sensors, and other components of the industry. Due to the tremendous increase in IoT, the amount of machine data has increased exponentially. The sensor data gathered from medical devices, traffic, agriculture, satellites, games, and smart grids are delivering high velocity data with high volume and variety. System log, call log, web log, and application log data also come under this category. Connecting Bluetooth devices, connecting Wi-Fi, file delete and create information, new software installation, application opening, and deleting the content from the recycle bins are termed as system log data. Call log data is the data obtained from telephone calls, such as call duration, call recordings, use of network, call starting and ending, duration, and call billing information. The user online activities, data transmission rate, and time stamping that are reported are considered as the web log data. All activities related to software are considered as application data, which includes software downloads, user id, and system files. These are e data that are automatically generated rather than manually entered. Machine data is beneficial for real-time data analysis, modelling, and making decisions.

6.3.1.3 Transactional data

These are the data generated from online transactions like invoices, online bills, order placements, tracking information, or delivery receipts. Due to the growth in e-commerce applications, bulk amounts of transactional data are available. It is a challenging task for the data analysts to use this kind of data. The transactional data may be either analytical data or master data. Peak data transaction rate, transaction time, and the nature of transaction can be analyzed by using this data. Proper usage of transactional data can reduce transaction failures, online threads, trouble shooting, and cost to service.

In fact, the big data is structured, semi-structured or unstructured data obtained from relational database, JSON files, and website, simultaneously. Most of the data are unstructured and homogeneous in nature, and thus, the data storage and data management also play an important role in BDE. Extraction of big data from data source faces critical issues in data integrity, authenticity, and security. In the next phase, the extracted data is transformed and loaded onto the database.

6.3.2 Data storage

It is the second layer of the BDE. The data is cleaned and transformed before storing it in the data storage medium. The data should be structured and formatted to fit into the destination system and application. The purpose of big data transformation is to make the data easy for handling by computers as well as humans. It is an expensive and intensive process in which the cost depends on the software, infrastructure, as well as the tools used. During transformation, the appropriate indexing, fields, and records are filtered to restructure the data. Before data propagation, the confidential and secured data are stored in an encrypted manner.

The common big data storage mediums are cloud storage, Data Lake, and data warehouses.

6.3.2.1 Data warehouse

In data warehouses, the formatted data is only stored and it is difficult to integrate the new raw data into the data warehouse. It is useful to store large collections of data for business purposes. The degree of data quality and data integrity is high if it is stored in a data warehouse. The data warehouse is designed to store big data for a specific application. The big data obtained from various sources undergo three steps: extract, transform, and load (ETL) before being stored in the data warehouse. The data mart is a smaller version of the data warehouse and contains a subset of data as compared to data warehouse, which is suitable for specific application as well as user. The schematic view of a data warehouse for big data storage is shown in Figure 6.3.

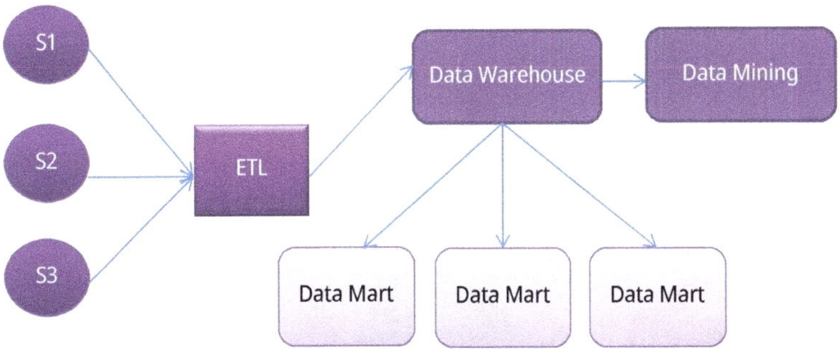

Figure 6.3: Schematic view of data warehouse.

6.3.2.2 Data Lake

It is also the commonly used storage medium for big, raw and granular data, which can be obtained from various resources. Identifiers and metadata indicators are used to access the raw data from the data lake. This raw data is more useful for data scientists rather than for business analysis.

Data Lakes are configured using inexpensive software. A Data Lake is formed by connecting various clusters and hence, the initial storage capacity is not a big problem. Most often, the clusters exist on the Cloud to support the distributive applications. The term Schema on Read represents the working of Data Lake, which indicates that the data is not fitted into any of the specific structures, initially. Based on the application, the data is fitted into the required schema, which saves wanted time required for ELT process.

6.3.2.3 Cloud storage

This storage medium is specifically used for sharable data connected through the internet. Boundless amount of data can be stored on the Cloud, and the data can be accessed from anywhere, thus increasing the accessibility of data. It is a secure mode of data storage, where credentials are required to access the data. Business productivity and efficiency are increased in cloud storage because of the remote accessing capabilities. Google Cloud and iCloud are the major real-time data storage mediums. Traditional storage protocols are used for big data storage on the Cloud. Powerful information management from remote accessing without deployment cost and hardware cost is the key feature of cloud data storage. The cloud storage paradigm for big data is pictured in Figure 6.4. Various data storage technologies are integrated to improve the reliability and effectiveness of big data storage.

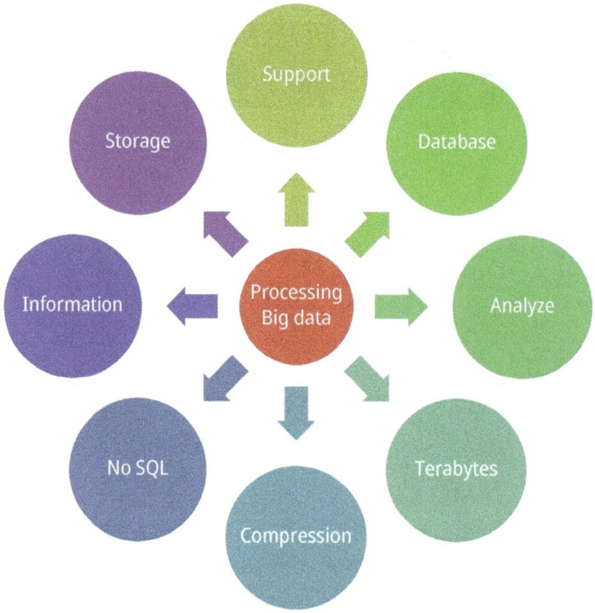

Figure 6.4: Cloud storage paradigm.

6.3.3 Data analysis

It is commonly referred to as a knowledge discovery subsystem that is required to process the data, like statistical and predictive analysis for specific application. It is essential to select the appropriate analysis tool, based on the business needs of the targeted domain. Depending on the problem domain and class of problem, the various analytical tools are selected; it also relies on the platform like web, mobile, text, video, network, and so on. Different kinds of analysis are needed for health care, e-commerce, production, security, and other real-time applications. In this phase, the hidden patterns, market trends, data correlations, and significant factors are analyzed for better decision-making, with less fraud and anomalies. Big data analytics is divided into four categories, namely diagnostic, descriptive, predictive, and prescriptive.

6.3.4 Data visualization

Data visualization increases the decision-making and data analytics capabilities by converting the massive data into a visual insight. Since the selection of visualization tool and method is a challenge, the most suitable visualization techniques for the industry are identified by Big data consultants. Some of the visualization techniques used in the real-time data analytics environment are listed below.

6.3.4.1 Kernel Density map

The nonparametric values and their distribution over the population is given by kernel distribution function. The Kernel Density map plots the kernel distribution function along with the frequency.

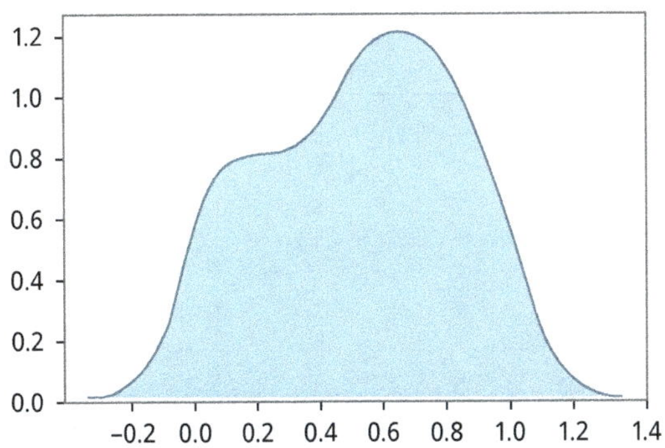

Figure 6.5: Kernel density map.

Figure 6.5 represents a map for kernel density. Gaussian function is used to map the kernel distribution function. From the shape of the distribution, the variance in the big data will be identified.

6.3.4.2 Box and whisker plot

A large amount of data distribution and the outliers can be easily identified by this kind of plot. This is the kind of statistical plot from which we will identify five different statistical measures. The L-estimators of the dataset – interquartile range, midhinge, range, mid-range, and trimean are visualized by the box plot. The box plot is mainly used to find the outliers in the dataset. Outliers influence the performance of the dataset, and they will be removed from the dataset to increase the performance of the model. The min value, max value, the first quartile Q1, third quartile Q3, median Q2, and the outlier for a dataset are shown in Figure 6.6. The interquartile range is calculated by subtracting the values of Q3 and Q1. The data spread, locality, and skewness of data are also visualized using the box plot.

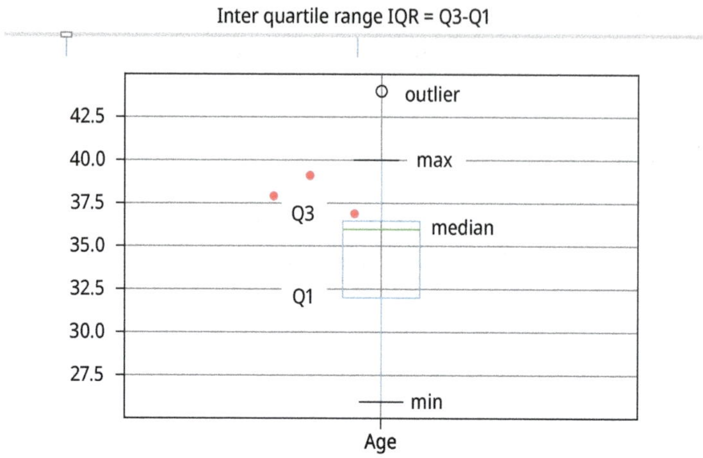

Figure 6.6: Box plot.

6.3.4.3 Histogram

The distribution of the numerical data in big data storage may be represented by a histogram in Figure 6.7. The frequency of data distribution in a defined range is visualized by the powerful data visualization tool, histogram. Here, the large range of data is compressed with in a limited bin for easy understanding and further analytics. There are a variety of histogram representations including probability distributed, right-skewed, left- skewed, random distribution, and bimodal histograms.

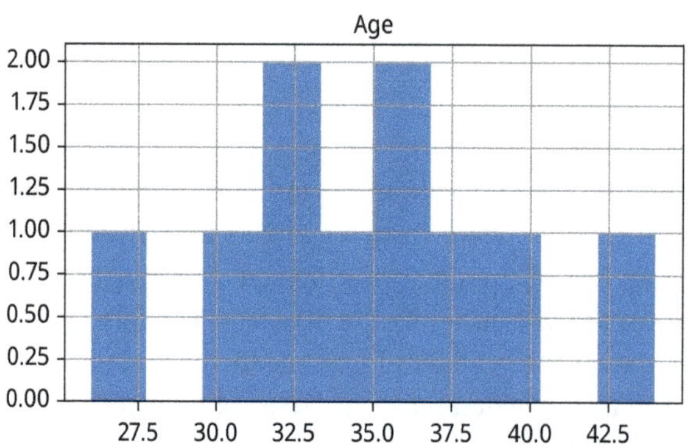

Figure 6.7: Histogram.

6.3.4.4 Word cloud

World cloud is one of the major visualization techniques especially used for social media analysis mentioned in Figure 6.8. It represents the frequency of words in big data text such that the size and width of the word is related to the frequency of the word in the particular text file. It is applicable mainly for unstructured data like the text in newspapers, web, and so on.

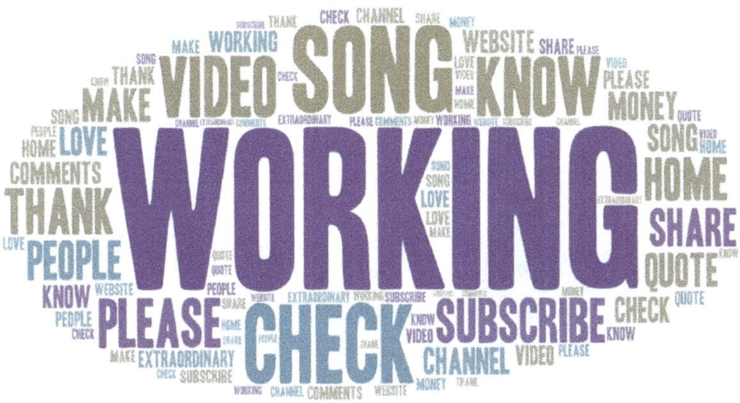

Figure 6.8: Word cloud.

6.3.4.5 Bar, pie, and line charts

These are some of the traditional visualization techniques applicable for small as well as large quantity of data. These plots are suitable for comparing the range of attributes in a dataset with one another. Generally, different colors and textures are used to differentiate the various attributes. A variety of bar and pie charts and line plots shown in Figure 6.9 are adopted for different big data applications.

Other visualization techniques such as heat map, correlation matrix, and scatter plots are used to show the relationship between the attributes in the data storage.

6.4 Architecture for big data analytics

Data from many sources will be loaded into the analytics program, which will then seek for trends, assess those patterns, and present the results in a way that is understandable to people. Analytics software searches through data sets that are too vast and diverse to be efficiently examined manually for hidden correlations, using mathematics, statistics, probability, and prediction models. To achieve a certain goal, the

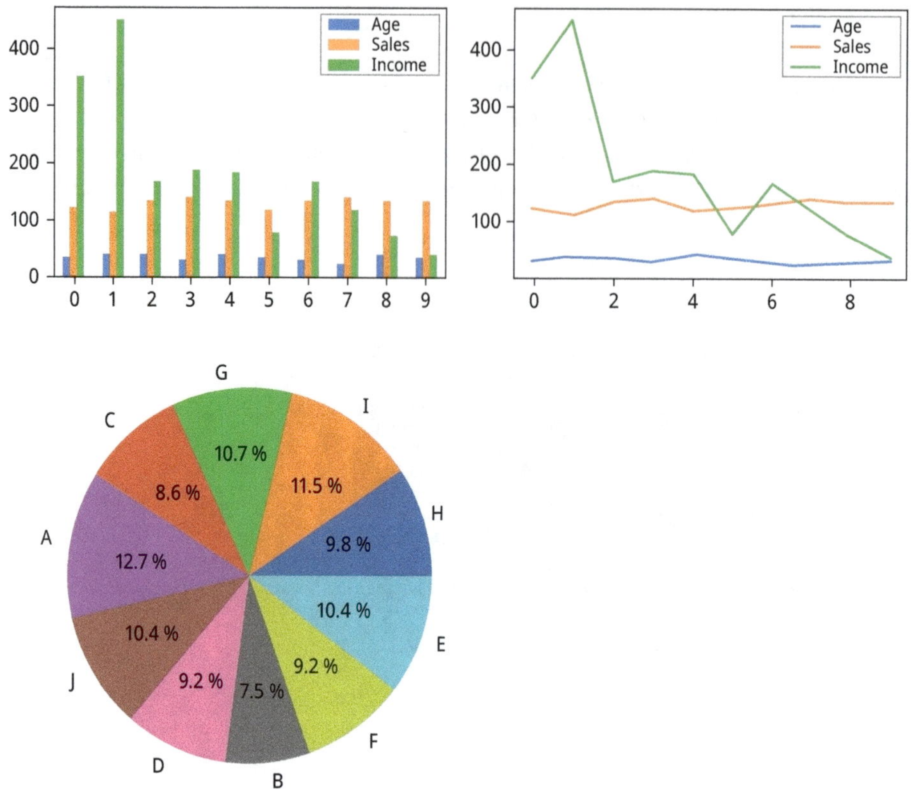

Figure 6.9: Bar, pie, and line chart.

best analytics solutions now on the market integrate a variety of software features with cutting-edge technologies like Machine Learning and pattern recognition.

By definition, big data is huge, and the processing is divided and carried out across several nodes. Although the ideas behind distributed processing have been around for a while, it is only recently that businesses have begun to employ them to analyze very big data sets in order to gain insight and make wise decisions. Moreover, the availability of open-source cloud platforms like Hadoop/MapReduce has promoted the use of big data analytics across a variety of industries. Traditional corporate analytics solutions have evolved to be more clear and user-friendly. On the other side, big data analytics tools are incredibly complicated, need a lot of programming, and call for the use of a range of abilities. Figure 6.10 represents the architecture of Big Data.

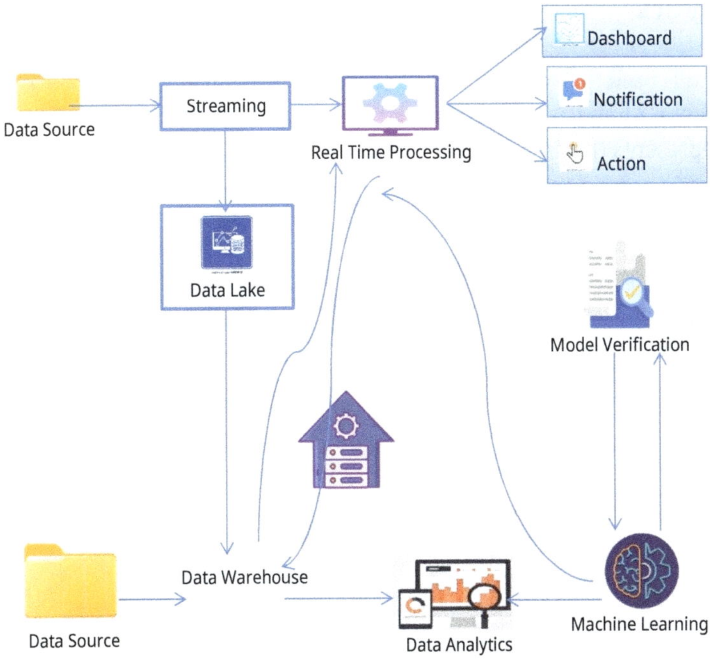

Figure 6.10: Architecture of big data.

6.4.1 Data sources

Big data architecture depends on the data source and type of data. The data sources are owned by third-party data providers. Relational Database Management Systems (RDBMS), data warehouses, cloud-based data warehouses, SaaS applications, real-time data from business servers and sensors like IoT devices, third-party data providers, and static files like Windows logs are just a few examples of data sources. Both batch processing and real-time processing of the managed data are options.

6.4.2 Data warehouse

The architecture's second layer that receives data for big data is the storage layer. It offers the infrastructure for archiving historical data, formats it for storage in that format, and stores it there. For instance, only Hadoop Distributed File System (HDFS) and RDBMS are used to store structured data for batch processing. The data is typically stored in the Data Lake in accordance with the needs of the system.

6.4.3 Data analytics

Data analytics, which is used to provide business projections, involves detecting correlations, hidden patterns, unnoticed trends, and other information from massive databases. Figure 6.11 represents the data analytics steps.

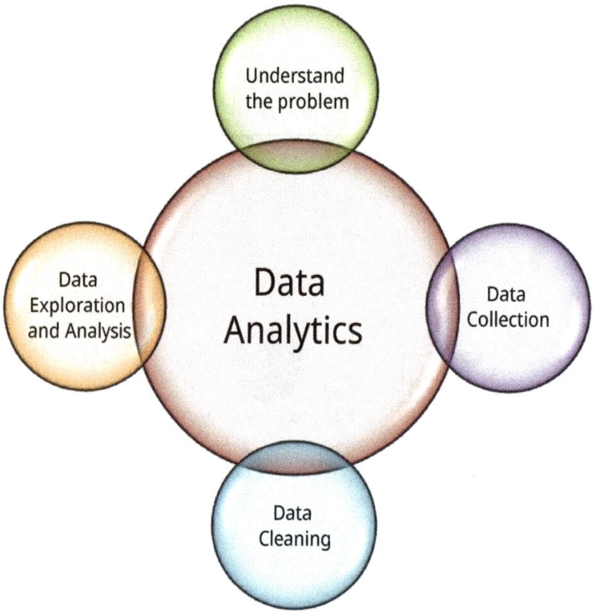

Figure 6.11: Steps involved in data analytics.

Understand the problem: The first phase in the analytics process involves comprehending the company problems, identifying the corporate goals, and planning a profitable solution. E-commerce businesses frequently struggle with problems including anticipating product returns, providing pertinent product recommendations, cancelling orders, spotting fraud, maximizing truck routing, etc.

Data collection: To solve the issues that your company faces, you must next gather transactional business data and customer-related information from the last several years. The data may include details such as how many total units of a product were sold, how much money was made from sales and profits, and when the order was placed. The future of a firm is greatly influenced by historical data.

Data cleaning: Your entire collection of data will now frequently be disorganized and contain undesired missing values. Such information is neither relevant nor appropriate for data analysis. Therefore, in order to prepare the data for analysis, you must clean it to remove unnecessary, redundant, and missing variables.

Data exploration and analysis: Executing exploratory data analysis is the next crucial step after collecting the appropriate data. To evaluate, display, and forecast future effects from this data, you need to employ business intelligence tools, data mining techniques, and predictive modelling. By using these techniques, you may determine the influence and connection between a specific characteristic and other variables.

– Descriptive analytics
The current state of the prescribed business is identified from the available data. It also describes the current statistics of the available data and enables prediction with the existing historical information. In descriptive data analytics, the correlation between the dataset variables and the relationship between the response variables and predominant input variables are also analyzed with suitable tools. For example, in the college admission process, if the college wants to identify the portal through which the number of inquiries is high, it can be identified by the historical data collected from the traffic. Descriptive analysis is initially the required analytical metric; the data required for the metric is also identified and then, the corresponding features are extracted from the big data using the appropriate tool and then analyzed. Hadoop, MongoDB, Cassandra, and Spark are some tools used for descriptive data analysis in big data environments.

– Diagnostic analytics
This kind of analysis is carried out to identify the cause of uncertainties in decision-making. The better assumptions to meet some criteria for achieving the projected metrics of the business are also analyzed in these analytics only. In-depth knowledge about the occurrence of failure in the prescribed project is identified in the diagnostic analysis. This analytics concentrates on the keyword "why"– why the production was reduced, why the profit reduced, why the failure occurred, and so on. Hypothesis testing, correlation-cause analysis, and diagnostic regression are some of the techniques available for big data diagnostic analytics. In-depth knowledge about the program is required for diagnostic analysis. Power BI, ThoughtSpot, Google Looker, Tableau, and Qlik Sense are some of the tools that are useful for this purpose. If the admissions are down even though there is demand for that course, the institution should identify the cause for lack of knowledge – online payment or server issues or less knowledge in that course. This type of analysis is identified as diagnostic analytics.

– Predictive analytics
The future of the problem is predicted with the help of past and present data by predictive analytics. An expert in the field is required for forecasting the predefined metrics with relevant information. The built-in models, regression models, Machine Learning, and Deep Learning algorithms are adopted for predictive analytics purposes. In regression analysis, the output and predominated input variables are correlated by the mathematical equations and the correlation coefficients are calculated for the prescribed solution, first. Then, the calculated coefficient is used for the

prediction of the metric. Driverless AI, IBM Watson Studio, Microsoft Azure Machine Learning, and SAP Predictive Analytics are some of the emerging predictive analytics tools available for business as well as research. In an institution, the scope of the course in the forthcoming academic year can be predicted with the past and current admission information.

– Prescriptive analytics
It is used to identify optimal decision-making by tuning the decision-making parameters. This analysis combined the feature of descriptive and predictive analytics to optimize the tuning parameters. Automated analytics is the advancement phase of prescriptive analytics. In the finance section risk, business and financial loss can be overcome using such analytics. RapidMiner Studio, Dataiku, Altair Datawatch Knowledge, and Halo Software By Logility are some of the commonly used prescriptive analytic tools. For example, to improve the admissions in the future, if the admission scenario and the procedures are also changed, it is an example of Prescriptive Analytics.

6.5 Applications

Real-time big data is supportive in all applications, because every field deals with large amounts of data. Therefore, there is a need to process big data in all fields. Some of the sectors supported by big data are healthcare, education, manufacturing, government, insurance industry, transportation, etc.

6.5.1 Healthcare sector

To enable physicians to use evidence-based medicine as opposed to requiring a battery of laboratory testing for each patient who enters a hospital, a cell phone app was used by a few organizations like Beth Israel, to collect data from millions of patients. Although battery tests are generally ineffective and expensive, they may be effective. The University of Florida has integrated Google Maps and open-source public health data to provide visual data that makes it easier to identify and effectively explore medical data used to monitor the progression of chronic disease. Obamacare has also used big data in a variety of ways. A few big data service companies in this industry are Recombinant Data, Humedica, Explorys, and Cerner.

6.5.2 Education

Big data is heavily used in higher education. Students receive data and subject resources via Google Class Room (GCR). An Australian organization with over 26,000 students has built a learning and management system that monitors students' logins, the length of time they spend on different system sites, and their general development over time. Big data is also used in a different application in education to evaluate instructors' performance, in order to create a favorable learning environment for both students and teachers. The number of students, the subject matter, the demographics, the aspirations, the behavioral classification, and a variety of other factors can all affect how effective a teacher is.

6.5.3 Manufacturing and natural resources

Predictive modelling is made possible by big data, which can be used to ingest and integrate enormous volumes of data from geospatial, graphical, textual, and temporal sources. This aids decision-making in the natural resources industry. Big data analytics has some fascinating uses, including the interpretation of seismic data and reservoir classification. Big data has been used, among other benefits, to gain a competitive edge and solve current production issues.

6.5.4 Big data in government

Numerous public services, such as fraud detection, health-related research, financial market analysis, and environmental protection, use big data.

Here are some further, more specific examples:

- The Social Security Administration (SSA) is employing big data to analyze a sizable number of social disability claims that have been submitted and received as unstructured data. The analytics are used to expedite decision-making and spot spurious or fraudulent claims by quickly and accurately estimating medical data.
- The Food and Drug Administration (FDA) is investigating and identifying disease and illness patterns linked to food using big data. This makes it possible to react more quickly, which has led to quicker treatment and fewer fatalities.

6.5.5 Insurance industry

By analyzing and forecasting consumer behavior using data from social media, GPS-enabled gadgets, and CCTV footage, big data also enables insurance businesses to better retain their customers. Since huge volumes of data are evaluated primarily during the underwriting stage, predictive analytics from Big Data has been engaged to provide speedier service when it comes to claims management. Additionally, fraud detection has been improved. Real-time claims monitoring throughout the claims cycle has been used to generate insights through the usage of vast amounts of data from digital channels and social media. Sprint, Qualcomm, Octo Telematics, and The Climate Corp. are some of the big data suppliers in this sector.

6.5.6 Transportation

Massive data from location-based social networks and fast data from telecommunications have recently had an impact on travel patterns. Unfortunately, research on understanding travel behavior has not progressed as quickly. Models of transportation demand are frequently still reliant on vague new social media platforms.

Big data has been used by organizations, governments, and individuals in the following ways:

- The use of big data in the transportation industry by private sector for competitive advantage, revenue management, technological advancements, and logistics
- Big data is used by individuals for a variety of purposes, such as route planning to conserve time and fuel, travel arrangements for tourism, etc.

6.5.7 Energy and utility industry

With smart meter readers, data can now be collected around each 15 min as opposed to once a day with traditional meter readers. Comprehensive data is being used to analyze utility consumption more effectively, which improves customer service and utility use management. Improved asset and staff management, which help spot errors and repair them as soon as possible before total failure happens, is another benefit of using big data in utility organizations. Alstom Siemens, ABB, and Cloudera are a few big data service providers in this industry.

6.6 Challenges

Real-time big data applications represent challenges in terms of design, implementation, and operation.

– Real-time event transfer

A processing location should receive real-time transmissions of all recent events from remote applications. Events can be transmitted from their dispersed sources as aggregated events, filtered events, or raw events. All created existing row, filtered, and aggregated events can be sent to a central processing point or dispersed intermediate processing points for preprocessing or additional filtering and aggregation, before being sent to the main decision-making unit.

– **Real-time situation discovery**
The goal of this stage is to identify anomalies in real-time commercial or operational situations and current events. Situation discovery techniques allow for the definition and real-time modification of the policies that are used to filter and aggregate event transfer. To address the observable organizational or corporate issues, it might also stimulate analytical or decision-making processes.

– **Real-time analytics**
It comprises using real-time analytical services to pinpoint the underlying causes of operational and business problems and outliers. Real-time analytical processing can make use of one, several, or many integrated analytical services. These in-the-moment analytical services ought to be able to forecast performance and pinpoint the dangers brought on by evolving operational and commercial contexts.

– **Real-time decision-making**
It is possible to choose the best alternative for enhancing current business operations or profitability and choose the most appropriate steps for a reaction to the business or operational environment, based on real-time analytics data. The dynamic and intelligent response to shifting business and operational conditions is made possible with the help of decision configurations.

– **Real-time responses**
This requires starting, carrying out, and keeping track of an activity described by a real-time decision-making procedure. For instance, if business agrees to purchase a certain number of shares, the real-time response system will initiate a stock purchase transaction, send the transaction request, and track the execution of the transaction.

6.7 Conclusion

Storing of terabytes and petabytes of data in data warehouse is needed to be analyzed. Therefore it is necessary to analyze the data in real time. Within seconds, real-time analytics may solve problems and support decision-making. In the event of an issue, organizations must act fast. With the use of real-time analytics, organizations may take quick action and be proactive by capturing opportunities or averting issues before they arise. Real-time analytics uses statistical techniques to analyze the raw data. Real-time event transfer, situation detection, analytics, decision-making, and real-time reactions are among the problems that must be overcome in order to develop effective real-time big data applications. Therefore, to design the most effective and efficient application, we must choose the best answers to these problems and develop a real-time application that is dependable and able to meet the deadlines.

References

[1] P. Liang, Y. H. Liu, Research landscape of business intelligence and big data analytics: A bibliometrics study, Expert Syst. Appl., 111(128), 2018, 210, doi: 10.1016/j.eswa.2018.05.018..

[2] P. Akhtar, J. G. Frynas, K. Mellahi, S. Ullah, Big data-savvy teams' skills, big data-driven actions and business performance, Br. J. Manage., 30(2), 2019, 252–271.

[3] P. Akhtar, Z. Khan, S. Tarba, U. Jayawickrama, The internet of things, dynamic data and information processing capabilities and operational agility, Preprint, 2017, 1–32, doi: 10.1016/j.techfore.2017.04.023.Pervaiz.

[4] X. Wang, Y. He, Learning from uncertainty for big data: Future analytical challenges and strategies, IEEE Syst. Man Cybern. Mag., 2016, 2(2), 26–31.

[5] Jain A., The 5 Vs of big data. IBM Watson Health Perspectives, 2017. https://www.ibm.com/blogs/watson-health/the-5-vs-of-big-data/. Accessed 30 May 2017.

[6] E. K. Clemons, R. M. Dewan, R. J. Kauffman, T. A. Weber, Understanding the information-based transformation of strategy and society, J. Manage. Inf. Syst., 34(2), 2017, 425–456, doi: 10.1080/07421222.2017.1334474.

[7] S. Erevelles, N. Fukawa, L. Swayne, Big data consumer analytics and the transformation of marketing, J. Bus. Res., 69(2), 2016, 897–904, doi: 10.1016/j.jbusres.2015.07.001.

[8] M. B. Forbes, How much data do we create every day? 2018. https://www.forbes.com/sites/bernardmarr/2018/05/21/how-much-data-do-we-create-every-day-the-mind-blowing-stats-everyone-should-read/#4146a89b60ba.

[9] D. Saidulu, R. Sasikala, Machine learning and statistical approaches for Big Data: Issues, challenges and research directions, Int. J. Appl. Eng. Res., 2017, 12(21), 11691–11699.

[10] J. Archenaa, E. A. Mary Anita, A survey of big data analytics in healthcare and government, 2nd International symposium on big data and cloud computing (ISBCC15), Proc. Comput. Sci., 50, 2015, 408–413.

[11] J. Zhang, Y. Chen, T. Li, “Opportunities of Innovation under Challenges of Big Data,” 10th International Conference on Fuzzy Systems and Knowledge Discovery (FSKD), IEEE, 2013.

[12] N. Mohamed, J. Al-Jaroodi, “Real-Time Big Data Analytics: Applications and Challenges,” In 2014 international conference on high performance computing & simulation (HPCS), pp. 305–310. IEEE, 2014.

[13] D. García-Gil, J. Luengo, S. García, F. Herrera, Enabling smart data: Noise filtering in big data classification, Inf. Sci., 479(2019), 2017, 135–152, doi: 10.1007/s10815-006-9022-8.

[14] W. A. Günther, M. H. Rezazade Mehrizi, M. Huysman, F. Feldberg, Debating big data: A literature review on realizing value from big data, J. Strategic Inf. Syst., 26(3), 2017, 191–209, doi: 10.1016/j.jsis.2017.07.003.

[15] B. Kitchens, D. Dobolyi, J. Li, A. Abbasi, Advanced customer analytics: Strategic value through integration of relationship-oriented big data, J. Manage. Inf. Syst., 35(2), 2018, 540–574, doi: 10.1080/07421222.2018.1451957.

[16] S. R. Swarna, A. Kumar, P. Dixit, T. V. M. Sairam, "Parkinson’s Disease Prediction using Adaptive Quantum Computing," 2021 Third International Conference on Intelligent Communication Technologies and Virtual Mobile Networks (ICICV), 2021, pp. 1396–1401, doi: 10.1109/ICICV50876.2021.9388628.

[17] G. Sasubilli, A. Kumar, "Machine Learning and Big Data Implementation on Health Care data," 4th International Conference on Intelligent Computing and Control Systems (ICICCS), 2020, pp. 859–864, doi: 10.1109/ICICCS48265.2020.9120906.
[18] S. M. Sasubilli, A. Kumar, V. Dutt, "Machine Learning Implementation on Medical Domain to Identify Disease Insights using TMS," 2020 International Conference on Advances in Computing and Communication Engineering (ICACCE), 2020, pp. 1–4, doi: 10.1109/ICACCE49060.2020.9154960.

Sruthi Anand, Granty Regina Elwin J., R. Priyadharsini, K. Ranjeetha Priya, D. Prabha

Chapter 7 Quantum processors/networks/sensors

Abstract: The concept of quantum computing was introduced to make processing faster so that processes can run in parallel. It harnesses the law of quantum mechanics that aids in solving highly complex problems. The law incorporates the concept of superposition and entanglement, which are the primary properties of quantum. The qubits are those involved in quantum, with the feasibility that a bit can represent any of the two states, unlike classical computers. With the advent of the new technology, it is important to enhance the other units that support in building up the technology of the future. This chapter proposes the different aspects involved in quantum computing, such as quantum sensors, the processing of quantum, the networks established, etc. It also emphasizes on the importance of moving from the classical computers to quantum, with comparisons. The focus is on knowing that there is the possibility of quantum computing in solving problems that cannot be solved in polynomial time. All these aid in uplifting the classic computing power to quantum. As it is understood that this technology is still in the research stage, the chapter provides information on what it is to actually build a quantum system. Having a brief idea on the different units, it also identifies the different applications of quantum with a few case studies.

7.1 Introduction to quantum computing

Quantum Computing is an emerging field in computing technology that incorporates quantum mechanics, which helps in solving complex problems. A quantum computer is a kind of processor that makes use of the smallest physical particles, such as atoms and molecules. It possesses the abilities of a computer but with added functionalities and abilities. Quantum computing works and operates on the idea of quantum theory,

Sruthi Anand, Department of Information Technology, Sri Krishna College of Engineering and Technology, Coimbatore, Tamil Nadu, India, e-mail: sruthianand@skcet.ac.in
Granty Regina Elwin J., Department of Information Technology, Sri Krishna College of Engineering and Technology, Coimbatore, Tamil Nadu, India, e-mail: grantyreginaelwinj@skcet.ac.in
R. Priyadharsini, Department of Science and Humanities, Sri Krishna College of Engineering and Technology, Coimbatore, Tamil Nadu, India, e-mail: priyadharsini@skcet.ac.in
K. Ranjeetha Priya, Department of Computer Science and Engineering, Sri Ramakrishna Engineering College, Coimbatore, Tamil Nadu, India, e-mail: ranjeethapriya.k@srec.ac.in
D. Prabha, Department of Computer Science and Engineering, Sri Krishna College of Engineering and Technology, Coimbatore, Tamil Nadu, India, e-mail: prabha@skcet.ac.in

https://doi.org/10.1515/9783110791402-007

which explicitly focuses and explains the nature of matter and energy at the atomic and subatomic levels [1].

Traditional computers operate and work on bits, whereas quantum computers work on quantum bits, called qubits. Qubits are short small objects such as electrons or photons. The classical computers revolve and function around on or off states, whereas a quantum works on mechanical spin states. In quantum computing, the qubit is the basic unit of information. Quantum computers execute their calculations pertaining to the object's state probability, even before it can be measured. This proves the fact that they can process huge data exponentially, when compared to the classical computers. As elaborated, the quantum state is defined as a mixed superposition, the state that can be compared when a coin is flipped even before it can land.

With classical computers and supercomputers, quantum computers have two major fundamentals: 1) quantum entanglement and 2) the superposition principle. The supercomputers use state-of-the-art processors for computing, wherein quantum computing aids in building new solutions for highly complex and unsolvable tasks. The classical computers even have thousands of GPU and CPU cores to improve the processing speed, but are still incapable of solving a few problems. There are occasions when even supercomputers get stumped, because of a complex problem [2]. Using quantum theory, it is even possible that the atomic particles may lie in more than one state at the same time, which might not be possible in the ordinary states of a computer.

7.2 Comparisons between a classical computer and quantum computer

Quantum computers can be used to solve complex problems such as diagnosing and running analysis for few chemical or drug trials, which cannot be fully completed over a particular period of time. Hence, they are highly complex and expensive.

Increasing the number of transistors in classical computers may expand the power linearly, whereas quantum computers grow exponentially.

On building quantum computers, complex problems would run faster than a normal activity that can be handled by a classical computer.

With the ability of these quantum computers, it might be possible to even formulate and work with drugs for cancer. They can also be employed in missiles and aircrafts.

With the potential of a qubit, it is understood that the quantum machines are processed simultaneously taking the capability of multiple classical computers, and not sequentially like the normal systems.

The primary difference between the classical and quantum computers lies in the way the data is processed. The qubits can make more powerful calculations for certain problems, as they have the ability to consider different states at the same time. Traditional computers use the logic of the binary states 0s and 1s, while the quantum takes the leverage of light and matter at the fundamental level.

Even nowadays, when a computer is designed to solve complex problems, it is done with the help of supercomputers. The supercomputers are a large group of interconnected systems that can perform parallel calculations at the same time, using the processors and powers of many independent systems [3].

Considering the task of analyzing the protein sequences and sorting them, it is predicted that the growing patterns of amino acids form a protein structure. This chain can be identified by the brute force algorithm using classical computers, wherein the system becomes more complex when the protein structure grows longer. And this could potentially be solved by quantum computers analyzing the protein structure that is folded in any of the trillion ways.

The quantum provides a solution to such problems that are capable of creating multidimensional space by linking different patterns to data points. For the problem discussed, the solution could be identifying all possible combinations of folds that require the least energy to be produced.

Properties of Quantum theory:

1) Superposition
2) Entanglement

7.2.1 Superposition

Unlike the normal computers that can reside in either of the two states, quantum can lie in any of the states as either 0, or 1, or even as a combination of the two.

The qubits can represent various combinations at the same time, making them reside in multiple states. Hence, a quantum computer with many qubits can simultaneously reach and achieve the potential outcomes. The result of that is measured by the qubits and once it is done, it is retained back to any of the binary states.

7.2.2 Entanglement

It is the process of taking the members of a pair of a qubit to appear as a single state.

It is the property of quantum mechanics, where the effect on one particular state can affect the other, although the particles are far apart or separated by long distances.

By this property, if the state of a qubit is changed, it will automatically change the other member of the pair in a way that can be predicted. It is by this property that the power of quantum computers can be understood or conveyed.

Thus, this ability of quantum computers with the properties of superposition and entanglement, paves a way to provide solutions for problems that are difficult to be handled by classical computers [5].

7.3 Hype or reality

Quantum computing has emerged to help in solving highly complex problems. With the advent of quantum mechanics, researchers and scientists were hoping to carry data in parallel bits, unlike the classical ones. The qubits were actually thought of for handling huge complex problems simultaneously, with multiple parallel bits. It is hyped, and prototypes of quantum are able to control a small number of qubits, somewhere around 100. This was typically found using theoretical experiments and the concept of quantum was expected to do more than what it is in a classical system. Researches are still going on for one that can process more than one million qubits to build a perfect quantum computer that will satisfy the needs of faster processing. While researchers are still in the process of analyzing and establishing systems that would cater the needs of business, it is predicted that in a period of five years, it would eventually be deflated [6]. Hence, only after understanding and working on these bottlenecks, the practicality of quantum computing would be a boon to the technology world.

7.4 Quantum processing unit

The QPU is also known as the Quantum processor and this, similar to the CPU of classical computers, is the brain of a quantum computer. Computations and calculations are done more rapidly by the QPU by manipulating the characteristics of particles like photons and electrons. Unlike conventional digital computers that store and process information in the form of the binary bits (0 and 1), the QPUs use qubits / quantum bits to store information. Data are encoded as 0s or 1s or both, at the same time. The new unit of information, qubit, which can be in any superposition of both 0 and 1 at the same time, forms the core of Quantum computing. The property of superposition is one of the primary reasons that facilitate the QPU's capability of handling numerous combinations of states simultaneously [4].

Quantum mechanics is difficult to simulate because matter, at a quantum level, may exist in numerous configurations. Probability theory is insufficient to represent these configurations because the various quantum states may interfere with each other, at the same time. The foundation of quantum computing is these numerous quantum configurations. These quantum states represent information, and quantum gate operations compute this information. Qubits are capable of representing information in the quantum states of matter.

Value of the qubit may either be 0 or 1 or a quantum superposition of 0 and 1. The quantum state vector is used to represent the configuration of the one-qubit quantum system. This vector is a unit normal two-dimensional column vector. The vectors, 1 0 and 0 1, are the basis for the vector space that represents the state of the qubit. They are similar to 0 and 1 of the binary space. The various configurations of a qubit may be pictured using the Bloch Sphere. The vectors other than the basis vectors are got by various rotations and can be seen in the Bloch sphere. Figure 7.1 represents the Quantum Gates.

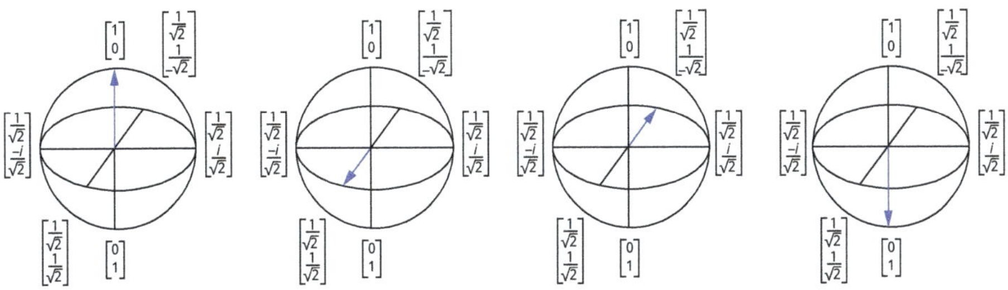

Figure 7.1: Quantum gates.

7.4.1 Quantum gates

All operations performed on a classical computer may ultimately be broken down into a series of Boolean operations, which, in turn, are implemented by the classical logic gates. Quantum gates are an extension of the classical gates.

7.4.2 Controlled NOT gate

This gate is a two-qubit quantum gate that accepts two inputs, the control input and target input, and generates two outputs. The CNOT gate flips the state of the second bit if the control bit's value is 1. Let the natural basis for the qubits be $(1\ \ 0)$ $(0\ \ 1)$ and let $|0\rangle$ and $|1\rangle$ denote them respectively. For a control qubit value of $|1\rangle$, the CNOT gate toggles the value of the target qubit from $|1\rangle$ to $|0\rangle$ or vice versa. For a control qubit value of $|0\rangle$, the target qubit remains unchanged. The following Figure 7.2 illustrates the working of the CNOT gate.

Input		Output	
Control Bit	**Target Bit**	**Control Bit**	**Target Bit**
$\|0\rangle$	$\|0\rangle$	$\|0\rangle$	$\|0\rangle$
$\|0\rangle$	$\|1\rangle$	$\|0\rangle$	$\|1\rangle$
$\|1\rangle$	$\|0\rangle$	$\|1\rangle$	$\|1\rangle$
$\|1\rangle$	$\|1\rangle$	$\|1\rangle$	$\|0\rangle$

Figure 7.2: Truth table: CNOT gate.

7.4.3 SWAP gate

The SWAP gate is a 2-input-2-output gate that swaps the values of its input qubits. The following Figure 7.3 illustrates the operation of the SWAP gate.

Input		Output	
a	b	a	b
$\|0\rangle$	$\|0\rangle$	$\|0\rangle$	$\|0\rangle$
$\|0\rangle$	$\|1\rangle$	$\|1\rangle$	$\|0\rangle$
$\|1\rangle$	$\|0\rangle$	$\|0\rangle$	$\|1\rangle$
$\|1\rangle$	$\|1\rangle$	$\|1\rangle$	$\|1\rangle$

Figure 7.3: Truth table of the SWAP gate.

7.4.4 One-qubit quantum gates

The gates, I and Z, represent the identity operation and these gates leave the qubits unchanged. The gates, *X* and *Y*, swap the states of the qubits. These gates are called Pauli gates and they perform the Pauli transformations. The Pauli X gate, also called the NOT gate, changes or switches the value of the qubit from one state to another. The Figure 7.4 represents the truth table of the one-qubit quantum gate.

Input	Output
$\|0\rangle$	$\|1\rangle$
$\|1\rangle$	$\|0\rangle$

Figure 7.4: Truth table of the one-qubit quantum gate.

The Pauli I gate is just the identity matrix and it leaves the qubits unchanged.

7.4.5 Quantum complexity

Quantum computers may not necessarily solve all problems faster than conventional computers. Certain problems that cannot be solved by conventional computers in polynomial time may be handled efficiently by quantum computers, and these problems are classified as Bounded-error Quantum Polynomials (BQP). Figure 7.5 illustrates the relationship between the complexity classes of problems. Integer factoring problem is an example of BQP.

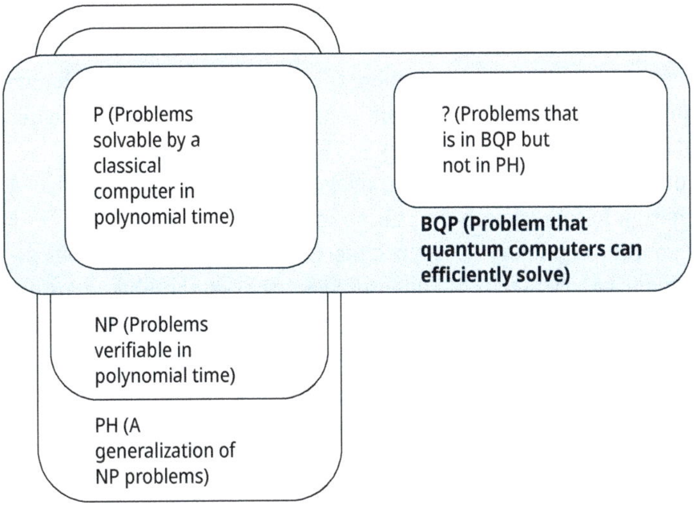

Figure 7.5: Relationship between complexity classes.

7.5 Quantum networking

In everyday life, digital information is transmitted and shared using the quantum network, which is most like the classical network. Quantum networks use qubits, which help encode information in a way that is entirely different from the classical way of thinking. The tricks used by qubits are fundamentally different from classical computing bits. The evolution of the quantum network is expected to bring a change in the entire world. The quantum processor can process quantum logic gates on a certain number of qubits.

Quantum networks transmit individual photons, which are usually referred to as quantum data. Quantum networks help enable and work with new applications like physics-based unhackable security that are more powerful by using the power inherent in the quantum states. They can solve tasks, which takes years to complete in classical computers, in fractions of seconds using quantum computers,.

7.5.1 Working of quantum networks

The quantum network works by means of transmitting encoded key over a fiber-optic cable with the shape of qubits among the endpoints. The qubits are usually polarized photons, which could tour effortlessly alongside fiber-optic cables. Any attempt to intercept the quantum key destroys the qubit's sensitive quantum country and the statistics it holds, by sending alerts to the endpoints that an intrusion occurred. The detectability of the intrusions is what guarantees the safety of the transmission.

The trick, or alternatively the challenge, of a quantum community is that quantum communications are restricted through the gap that photons can journey over fiber-optic cable, generally around 100 km [10]. To resolve this problem, Quantum Xchange has developed "relied-on-nodes," which function in an identical manner as a repeater does in a conventional communications community.

The key is decrypted into the original bits at every node. After this, it is placed back to the quantum nation for addition to the subsequent node.

The invention of quantum technology has helped increase the quantum capabilities – quantum computers, quantum cryptography, quantum-secure cryptography, quantum communication in the international front. Building a quantum network is a challenging task that is still in process. A few countries have moved towards the development of these networks in a faster stride. Now, the U.S. is displaying development as well, with Quantum Xchange's Phio community and related relied-on-node technology.

7.5.2 Qubit

Qubits are the constructing blocks of quantum data. Whereas a classical bit should be both a 0 and a 1, a qubit may be in a superposition of each zero and 1. In addition, qubits have the odd belongings that after measured, they probabilistically disintegrate to both a zero and a 1. For example, if we degree a qubit in an "identical superposition" of zero and 1, there can be a 50/50 shot of measuring both states. To top this, a couple of qubits can become "entangled."

7.5.3 Benefits of quantum networks

We have already set up that quantum networks are extremely steady and are impervious to any form of cyber hacking. It may not be possible to intercept encrypted messages. However, apart from the warranty of security, quantum net can transmit big volumes of statistics through extensive distances at a far quicker velocity than classical networks are known to. This can be progressive for apps and software, in addition for any updates they want to set up over the air.

The economic quarter will substantially benefit from the usage of quantum net, specifically in securing online banking transactions. Consumers will sense more secure and assured sharing private statistics and doing their banking and economic sports online, due to this promise of security. Other sectors that enjoy substantial advantage from the usage of quantum net consist of the general public and healthcare sectors [7–9]. A quicker and more secure net will assist those sectors, expedite their techniques, and make offerings promptly. Quantum computing will even permit groups beneath this quarter to remedy complicated issues and to conduct large-scale experiments and studies.

Quantum networks can permit primarily physics-based totally ultra steady communication, extra effective quantum computers, and higher quantum sensors. Just as the creators of the ARPANET could not have expected the various methods wherein we use the net today, there are possibly many extra packages of quantum networks that we cannot consider till we construct them.

7.5.4 Quantum networks in future

Quantum networks can permit primarily physics-based totally ultra steady communication. Once quantum net takes off, we can count on the beginning of an entire new industry. Of course, classical or everyday net will continue to exist and they may exist, aspect through aspect. While we will count on huge agencies to make use of quantum networks to guard the huge amount of precious information they have got in their possession, men or women purchasers, in all likelihood, will continue the use of classical net. This is not unexpected, considering that quantum net is a brand new era and could, in all likelihood, be priced high, in the beginning. In addition to sluggish adoption due to the price of overhauling contemporary traditional systems, there is, additionally, the reality that it takes time for humans to adopt new technology. This loss of urgency is likewise rooted inside the "if-it-ain't-broke-why-fix-it" mindset that purchasers often hold one to, whilst new technology is introduced. However, in time, quantum net would turn extra handy and extra cheap to a developing wide variety of humans. The longer it is used, the more common and mainstream it turns into.

7.5.4.1 Quantum sensors

A sensor is a device that produces an output signal for sensing a physical phenomenon. In other words, a sensor is a device or a machine that detects the changes in the environment and the information sent to other electronic devices, frequently a computer processor. An example of an early quantum sensor is an **avalanche photodiode (APD)**, which is used to detect entangled photons. Sensors are always used with other electronics in everyday objects, such as touch-sensitive elevator buttons and lamps that dim or brighten by touching the base, and in innumerable applications, of which most people are aware.

A quantum sensor is a device that works by the principles of quantum physics, which is based on manipulating at the submolecular level. The field of quantum sensing deals with the design and engineering of quantum sources and quantum measurements. Quantum sensors are built on continuously variable systems, i.e., quantum systems characterized by continuous degrees of freedom, such as position and momentum quadrature. Quantum sensors will also play a major role in discovering new quantum materials needed to build a large-scale quantum computer.

7.5.4.2 History of quantum sensing

Earlier, quantum sensors were found in 1600s, during the period of Galileo, when he used his own pulse to compare it with the swinging of the pendulum in the Cathedral of Pisa. He found that his heartbeat had a specific duration and it was exactly matching with one swing of the pendulum. The basic idea behind this study was the concept of tick rate and time series. The reason for this was based on the fact that atomic clocks were more accurate, having an error rate of only 1 s in 100 million years. Another important application of these quantum sensors are found in atom interferometers that exploit the characters of quantum mechanical nature, which includes both the properties of protons and neutrons. The working of the atom interferometers is based on the concept of interference between the two waves in a space. With the light pulses, each atom in the quantum has been identified with its momentum. Interferometry is mostly applied to telecommunication industry, navigations, imaging, and constructions. All these applications employ laser light to illuminate the atoms. The atom absorbs light from the laser and because of this, the momentum will increase, which makes the atom take a separate path from that it originally intended to follow. After the absorption of light is done and the recombination of the interference pattern of the atoms is done, the resultant difference of the gravitational strength of the two paths is taken. Another important application of the atom interferometers is their use in the military sector for global positioning systems(GPS), which did not turn out positive. They started failing in untapped locations. They also played a major role in the oil and gas industry for studying the seafloor.

7.5.4.3 Applications of quantum sensors beyond quantum physics

Quantum sensors are used in different types of applications due to their high sensitivity, also in critical operations. Quantum sensors are also used in the fields of renewable energy, nuclear energy, and geothermal energy. In nuclear power plants, these sensors are deployed to help in the output of the nuclear energy. In nuclear energy plant devices, detection of isotopes was done with atom interferometric sensors. In the area of manufacturing infrastructures, these sensors aid in detecting early faults in equipment, which can generate and obtain future performance diagnostic reports. Due to these approaches, it has replaced the traditional time-based maintenance practices with condition-based maintenance, which can reduce the overall cost and minimize equipment failures. The application of quantum sensors has been increasingly done in electricity transmission grids for monitoring and analyzing the performance. The core indicators in the field such as temperature loss, strain, and stress in electrical power lines can be used to determine the electricity loss. The electrical grid manufacturers make use of the input from the quantum sensors to improve the efficiencies of the grids.

7.5.4.4 Constructions

Quantum sensors are widely used in developing smart buildings recently. They are used in the construction industry to monitor the energy usage throughout the day, in each building.

Once the results are retrieved, it helps in the analysis and monitoring of the various functions of the buildings such as ventilation, heating, and lighting. This will help reduce the overall energy consumption of the building, which would contribute to consuming fewer resources.

7.5.4.5 Defects in diamond

Diamond quantum computing is a field of quantum computing that is based on the process called optically detected magnetic resonance with the (Nitrogen Vacancy) NV defect. It is observed when changes occur while measuring fluorescence after shining green light on an NV defect, or on an ensemble of them, while scanning an applied microwave field.

Color centers in diamond are magnetic quantum sensors that have gained a wide range of applicability over the last decade. Defects can be synthesized by injecting nitrogen, silicon, germanium, and other atoms into the diamond lattice.

7.5.4.6 Superconductors

The Superconducting Quantum Interference Device (SQUIDs) is one of the oldest quantum sensors built with superconducting interferometers. SQUIDs have been successfully used for materials characterization and clinical magnetoencephalography.

7.5.5 The future of quantum sensors

Quantum sensors have several advantages and they are used in many applications. This is due to the fact that the availability of the hardware is fair. The parameters are sensed outside the laboratory, which is a great barrier that is commonly faced by the advanced technologies. Hence, it creates a domain gap between the actual environments and the quantum physics research. In the field of metrology, quantum coherence is accompanied by the environment loss and noise, which creates an impact in the signal-to-noise ratio due to quantum coherence. A majority of these quantum sensor applications are deployed at the laboratory level and hence scaling them to an industry perspective is challenging. But the demand for the advanced quantum technologies and sensing has risen, which could drastically move form theoretical experiments and become mainstream commercial quantum systems.

7.6 Applications of quantum computing

7.6.1 Quantum computing

Quantum computing is a computation technique used to solve complex computational problems, which seem to be very complex to solve even by some traditional and supercomputers. The operations that are performed in this technique use the phenomena of quantum mechanics such as superposition, interference, and entanglement to solve these complex problems. The computation is normally done by using a special type of computers called quantum computers. These computers are very different from the classical computers in the way they look and perform. Quantum computers perform the operations at a high speed and consume very low energy.

7.6.2 Types of quantum computing

There are three different types of quantum computing, namely, Quantum Annealer, Analog Quantum Simulations, and Universal Quantum. These types are classified based on the problems they encounter. Figure 7.6 illustrates the types of quantum computing.

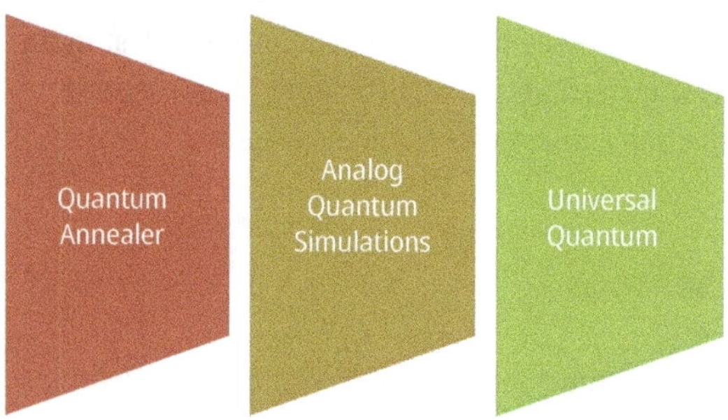

Figure 7.6: Types of quantum computing.

7.6.2.1 Quantum annealer

There could be many possible and feasible solutions to a complex problem. Among those solutions, the best optimal solution can be found by using quantum annealer. Initially, its states are represented as energy levels. These states are therefore reproduced by the superposition and entanglement properties, resulting in the lowest energy level, which, in turn, gives the optimal solution to the problem.

Analog quantum simulations

Analog quantum simulations help one to deal with physics problems that are difficult to be solved by the classical computer. Here, it stimulates more quantum interactions, which are controllable by the quantum computers. Besides solving the physics problems, it has a large application in sampling, quantum chemistry, materials science, etc.

Universal quantum

This type of computing mode is more powerful and it is also difficult to build because of the technical challenges involved in the process. This universal quantum computing method was introduced in quantum computers, so that the computer could perform all sorts of complex problems with a high speed and provide an accurate result. This is widely used in searching, secure computing, machine learning, cryptography, etc.

7.6.3 Applications

Quantum computing technique has been developed mainly to solve the complex problems that cannot be solved by using a classical computer. Besides the complexity, the quantum computers are developed in such a way that the processing time of the problem gets reduced because of the quantum algorithms used. This computing is widely used in many areas such as Artificial Intelligence and Machine Learning, drug design and development, cybersecurity and cryptography, computational chemistry, financial modelling, weather forecasting, logistics optimization, etc. The uses of certain applications are given in Figure 7.7 below:

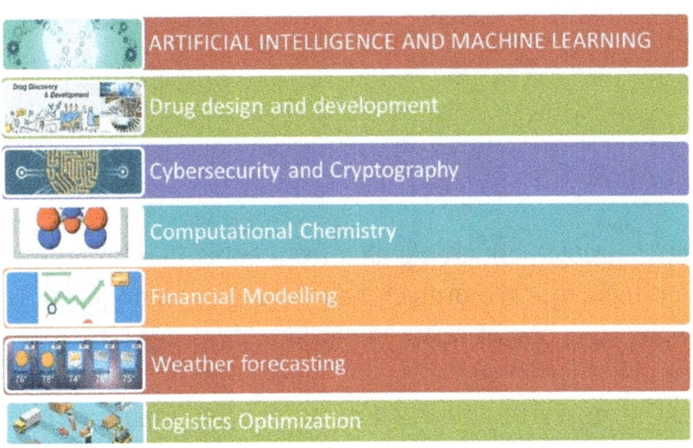

Figure 7.7: Applications of quantum computing.

7.6.3.1 Artificial intelligence and machine learning

Artificial Intelligence and Machine learning are both emerging technologies in the field of data science. They train the machines, process the knowledge in them, and model them to perform human related tasks as also some other complex tasks. For training and processing the model, different types of algorithms are used. When the complexity of the task increases, i.e., when a large amount of data is required to train the model, processing in the conventional computers takes more time, which delays the end result. Also, there will be a change in the accuracy of the data. Since the applications of Artificial intelligence and Machine learning are increasing over time, the task will also be more challenging than usual. With the use of quantum computers, the complexity of the task decreases. More data can be processed at any given time with high speed, providing more accurate results.

7.6.3.2 Drug design and development

One of the fast-growing Industries after the COVID pandemic is the pharmaceutical Industry. The pharma Industry is capable of producing a large number of medicinal drugs and vaccines to treat illnesses. It is known that drugs are produced by combining the atoms and molecules in a correct proportion. This industry uses many types of computational techniques to get the correct proportion of drugs. It can be done by using conventional computers to calculate the exact proportion of the molecules and atoms. When the compound involves a greater number of interactions between many different molecules, the conventional computer takes more time to give the result with cybersecurity, whereas the quantum computers have the ability to perform more tasks at the same time. Therefore, a greater number of interactions between the molecules will give us the exact proportion of the molecules to be taken in a very minimal time, also with greater accuracy. Apart from medicinal drugs, this computation also provides a way to form new products, to calculate the reaction rate, and to stimulate the catalytic process.

7.6.3.3 Cybersecurity and cryptography

Nowadays, it is very important to protect vulnerable data from being used to others. Each organization uses different techniques to protect their data. One among them is using cryptography. Cryptography uses one of many methods and encodes the plain text into a cipher text, which can be then decoded by using the public key. Cyber security operates in such a way that it provides a platform to secure the important and confidential data of the users. The encryption process is done by using conventional computers. The drawback is that these encryptions could be easily decoded by the hackers by some advanced technique. To overcome this problem, quantum computing is used. This technique not only solves the complex problem but also provides quantum encryption to secure the data. Quantum computing uses the Machine Learning algorithm to create a quantum encryption method, which is known as quantum cryptography. This encryption cannot be easily cracked by any other conventional computer, and thus, it secures the data.

7.6.3.4 Financial modeling

To invest in the stock market, one must be aware of some of the key factors in it. The key factors include market value, returns, risk factor, and so on. One must calculate all these before investing in the stock. To calculate them using a conventional computer, a simulation technique called Monte Carlo makes continuous runs on those computers. This process consumes more time than expected. With an increase in the complex calculation and the factors, one can use quantum computers to perform the calculations. This technique results in high quality of the solution obtained, and it is also reduces the time taken to perform the calculations. With the development of algorithms in quantum computers, more benefits in finance can be seen.

7.6.3.5 Weather forecasting

Weather forecasting is done by the meteorologists to predict the change in weather and to find out the accurate climatic change. The forecasting depends on certain factors, which include air, temperature, pressure and humidity. Besides these factors, conventional computers take more time to predict the weather and come up with very low accuracy. But the quantum computer has the ability to read through a large amount of data and predict the weather with high accuracy in very little time. Also, the prediction will never be false. Besides predicting the weather, these quantum computers are very helpful to the meteorologist in developing climatic models to get more detailed information about the weather, which helps in the migration process.

7.6.3.6 Logistics optimization

Data analysis and Machine learning models pave the way for a logistics workflow for many industries and other organizations. These models give the optimal solution to find the best route in traffic, or the airway route, or the best decision to be made in order to book profits. To achieve this, all the possible calculations and ways must be done repeatedly to find out the optimal way. It is possible to do the computational work even in classical computers. But when the dataset size and the complexity in the problem increase, it becomes difficult to proceed with this. Therefore, to overcome this complexity, quantum computers are used. It mainly depends on two computational approaches to solve this problem namely, quantum annealing and universal quantum. Quantum annealing is mainly used to find the optimal solution to a complex problem, whereas universal quantum solves types of the computational problems, which in turn, results in a great outcome.

Quantum computing technique is still in the developing process. However, once the full processing work is done, it will be widely helpful in all areas because of its high-speed computation and accuracy. Besides the application and the uses, one must also know the threats involved in this computing method. Researchers are working on them to make quantum computing a fruitful method.

7.7 Conclusion

The chapter covers the various ideas regarding quantum units like sensors, networks, and the processing unit. These types of quantum help revolutionize computation by making it possible to solve highly complex problems. While treating and understanding quantum computers for everyday usability, they have been advancing only in theory. It is focused on the study of particles at the sub atomic level. As discussed, it can create computers with power that can never be matched by classical systems. Although the technology is developing at a low pace, it is expected that it can build fault-tolerant systems by 2030. And this would benefit a number of organizations in various fields in the near future. Apart from its exceptional processing capability, it can also optimize the storage facility that is a constraint today. And because of these capabilities, the technology expects to identify and resolve major issues arising in the field of drug discovery and manufacturing, finance, cyber security, climatic change, etc. Although there are a number of advantages in various fields, it is necessary that there are guidelines set in order to prevent harm that can be caused due to it. It also points to the need to educate people and organizations on the business risks involved in it. Thus, the future of quantum is enormous, if it is manipulated and treated in the right sense. They will definitely not replace the classical computers; there are still many areas where classical computers would always remain a better choice.

References

[1] https://www.ibm.com/topics/quantumcomputing#:~:text=Quantum%20computing%20is%20a%20rapidly,available%20to%20thousands%20of%20developers.

[2] https://www.investopedia.com/terms/q/quantum-computing.asp.

[3] https://medium.com/@markus.c.braun/a-brief-history-of-quantum-computing-a5babea5d0bd.

[4] https://www.technologyreview.com/ 2019/01/29/66141/what-is-quantum-computing/#:~:text=In%20both%20cases%2C%20the%20goal,and%20another%20is%20called%20entanglement.

[5] https://plato.stanford.edu/entries/qt-quantcomp/#Qubi.

[6] https://docs.microsoft.com/en-us/azure/quantum/concepts-the-qubit.

[7] S. M. Sasubilli, A. Kumar, V. Dutt, "Improving Health Care by Help of Internet of Things and Big data Analytics and Cloud Computing," 2020 International Conference on Advances in Computing and Communication Engineering (ICACCE), 2020, pp. 1–4, doi: 10.1109/ICACCE49060.2020.9155042.

[8] P. K. Kotturu, A. Kumar, "Comparative Study on Machine Learning models for Early Diagnose of Alzheimer's Disease: Multi Correlation Method," 2020 5th International Conference on Communication and Electronics Systems (ICCES), 2020, pp. 778–783, doi: 10.1109/ICCES48766.2020.9137872.

[9] S. R. Swarna, A. Kumar, P. Dixit, T. V. M. Sairam, "Parkinson's Disease Prediction using Adaptive Quantum Computing," 2021 Third International Conference on Intelligent Communication Technologies and Virtual Mobile Networks (ICICV), 2021, pp. 1396–1401, doi: 10.1109/ICICV50876.2021.9388628.

[10] https://www.infineon.com/cms/en/discoveries/quantum-computing.

G. Ignisha Rajathi, R. Vedhapriyavadhana, R. Niranjana, N. Pooranam, R. Johny Elton

Chapter 8
Quantum computing in automata theory

Abstract: In the contemporary era we dwell in, every now and then, new technologies are developed by master brains such as researchers and scientists. This also paves the way to discover replacements to classical computers. One such progression among the great inventions is the essence of quantum computing. They are brought to light for the purpose of solving problems and scenarios solved by classical computers but in a more effective way and in the least span of time. The study of quantum computing focuses on finding new ways to compute by using quantum physics phenomena. A qubit, in contrast to a typical computer bit, can be either 0 or 1, or it can be a superposition of both 0 and 1. Quantum theory commemorates with automata theory, thereby formulating quantum finite automata. Quantum finite automaton, or basically a deterministic finite automaton, is one of the simplest models of computation, which is an input-driven model, which takes an input, and according to the input, we just read the machine, and it changes its state. This proves that any quantum finite automata are completely capable of recognizing languages like Tally with given cut point too. Such languages recognized by a cut point, by quantum finite automata, are called stochastic languages. It further collaborates with the chemical industry with its qubit power over molecular collide, interactive models, and collaborative string processing using nonregular languages.

8.1 Introduction

Artificial intelligence and quantum computing are both game-changing technologies. To make major progress with artificial intelligence, quantum computing is probably necessary. Artificial intelligence [1, 2] is constrained by the computational power of conventional computers, producing useful applications on it. Artificial Intelligence (AI) may benefit from a computational boost from quantum computing, allowing it to

G. Ignisha Rajathi, Computer Science and Business Systems, Sri Krishna College of Engineering and Technology, Coimbatore, Tamil Nadu, India, e-mail: ignishaelton@gmail.com
R. Vedhapriyavadhana, School of Computer Science Engineering, Vellore Institute of Technology, Chennai, Tamil Nadu, India, e-mail: vedhapriya.vadhana@gmail.com
R. Niranjana, Electronics and Communication Engineering, Sri Krishna College of Engineering and Technology, Coimbatore, Tamil Nadu, India, e-mail: niranbme@gmail.com
N. Pooranam, Computer Science and Engineering, Sri Krishna College of Engineering and Technology, Coimbatore, Tamil Nadu, India, e-mail: pooranam.n@gmail.com
R. Johny Elton, Indsoft Technologies, Tirunelveli, Tamil Nadu, India, e-mail: erjohnyelton@gmail.com

https://doi.org/10.1515/9783110791402-008

handle more challenging issues. Quantum AI refers to the use of quantum computing to calculate machine learning (ML) algorithms. Quantum AI can assist in achieving results that are not attainable with classical computers because of the computational benefits of quantum computing. A universal model known as quantum mechanics is built on ideas that are not found in everyday experience. For quantum computing to process data, a quantum model of the data is required. Hybrid quantum-classical models are also necessary for the quantum computer's error correction and efficient operation.

8.1.1 Quantum computing

In general, quantum computing [3] helps researchers by reviewing its features and qualities. Quantum information can be viewed as data packets that are stored in qubits for the purposes of computing. The features that render quantum data [4] important, such as superposition and entanglement, also make it difficult to observe and store. Additionally, ML must be used in the process of appropriately evaluating and interpreting this data due to quantum noisy data. It is very likely to acquire useless data in hybrid quantum-classical models when simply employing quantum processors to obtain quantum data. As a result, when a hybrid model is supported by faster data processing components like the CPU and GPU that are frequently used in traditional computers, a hybrid model is created.

8.1.1.1 Quantum algorithm

A quantum algorithm [5] is a series of actions that results in a problem's resolution. One needs to employ particular instruction sets that the device is built to handle in order to carry out these processes on a device. In contrast to classical computing, quantum computing presents new instruction sets, which are based on a completely different idea of execution. Quantum algorithms are designed to accelerate the solution by using quantum states like superposition and entanglement.

Quantum computers provide programmers a new tool for particular applications, since they are made to accomplish jobs considerably more precisely and effectively than traditional computers. There is no question that conventional computers will continue to exist in the near future, and quantum computers will be needed to support their specific capabilities, such as systems optimization. AI and quantum computing are both game-changing technologies, and for artificial intelligence to make substantial progress, quantum computing is a requirement. AI is constrained by the computational power of conventional computers, producing useful applications on computers. AI may benefit from a processing gain from quantum computing, allowing it to handle more challenging issues in a variety of commercial and scientific domains. The development of computer technology based on the concepts of quantum theory is the subject of the branch of study known as quantum computing. The ability of the quantum computer to be in various states and to carry out tasks using all potential permutations gives an immense processing power, according to the laws of quantum mechanics.

8.1.1.2 Quantum influence

For decades, there have been some restrictions on solving complicated issues. The researchers require more than just software; they also need a powerful hardware. Researchers are looking for a better method for processing and extracting the enormous volume of data due to various limits in software optimization. Some ways that quantum computing may influence future developments in artificial intelligence [6, 7] are handling large volume of datasets, finding quick solutions to challenging problems, improving models, integration of multiple datasets, preventing fraud detection, and overcoming technical challenges.

8.1.1.3 Study of quantum computing

The field of study known as "quantum computing" is devoted to creating computer systems that fit to the ideas of quantum theory [8, 9]. The capacity of the quantum computer to be in various states and to carry out tasks using all conceivable permutations at once would allow it to operate according to the laws of quantum physics and gain vast processing capability. Quantum computing's optimized and reliable AI may speed up years of investigation and provide technological advancements. Neuromorphic cognitive models, reasoning under uncertainty, and adaptive ML, are some of the main problems that modern AI is now addressing. Quantum AI is one of the most likely next-generation solutions.

8.1.1.4 Quantum parallelism

There are several advantages in quantum computing compared to conventional computing. One of the main benefits of quantum computing is that it can answer problems much more quickly than conventional computers. Due to a phenomenon called "quantum parallelism," which combines the traits of a particle and a wave, a quantum computer is able to simultaneously study a wide range of potential solutions. Quantum computers are perfect for use in Machine Learning and Artificial Intelligence applications because they can operate in difficult situations and are noise-isolating. Another advantage is that quantum computers can outperform weaknesses in traditional computing systems.

8.1.1.5 Artificial Intelligence and quantum computing

AI [10, 11] is quickly replacing human intelligence as the dominant force in the technological world. It has been projected that in the near future, machines powered by AI would be able to perform tasks that are beyond the capabilities of their human counterparts. One of the most recent developments in AI is quantum computing, which has attracted a lot of attention in terms of research and development. A great degree of progress has also been made in AI during the past few years. It has the capacity to create lifelike 3D images and films. The acceptance of quantum computing has begun, which has given rise to quantum AI. Nowadays, quantum computers are used in AI. Their full integration will lead to the technological revolution of the century. The ability to break down linguistic boundaries is one of the most fascinating predictions for quantum computing in AI. The only language that AI models can now understand is the one that was used to train them. As a result, if we want AI to learn a new language, we would have to start from scratch. However, quantum computing can help AI models get around linguistic barriers. The models were simply trained and translated into another.

8.1.1.6 Quantum trends

Quantum computing is one of the five major technological trends for 2019 according to Gartner [12], and it has the potential to completely transform the IT sector. According to Gartner, quantum computing has enormous potential for ML and AI applications. Due to the computational constraints imposed by traditional computing architectures, data scientists are now unable to utilize important opportunities. Since quantum computing would be able to handle much more data much more quickly, it could enable faster structured predictions in the areas of unsupervised learning, semisupervised learning, and deep learning. Particularly optimization-related use cases might gain from this. AI will be able to concurrently interpret and comprehend several languages. This will make it possible to create a multilingual, international AI. Another exciting forecast for the future of AI is the potential to use quantum computing to create models with enhanced decision-making capabilities: Using larger datasets to train models will be achievable with quantum computing. Financial models usually have a high rate of error because of the limited amount of data used to train them. AI will be able to make more accurate conclusions, consequently. Quantum computing is the prominent technology, which is used in various applications in the real world other than AI. It is used in cybersecurity & cryptography [13], weather forecasting, industry sector [14], healthcare, chemical, and manufacturing and industrial design [15].

8.1.1.7 Quantum principles

The model for quantum cryptography is often constructed using the fundamental principles of quantum physics [16]. The quantum future and some encryption techniques, such Advanced Encryption Standard (AES), RSA (Rivest-Shamir-Adleman), and Dara Encryption Standard (DES), are providing workable solutions to the numerous discrete logarithmic difficulties and secure data transmission. Quantum computing is used to speed up the weather forecasting process, because it takes a lot of time to process large number of data. Due to global climate change, it is mostly used to forecast the weather. In manufacturing design [17], effective procedures and designs are required to manufacture high-quality products. Industrial designers must take into account a number of aspects in order to produce a functional product. This is essential in the equipment, transportation, and electronics industries. Designers may need several prototypes, for instance, when building a high-speed jet. It can also help manufacturers troubleshoot more effectively. A quantum computer can be used to enter information regarding machine failure and help pinpoint the problem areas. The rest of this chapter describes coalescence of quantum finite automata (QFA), role of QFA, and future of quantum computing.

An interesting discussion on the two great giant knowledges such as quantum computing and automata theory has been depicted in this chapter as a coalescence of quantum finite automata. Further, the magnanimity of these two theories are deliberated individually in the introduction and its arena further developed towards quantum finite automata.

8.1.2 Automata theory

In automata theory, the formal language that we usually learn is modelled or depicted by using graph models. We have a start state, we have transitions, and then, we have a finalized state. Hence, a basic deterministic finite automaton is given by A which consists of a set of states $Q1$, finite alphabets Σ, a transition function δ, a start state $Q0$, and a set of final states F. F is a subset of Q and $q0$ is an element of Q. Sigma represents a finite set of symbols:

$$A = \left\{Q, \sum, \delta, q0, F\right\}$$

where

$$F \in Q \text{ and } q0 \in Q$$

When the machine is in state q, and it reads a symbol, it changes state to Q', as shown in Figure 8.1. Thus, it is simply an input driven machine and it is a very simple model of computation. But there are some variants of finite automata, which are called finite state machines. The same can be conducted by adding additional tapes, giving additional movements, and observation can be done post this. This is what defines one-way finite

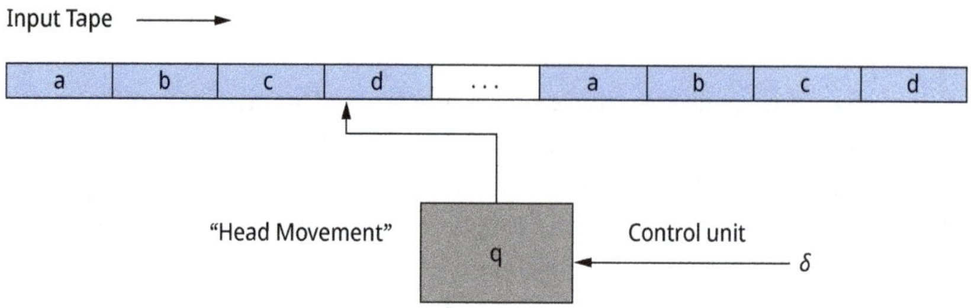

Figure 8.1: One-way finite automata.

automata. This is the same as deterministic finite automaton (DFA), except that we now have an input tape, we have a read-only input tape, and the head can move in a single direction from left to right. There is a generalization of this model to DFA, where we allow the tape head to move in both directions. Hence, the transition function for the two DFA is defined as Q times Σ. So, it can stay or take the choice of movement to its left or right as shown in Figure 8.1.

We also have a nondeterministic finite automaton. The transition function, a nondeterministic transition function, is expressed as Q times Σ and it maps to the power set of Q. Thus, it can map to two different states with the same input. But we do not allow this in a DFA and so a confusion might arise that this method is way simpler than Turing machines. But the thing is that QFA, the quantum counterpart of finite state machines, is way more established than quantum Turing machines.

It is the effortlessness nature that QFA poses that acts as a good teaching-learning basis for absorbing the concepts of quantum mechanics in computer theory. The designs that were primarily structured during the inception of quantum theory in computation displayed a lot threats and loose ends. This was because they did not leverage the complete capability of the quantum mechanics. In fact, they were not even able to display the results that the conventional systems were able to provide at that time. Their true potential was not yet tapped. Say et al., in their paper, discussed about quite a few quanta finite automata algorithms which have been proved to exhibit far more efficiency than the conventional systems [18].

8.2 Coalescence of quantum finite automata

Quantum computing and Automata Theory are intertwined in ways in which the quantum mechanics overpowers the classical methods. Here, a multinumber of algorithms have been used, and they are all based on a simple single component. The quantumness of such a solitary quantum element-based machine can best be indicated as a point on a unit circle, and any necessary operations performed on it can be interpreted as a rotation or reflection of that particular point.

If λ is the angle of rotation, then the counterclockwise rotation component could be represented as C_λ, which is given by:

$$C_\lambda = \begin{pmatrix} \cos\lambda & -\sin\lambda \\ \sin\lambda & \cos\lambda \end{pmatrix}$$

This particular research work has proved that QFA can distinguish a good number of tally languages with cut point, which outnumbers the classical model capability. Towards the final stage of performing computations, the qubit involved is quantified, and if the initial state is observed, then the system accepts the input. The probability of acceptance is also an important factor in these computations.

A quantum finite automaton forms a connection between each single string and a number in a row array of 0 and 1. The strings to be analyzed for language recognition can be fragmented into three groups each, on selection of a particular cut point, α, within the abovesaid interval. The strings to be analyzed that are putative based on the probability being more than the cut point α, proves that the given string belongs to a language that is recognized. This is represented as a system LG, which is a function of N, the recognizing entity and α.

$$\mathrm{LG}(N, \alpha) = \left\{\mu \epsilon \sum * |fN(\mu) > \alpha\right\}$$

This proves that any QFA are completely capable of recognizing languages like Tally with given cut point too. Such languages recognized by a cut point by QFA are called stochastic languages. It further collaborates with the chemical industry with its qubit power over molecular collision, interactive models, and collaborative string processing using nonregular languages.

8.3 Role of quantum finite automata: case study

8.3.1 Chemical industry

While classical computers are based on mechanical structures working through the language of mathematics, quantum computers leverage the strange rules of the quantum world. They rely on two phenomena that happen beyond the atomic level, namely, superposition and quantum entanglement. Repeatedly measuring a qubit will uncover the probability that it is in a particular position. This is important as this is how the universe works too. Nature is also quantum. Our brains and the drugs we are searching for work in probabilities, so they can only be accurately simulated on quantum computers. Quantum computers work in synergy with the quantum mechanics of the universe and they leverage its capabilities. With more qubits, the processing power grows exponentially. Due to quantum entanglement, two separate particles can be at extreme opposite ends of existence and still be entangled, that is, changing one of them would immediately change the other, without any physical connection between them. This is how quantum computers will be able to work. The trick is to entangle qubits to make them work together and with that, we can exponentially grow the processing power with every added qubit.

When two molecules collide, an atom or atoms can transfer from one molecule to another to produce a new molecule. This particular process is called a chemical reaction. And, it is a really important process, which occurs everywhere, in biology, in the laboratory, in the industry, etc. There are a few things that are more important than chemical reactions and it is important, as a chemist, to understand, the very basic mechanism behind every chemical reaction, like, when molecules collide, which are the atoms that are going to transfer and where, what are the products, and also, how fast this chemical reaction can occur. Understanding the speed or rate of various chemical reactions and the factors influencing them comes under the topic chemical kinetics, and many scientists spend a lot of time trying to study this. It is quite a complicated and expensive process. Nowadays, in the modern world, there is another way to find the rates of chemical reaction, without having to even conduct an experiment, by using both theory and computers.

The most crucial fact is that when one molecule collides with another, there exists a potential energy between them, which initially goes up to a maximum and comes down the other side. If the molecules have enough speed or kinetic energy above the level of barrier of the reaction, the barrier will fade away, and reaction happens very quickly. On the contrary, most often, the molecules do not have enough kinetic energy to overcome the barrier. Even then, there is a probability for the chemical reaction to occur. This happens due to the capability of the atoms to tunnel through the barrier, rather than over it, and go from one side to the other. This is a remarkable effect that arises due to quantum theory. In practice though, that does not happen because the probability of tunnelling through is so very low. Thus, a very important part of modern quantum chemistry is to calculate the tunnelling probabilities for passing through the barrier, from one side to another.

Schrodinger's equation could be applied for the tunnelling of atoms through the barrier, in order to calculate the tunnelling probability. This process is very temperature-dependent. The equation is solved for low collision energies, high collision energies, and normal collision energies. An average of all these calculations is taken to get the rate of reactions.

Conventional quantum physics and conventional automation models have been mutually applied on each other in order to develop unimaginably efficient simulation and prediction models. Thus, QFA models could be the perfect choice in the chemical section, in order to compute scenarios of different chemical reactions and determine their rates. Bhatia and Zheng came up with mock-ups of chemical reactions applying the models of formal version of the languages, signified by dual-path QFA [19]. The sole purpose of this trial was to identify rates of chemical reactions without actually carrying out the experiment with the chemicals.

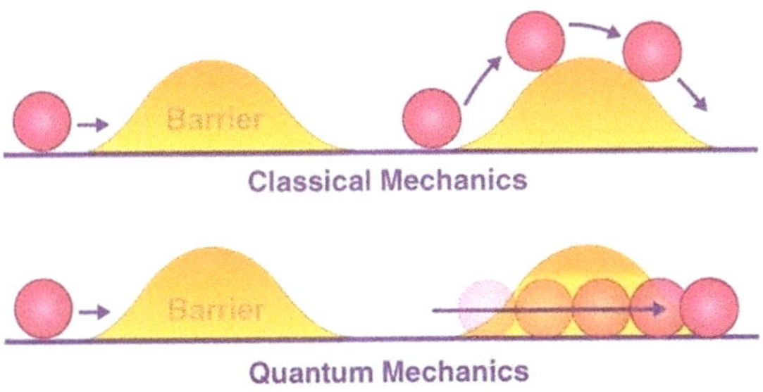

Figure 8.2: Difference in chemical kinetics in case of classic model and quantum model.

The dual pathway quantum finite automaton is implemented in such a way that quantum leaps are applied by getting the input variables form original source. Both pathways are quantum counterpart of each other. The source is allowed to move in either left or right directions, or it can also remain stationary.

Quantum modelling of the chemical reactions is carried out by first implementing language recognition. This comprises three main phases: computation box where the actual computation takes place, a source translator, and finally, a complex system, which closely watches the response of the chemical automation process. Three different categories of chemical reactions were discussed here. The first is the chemical reaction consisting of regular language. The second is the chemical reaction consisting of context-free language. The third category is the chemical reaction consisting of context-sensitive language, as shown in Figure 8.2. The dual way QFA approach is successful as an efficient method to model balanced chemical reaction, with a give-or-take limit of one-sided error, made to freeze along the linear time axis.

8.3.2 Interactive proof system

Interactive proof systems begin and follow on probabilistic Turing machines, probabilistic computations, and its associated class, backpropagation algorithms. In some sense, it is a probabilistic version of np. This model is a very important one. It involves a great deal of research in complexity and also, most of its connection is with cryptography. Some part of its genesis also comes from cryptography. Applications where multiple parties are communicating in order to achieve certain goals can benefit a lot from interactive proof systems.

An example of an interactive proof system is the graph isomorphism problem, which is a problem of testing whether two graphs are isomorphic. Isomorphic graphs are really just the same graphs, with one of them being relabeled or permuted, so that they may look superficially different, but they may appear with a different sequence of labels or the nodes may appear in a different order.

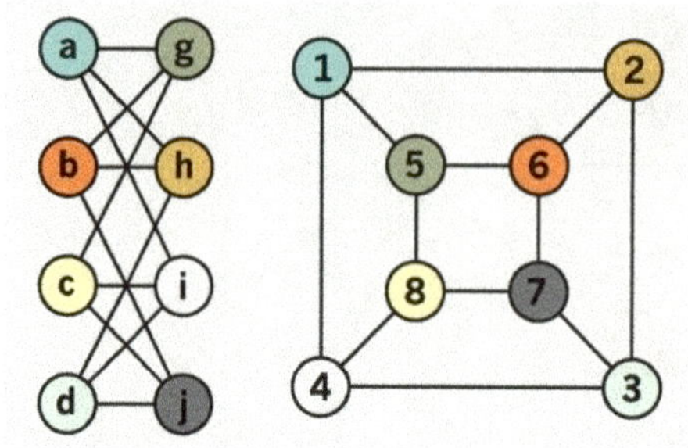

Figure 8.3: Isomerism of two graphs –an informal model of an interactive proof system.

Thus, here, we have associated computational problem, called iso, which is given a pair of graphs. To test the isomerism of the graphs in polynomial time is an unsolved problem. This is a problem on which a lot of research work is still being done in order to get a polynomial time algorithm. This is where interactive proof systems play a crucial role in offering a solution in a more convenient way, in polynomial space. There are two interacting parties – one is the Verifier(V), which is a probabilistic polynomial in time, and the other is Prover(P), which has unlimited computational power.

Both of them get to see the input, and the whole computation in the verifier space seems to be polynomial. The only thing that is not polynomial is the prover's work, which is unlimited. Probabilities of the verifier that will be accepted by the prover will have to be computed. These complex probabilistic computations were backed up with efficiency, with the entry of quantum mechanics into cryptography. The idea of performing a task efficiently on a quantum device that is believed to be intractable is impossible for classical computers. Due to application of quantum mechanics in computing, various advantages surfaced while doing the experimentation. Sometimes, even if we have noise in our system, it would still be intractable to sample from that.

We are basically interested to know how hard nonlocal games are. This has been a problem that has puzzled us for a very long time. Nonlocal games are simple models that have connections to represent bell inequalities. This is a classic example of an interactive proof system but with multiprover proofs to be dealt with. Research is being conducted on the compression of quantum automata in such multiprover interactive proofs, as in Figures 8.3 and 8.4.

Over the decades, since the introduction of quantum mechanics and application of QFA in multiprover systems, a lot of work has been and is still being done, to mathematically model such quantum mechanical models in various multiple ways. Harumichi Nishimura et al. [20] have done extensive work in applying the possibilities of QFA to the transactions happening between a verifier and a prover, in a typical interactive proof system.

First, they tested the model with single pathway QFA model, which was built to oppose strongly evident provers. From this, they were able to abstract the containment

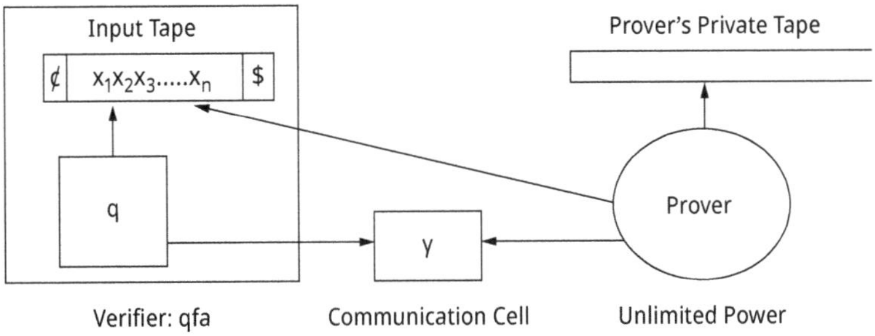

Figure 8.4: Block diagram of a quantum interactive proof system with QFA verifier.

interactive proof by combining a few theorems, propositions, and lemmas that are relevant to the model.

Second, they tested the model with a dual pathway QFA model, which was built to oppose strongly evident provers. Here too, they applied certain theorems, propositions, and lemmas relevant to the model.

The most essential part of designing a model is measurement of the results that could be obtained. Measurement in a quantum finite automaton-applied design is obviously to excerpt the target quantum state. It is already a known fact that QFA has limited memory. Hence, the number of iterations to carry out the measurements becomes restricted due to memory deficiency.

The prover's ability to extract information from the verifier is the strength that defines prover. The quantity of the communications occurring between the verifier and prover evidently increases the QFA to be well-versed in certain languages. In this work, the authors have also given six tracks to take forward this work for future research, which would be fruitful in improvising the QFA model design. They are effects of altering the verifier's capability, restraining the power tactic, collaborating through a conventional channel, using preceding entanglement, implementing multiple prover environment, and making knowledge-based interactions.

8.3.3 Nonregular languages

Automata is a data stream processing algorithm, and quantum automata is the mechanics that can be employed to optimize the size of memory, compared to classical automata. Consider a quantum automaton that can more than sufficiently recognize a unary language, MODp. It can be considered as a language consisting of words of length p, where p is a prime that is known one-way classical automata. MODp can be recognized by the QFA with O(log p) states, using fingerprinting technique. Soon, constant O depends on required probability of correct answer.

For x Є MODp, the answer is always correct with probability 1. According to Ambainis and Nahimovs, for any $\varepsilon > 0$, there is a QFA with $4\frac{\log 2p}{\varepsilon}$ states, recognizing MODp with probability of at least $1 - \varepsilon$, where ε is the probability of error.

We have a quantum register of qubits. One of them is the target qubit and others are called the control qubits, which keep it entangled uniformly. Then, we obtain a uniform distribution or superposition of this control qubits, by applying Hadamard transform. In this way, we will be able to obtain the quantum states, where $d = O$ (log p).

On reading the input symbol, the automaton algorithm rotates the target qubit on an angle $\alpha = \frac{2\pi k_i}{p}$, for each value of i Є {0 . . . d}. For reducing probability of error, we use a series of rotations on different angles and for each quantum state i, we have the angle α. So the target qubit is rotated around this angle, while reading a single symbol, for e.g., "*a*". After reading all n symbols, the resulting angle is $\frac{2\pi k_i n}{p}$. If n mod p is 0, or in other words, if n is divisible by p, then target qubit will reach state 0 with probability 1. If n mod p is not zero, then the target qubit will be in some nonzero state, with some positive probability. For minimizing the probability of error, we can use repeating of the probability. It calls for a probability boost up. In case of streaming data, we are not able to repeat the algorithm for all input symbols. So, instead of this, we use several rotations of the target qubit at different angles. This also helps in minimizing the probability of error. So, we have d parameters, each angle determining a state, and we apply specific rotation on the target qubit.

Now, the confusion of how to differentiate these states arises. For this, particular issue, we transform the state I, to get state $|\langle 1 \ldots .11\rangle$ and apply control rotation on an angle α_i. The second matter of concern is how to implement such a transformation in qiskit, as it is a multi-controlled qubit rotation operation. For this purpose, we use circuit decomposition for controlled rotation operation, where U is the rotation of angle and V is the rotation of half of the angle.

Consider U is rotation on some λ and V is the rotation on $\lambda/2$; then $U = V^2$.

In order to implement control operation with many control qubits, consider the scenario where there are multiple control qubits in a particular gate; we can use decomposition of that gate. The problem of using n control qubits can be solved using $n - 1$ control qubits. This could be implemented recursively, by creating a recursive function for n control qubits.

In order to optimize this computation, we can use x-gates to switch from one state to another as also specific number of reordering of states. A lot of available coding mechanisms could be used like gray code system, which proves to be very useful in this case. The coding is selecting in such a way that consecutive states differ in binary value only by one bit position. This allowed us to use less umber of x-gates in a circuit, thus providing a good step for optimization.

Single pathway lines exhibit what is maintained in relation and dual pathway lines exhibit the connectivity based on equivalence. The connectivity within the quantum computed model clearly proves that the non-regular languages recognized by

MO-1 quantum finite automaton, are actually maintained within those recognized by MM-1 quantum finite automaton. MO-1 g quantum finite automaton was able to simulate deterministic and probabilistic quantum finite automaton. This small group was able to come up with not only very good recognition of regular languages but also a few non-regular languages in polynomial space.

Amano and Iwama [21] illuminated on the potential capability of 1.5 quantum finite automaton, demonstrating that it was not sufficient to solve the issue of emptiness. Nakanishi et al. [22], demonstrated that 1.5 quantum finite automaton was capable of recognizing a particular non-regular language L, with a probability which was below the acceptable borderline of 2 or 3, whereas the probability was higher for its classical model counterpart. Yakaryilmaz [23], made a formal confirmation of the above mentioned theoretical assumption that 1.5 quantum finite automaton was, indeed, able to recognize and accept non-regular stochastic languages.

The dual pathway quantum finite automaton(2QFA) was evidentially vouched for being far more superior in computation and recognition, than its classical model counterpart, by Kondacs et al. [24]. It was also experimentally proposed that it has the capacity to accept and recognize non-regular languages better than single pathway quantum finite automaton, but with the exception of single-pathway error in linear space of time domain. The entire effect of all these lies under the banner of the emergence of AI [25–32] being reliable and secure in all fields such as Medical Science, Structural Science, Literal sources, etc.

8.4 Conclusion

Using the principles of quantum physics, a new paradigm of computation has emerged. For tasks like factoring, quantum computers may be enormously faster than traditional ones. More constrained models, including quantum variations of finite automata, have also been researched in addition to full-scale quantum computers. In this chapter, we examine many models of quantum finite automata (QFA) and their characteristics in this chapter. QFA integrate quantum computing with the theory of finite automata. The study of QFA has covered a wide range of models and features. This chapter's major goals are to discuss the interactions between quantum theory and AI and to look at some (possible) uses of quantum computation in AI. There are some situations where quantum models are as powerful as classical models or share characteristics with them (for instance, the ability of 1QFA to identify languages with bounded or unbounded error). A brief explanation of quantum computation is given to the readers who are unfamiliar with it, so that they can understand the power of quantum computation and apply these techniques for their problem statements. Although some portions of the map are very poor and others are missing, this chapter will serve as a useful road map for AI researchers who wish to investigate more and deeper linkages between AI and quantum computation as well as quantum Automata Theory.

References

[1] N. Abdelgaber, C. Nikolopoulos, "Overview on Quantum Computing and its Applications in Artificial Intelligence," IEEE International Conference on Artificial Intelligence and Knowledge Engineering, 2020.

[2] J. Barzen, F. Leymann, Quantum humanities: A vision for quantum computing in digital humanities, SICS Softw.-Inensiv. Cyber-Phys. Syst. 35, 153–158 (2020). https://doi.org/10.1007/s00450-019-00419-4.

[3] R. Feynman, Simulating physics with computers, Int. J. Theor. Phys., 21(6), 1982, 467–488.

[4] https://research.aimultiple.com/quantum-ai/.

[5] https://research.aimultiple.com/quantum-ai/#:~:text=Quantum%20AI%20is%20the%20use%20of%20quantum%20computing,are%20not%20possible%20to%20achieve%20with%20classical%20computers.

[6] https://www.bbvaopenmind.com/en/technology/digital-world/quantum-computing-and-ai/. Accessed 12 August 12 2022.

[7] D. Aerts, M. Czachor, Quantum aspects of semantic analysis and symbolic artificial intelligence, J. Phys. A Math. Gen., 37, 2004, 123–132.

[8] P. J. Coles, S. Eidenbenz, S. Pakin, A. Adedoyin, J. Ambrosiano, P. Anisimov, D. Gunter, Quantum algorithm implementations for beginners, ACM Trans. Quant. Comput., 3. Doi: 10.1145/3517340. 2018, 1–92.

[9] M. Ying, Quantum computation, quantum theory and AI, Artif. Intell., 174(2), 2010, 162–176.

[10] https://www.e2enetworks.com/blog/what-quantum-computing-will-mean-for-the-future-of-artificial-intelligence.

[11] E. Bernstein, U. Vazirani, Quantum complexity theory, SIAM J. Comput., 26, 1997, 1411–1473.

[12] https://studiopulpit.com/2020/02/20/quantum-computing-and-artificial-intelligence-ai/.

[13] S. Kute, C. G. Desai, Quantum cryptography: A review, Ind. J. Sci. Technol., 10(3), 2017, 1–5.

[14] F. Bova, A. Goldfarb, R. G. Melko, Commercial applications of quantum computing, EPJ Quantum Technol., 8, 2021.

[15] A. Bayerstadler, et al., Industry quantum computing applications, Quantum Technology and Application Consortium – QUTAC., Bayerstadler, A., Becquin, G. et al. Industry quantum computing applications. EPJ Quantum Technol. 8, 25, 2021, 1–17.

[16] https://analyticsindiamag.com/top-applications-of-quantum-computing-everyone-should-know-about/. Accessed 15 August 2022.

[17] https://www.themanufacturer.com/articles/the-potential-of-quantum-computing-for-manufacturing/. Accessed on 29.05.2023

[18] C. Say, A. Yakaryilmaz, Quantum Finite Automata: A Modern Introduction, in: Cristian S. Claude Et al (Eds), Computing with New Resources, 2014, doi: 10.1007/978-3-319-13350-8_16.

[19] A. S. Bhatia, S. Zheng, A quantum finite automata approach to modeling the chemical reactions, Front. Phys., 10 November 2020, Sec. Physical Chemistry and Chemical Physics https://doi.org/10.3389/fphy.2020.547370.

[20] H. Nishimura, T. Yamakami, An Application of Quantum Finite Automata to Interactive Proof Systems (Extended Abstract), in: M. Domaratzki, A. Okhotin, K. Salomaa, S. Yu (Eds.), Implementation and Application of Automata, CIAA 2004. Lecture Notes in Computer Science, vol. 3317, Springer, Berlin, Heidelberg, 2005. https://doi.org/10.1007/978-3-540-30500-2_21.

[21] M. Amano, K. Iwama. "Undecidability on quantum finite automata," *STOC '99*, Proceedings of the Thirty First Annual ACM Symposium on Theory of Computing, ACM, 1999, pp. 368–375.

[22] K. H. M. Nakanishi, T. Indoh, K. Hamaguchi, T. Kashiwabara, On the power of quantum pushdown automata with a classical stack and 1.5-way quantum finite automata, Tech. rep., Nara Institute of Science and Technology, 2001.

[23] A. Yakaryilmaz, A. C. Say, Languages Recognized with Unbounded Error by Quantum Finite Automata, In: Frid, A., Morozov, A., Rybalchenko, A., Wagner, K.W. (eds) Computer Science - Theory and Applications. CSR 2009. Lecture Notes in Computer Science, vol 5675. Springer, Berlin, Heidelberg. https://doi.org/10.1007/978-3-642-03351-3_33

[24] A. Kondacs, J. Watrous, "On the power of quantum finite state automata." Proceedings 38th Annual Symposium on Foundations of Computer Science (1997): 66–75.

[25] G. I. Rajathi, Dr. G. W. Jiji, "Chronic Liver Disease Classification using Hybrid Whale Optimization with Simulated Annealing and Ensemble Classifier," Symmetry, MDPI, January 2019, (ISSN 2073–8994). (Annexure I) (Impact Factor: 2.143). https://www.mdpi.com/2073-8994/11/1/33.

[26] G. I. Rajathi, Dr. G. W. Jiji, A novel automatic liver segmentation by level set method over real-time sensory computed tomography, Wireless Pers. Commun., 109(3), 2019, 1987–2010, Springer, (ISSN 0929–6212) December 2019, (Annexure I) (Impact Factor: 0.959), doi: 10.1007/s11277-019-06664-9.

[27] G. Sasubilli, A. Kumar, "Machine Learning and Big Data Implementation on Health Care data," 4th International Conference on Intelligent Computing and Control Systems (ICICCS), 2020, pp. 859–864, doi: 10.1109/ICICCS48265.2020.9120906.

[28] S. M. Sasubilli, A. Kumar, V. Dutt, "Machine Learning Implementation on Medical Domain to Identify Disease Insights using TMS," 2020 International Conference on Advances in Computing and Communication Engineering (ICACCE), 2020, pp. 1–4, doi: 10.1109/ICACCE49060.2020.9154960.

[29] S. M. Sasubilli, A. Kumar, V. Dutt, "Improving Health Care by Help of Internet of Things and Bigdata Analytics and Cloud Computing," 2020 International Conference on Advances in Computing and Communication Engineering (ICACCE), 2020, pp. 1–4, doi: 10.1109/ICACCE49060.2020.9155042.

[30] P. K. Kotturu, A. Kumar, "Comparative Study on Machine Learning models for Early Diagnose of Alzheimer's Disease: Multi Correlation Method," 2020 5th International Conference on Communication and Electronics Systems (ICCES), 2020, pp. 778–783, doi: 10.1109/ICCES48766.2020.9137872.

[31] S. Mary Fabiola, G. Ignisha Rajathi, R. Vedhapriyavadhana, L. R. Priya, Artificial intelligence to combat the sting of the pandemic on the psychological realms of human brain, Springer Nat., March 2022, doi: 10.1007/s42979-022-01038-8.

[32] G. I. Rajathi, R. VedhaPriyaVadhana, L. R. Priya, Robotic dustbin on wheels, Int J. Innov. Technol. Explor. Eng., 9(1), November 2019, ISSN No. 2278-3075.

N. Pooranam, D. Surendran, N. Karthikeyan, G. Ignisha Rajathi

Chapter 9
Quantum computing: future of artificial intelligence and its applications

Abstract: Quantum computing (QC) is the emerging technology for solving complex problems through the laws of quantum mechanism and optimized algorithms. To examine the possibilities of active cooperation across both disciplines of computer science, this article explains the core components of quantum computing and conveys some of the important features linked by quantum computing and Artificial Intelligence (AI). To make major progress with artificial intelligence, quantum computing is probably necessary. Quantum computing's training and processing power can be used to improve AI and machine learning (ML) algorithms at a reduced cost. AI can provide quantum computers with crucial error-correcting algorithms. Several AI algorithms have been successfully applied to quantum computers; they are constrained by the computational power of conventional computers despite developing useful applications. Artificial intelligence may benefit from quantum computing's computational boost in order to handle more challenging issues. This case study explains the importance of quantum computing, the future of QC in AI, and the different applications of quantum computing, such as AI and ML, cybersecurity and cryptography, and weather forecasting.

9.1 Introduction

Actual Quantum is a quickly developing technology that uses the principles of quantum mechanics to address issues that are too difficult for conventional computing. It is a subject of study committed to designing computer systems based on the ideas of quantum theory. It complies with the laws of quantum physics [1, 2], and gains a significant amount of processing capacity by having numerous states and being able to carry out tasks simultaneously using all feasible permutations. The majority of researchers are

N. Pooranam, CSE, Sri Krishna College of Engineering and Technology, Coimbatore, Tamil Nadu, India, e-mail: pooranam.n@gmail.com
D. Surendran, CSE, Karpagam College of Engineering, Coimbatore, Tamil Nadu, India, e-mail: d.surendran@gmail.com
N. Karthikeyan, School of Computer Science and Engineering, Vellore Institute of Technology, Chennai, Tamil Nadu, India, e-mail: narenkarthikeyan.mecse@gmail.com
G. Ignisha Rajathi, CSE, Sri Krishna College of Engineering and Technology, Coimbatore, Tamil Nadu, India, e-mail: ignishaelton@gmail.com

https://doi.org/10.1515/9783110791402-009

continuously searching for efficient superconducting quantum processors. The main keys for developing quantum computers is to build fast computing speed and capacity for storing data set that can change the world. Nowadays, most of the companies are doing research on Quantum Computing along with engineers and modern technology to find superfast quantum computers. IBM is one of the most and dominant one in the research work. It has launched the largest, highest performing fleet of quantum systems in the world. Significant figures of their achievement are; more than 400,000 user base run on one trillion circuits on the IBM-Quantum Computer; Quantum Volume achieves Folcon R10 processor and provides fast transaction and efficient storage. It has shown improvements over conventional computers in terms of speed, accuracy, and efficiency by utilizing innovative tools designed for particular tasks.

Classical computers [3] are required to support and make the system optimized. Even the advanced techniques in the field of quantum computer are designed and should be developed more for further consistency and to progress to the next level of the modern world. To build quantum computers, various military and national security organizations are sponsoring and funding research in the field of quantum computation. Large-scale quantum computers can answer problems significantly faster than classical computers using the best algorithms currently available.

9.2 A comparison of classical and quantum computing

In general, the concept of superposition is used in quantum computing to manipulate data [4]. Using theoretic computer science and utilizing qubits, quantum computers are different from conventional or digital computers. In conventional computers, at each point in time or for whatever number of bits, data processing must be done only in binary state. Before switching states, a transistor or capacitor must be in either 0 or 1; this is currently measured in billionths of a second. The rate at which these devices can be made to change states is currently constrained. The physical constraints of materials and the barrier for the application of the classical rules of physics are being approached as we go toward smaller and quicker circuits. Beyond this, the quantum realm dominates. The charge or polarization of a variety of elementary particles, including electrons and photons, can be employed in a QC to represent the numbers 0 and/or 1 in the quantum world. The term "quantum bit," or "qubit," refers to each of these particles. Quantum computing is built on the nature and behavior of these particles.

9.3 Elements of quantum computing

To address the issues of memory consumption, increasing the speed of the process, redesigning the algorithm build to solve the problems, quantum computer makes good sense to perform the majority of the processing on the classical machine by adding a quantum circuit and some sort of interface between the classical and the quantum logic. QC has some strong advantages. Multiplying two numbers is easy for a digital computer. The reverse process of finding the factoring is much harder. Factoring a 400-digit number would take years, even on the most advanced supercomputers. Peter Shor demonstrated in 1994 that a sizable and reliable quantum computer would be able to locate those factors considerably more quickly. Basic elements of QC [5] are described below.

9.3.1 Bits and qubits

Classical computers refer ones and zeros for manipulation whereas QC refers quantum bits or qubits. In a digital computer, the data is processed and stored in the format of 0s and 1s for either ON or OFF. For example, in a toss of a coin that results in a head or a tail, there are two distinct and distinguishable possibilities. Qubits, gates, and circuits are specified in QC [6, 7]. It effectively performs operations on qubits, which are identical to the bits used with classical or digital computers. It has the additional property of superposition of states. A classical register with two bits can only store two numbers at a time, whereas a quantum register with two qubits can simultaneously store four numbers in superposition. More numbers can be stored in a quantum register with 300 qubits than there are atoms in the universe. A problem occurs while reading the output in a superposition state. Instead of a bit used in classical and digital computer, data is encoded in a sequence of qubits that are utilized in quantum computers. The qubit is a superposition of both $|0|$ and $|1|$, which differs from bits.

$$|\Psi| = \alpha|0| + \beta|1| \tag{9.1}$$

Here α and β are complex numbers,

Similar to tossing a coin, a bit must be in either the first or the second state, whereas a qubit can be either 1 or 0. As a result, our measurement's outcome is either 0 with probability $|\alpha|^2$ or 1 with probability $|\beta|^2$.

9.3.2 Quantum gates

To build the QC, a unique circuit type known as a quantum gate is required (QG). It is necessary to construct computer circuits and logic gates. The major difference between the input and output states of quantum gates is the superposition of the $|0|$ and $|1|$ states. Instead of being in one quantum state, the Eigen selected states of quantum gates are a tangle of quantum states. It is considered by NOT gate to change the state $|0|$ to state $|1|$ for improved comprehension.

$$\alpha|0\rangle + \beta|1\rangle -> \alpha|1\rangle + \beta|0\rangle \tag{9.2}$$

A quantum computer can hold up to two extra bits of information due to the superposition level.

In general, the QG is classified into two types – single qubit gate and multiple qubit gate. A single qubit can be represented by a column vector, and a gate working on the qubit can be represented by a 2 × 2 matrix. For instance, the structure of a NOT gate's quantum equivalent is as follows:

$$\begin{bmatrix} 0 & 1 \\ 1 & 0 \end{bmatrix}$$

A real quantum gate must be reversible, requiring the usage of control lines in multi-qubit gates that are independent of unitary transformation. The CNOT gate scenario uses the conventional XOR with the input on the *b* line and the control line. A 4 × 4 matrix is used to represent the two-qubit gate as follows.

$$\begin{bmatrix} 1 & 0 & 0 & 0 \\ 0 & 1 & 0 & 0 \\ 0 & 0 & 0 & 1 \\ 0 & 0 & 1 & 0 \end{bmatrix}$$

9.3.3 The ket | >

The ket (| >) is a part of Dirac's notation; it is merely a vector's notation. The state of a qubit in C2 is a unit vector, and $\begin{bmatrix} \alpha \\ \beta \end{bmatrix}$ is a vector, mentioned as follows:

$$\alpha|0\rangle + \beta|1\rangle \tag{9.3}$$

And

$$|0\rangle = \begin{bmatrix} 1 \\ 0 \end{bmatrix} \quad |1\rangle = \begin{bmatrix} 0 \\ 1 \end{bmatrix} \tag{9.4}$$

9.3.4 Entangled states

Despite of how far the states are from one another, subatomic particles are related to one another because they are in an entangled state. When combined, they have an instant impact on measures. For computational needs, this effect is helpful. Consider the subsequent condition, which is not entangled

$$\frac{1}{\sqrt{2}}|00\rangle + \frac{1}{\sqrt{2}}|01\rangle \tag{9.5}$$

and is expanded into eq. (9.6)

$$\frac{1}{\sqrt{2}}|00\rangle + \frac{1}{\sqrt{2}}|01\rangle + 0|10\rangle + |11\rangle \tag{9.6}$$

The second qubit's state changes from zero to one, 100% of the time when the first qubit is measured (a partial measurement) to:

$$\frac{1}{\sqrt{2}}(|0\rangle + |1\rangle) \tag{9.7}$$

It gives the equal probability for 0 or 1.

9.3.5 Quantum circuits

A quantum circuit is a quantum state that contains one or more qubits with unitary operators. A fundamental quantum circuit is made up of several computations and measurements on the state of n-qubits. Each unitary operation can be represented by a $2n \times 2n$ matrix, as seen in Figure 9.1.

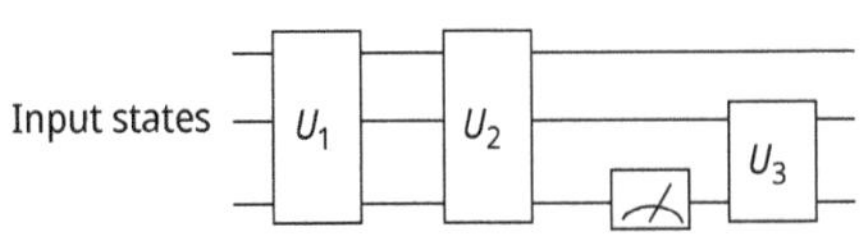

Figure 9.1: *N*-qubits quantum circuit.

The boxes contain either unitary quantum logic gates or are collections of gates, with each line being an abstract wire; the measurement symbol is "meter." All these – gates, wires, input, and output mechanisms – work together to create quantum algorithms. Quantum circuits can always be rearranged. When compared to conventional classical circuits, quantum circuits run only once, from left to right.

9.3.6 Quantum computers

In Figure 1, the registers V and W represent the input and output qubits of a quantum computer, respectively. It is possible to build up the input register as a superposition of states, for example, as a superposition of all the numbers between 0 and $2n$. The computer then performs a sequential computation of the function treated to all $2n$ integers. According to the Quantum Measurement Postulate (QMP), every bit from the output register is assigned a Boolean value when we measure W, based on the entangled wave of qubits that results.

9.4 Benefits of quantum computing

A traditional computer processes the data bit by bit whereas the QC processes it all simultaneously. But it does not give a suitable solution for all types of problems. Researchers know where it can be used and how its affects certain operations. Some of the areas where QC is required are as listed below.

9.4.1 Database processing

In the digital world, data is growing everyday due the digital media. Data storage [8] is a major challenge for researchers. Based on the user's query, the normal computer processes the data by checking every record one by one in the database. When the volume of the database increases, the normal computer does not give better results in terms of computation time. So, the quantum computer replaces the previous one for processing data efficiently. A quantum computer's great strength is to perform many operations simultaneously and give the best result by checking many records at once with quantum.

Some of the advantages of quantum computing in database are as follows:

- Processing the data simultaneously and better than 1,000 times the normal computer,
- Cost-effective,
- Computation power is higher when compared to the earlier computers,
- Best for simulation.

9.4.2 Security

Data security [9] is one of the most challenging parts for researchers who try to make communication secure from one end to another. This requires boosting the quantum technique for digital encryption solutions in which certain standard algorithms are used to convert the data into an encrypted format. The inverse process of the same algorithms helps to decrypt the data without any loss. All attempts are made to break the encrypted key of the possible solutions, though they take a lot of time. Even if broken by brute force, the quantum in security mechanism does not allow the situation, which is a possible scenario.

Some of the merits of QC in security mechanism are as follows:

- Generating the random number to enrich the encryption technique,
- Making the encryption technique stronger by referring different cryptographic combinations simultaneously, such as Advanced Encryption Standard (AES),
- Providing secure communication over the network [10] and cloud through standard set of protocols and advanced technologies,
- Providing fast and constant communication for real-time data transfer (example: receive the shorter keys to decrypt the commands from the pilot instead of the long key sequence).

9.4.3 Artificial Intelligence

Quantum computing with AI is used to perform a task accurately, fast, and efficiently compared to existing techniques such as convolutional technique, as shown in Figure 3. Researchers took the advanced tool for some specific application and problem solving. Quantum Computing [11] improves the speed and efficiency of some of the most complex optimization problems. For example, large factories may utilize machine learning to help maximize output by determining how each individual process and component could be used most efficiently. Quantum computers can assist by giving insights more quickly and correctly, resulting in streamlined output, decreased waste, and cheaper prices.

9.5 Quantum computing: define the future of AI

A new version of the TensorFlow framework [11, 12], named as TensorFlow Quantum (TFQ), has been launched. It is a quantum machine learning library and a hybrid of quantum-based classical ML models. Fast and accurate real-time problem solving is achieved by TensorFlow library-Quantum computing framework model. Quantum data state the process in TensorFlow Convolutional Neural Network, as shown in Figure 9.2. Here, the input is either empty or an X gate, which constructs the remaining model quantum's components, reads Z inference, and compares the label for the exited and nonexited state of "1" or "−1", respectively.

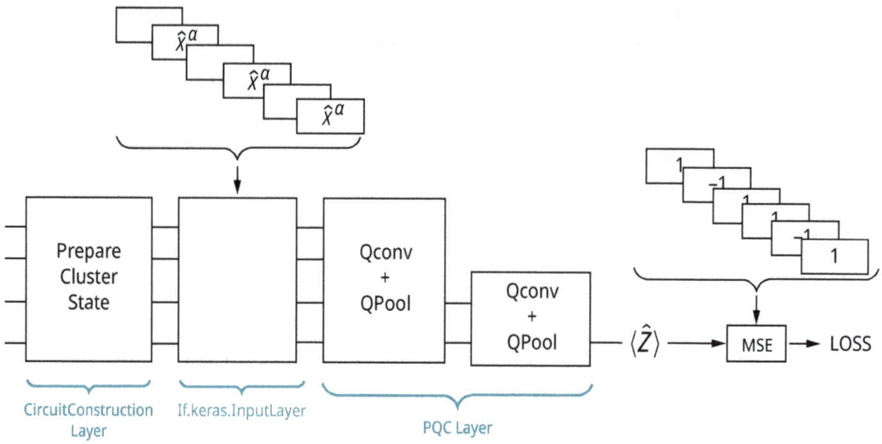

Figure 9.2: Block diagram of TensorFlow quantum model.

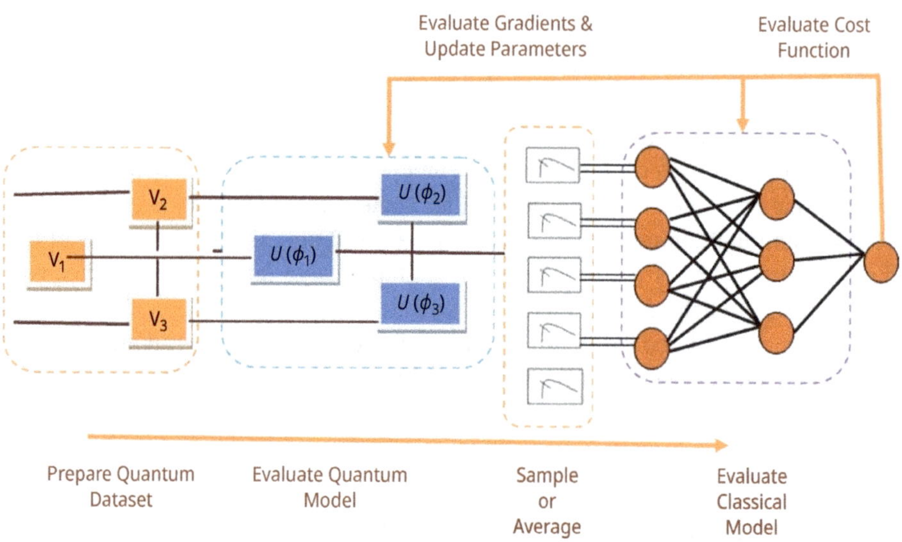

Figure 9.3: TensorFlow quantum model with AI.

TensorFlow builds hybrid quantum classical model through quantum data, and integrates quantum algorithms and logic designed circuits, as shown in Figure 9.3. Finally, it provides new APT with high performance.

TFQ is used in several applications, including supervised learning for quantum classification, simulating noisy quantum circuits, quantum control, and quantum approximate optimization. Some of the advanced quantum learning models of TFS are meta-learning, Hamiltonian learning, classification of quantum phase transitions, and reinforcement learning.

The steps involved to create an effective model for unsupervised tasks are as follows [13].

1. Convert quantum data to the quantum dataset – It is a multidimensional array data, referred to as “Quantum Tensors.”
2. Choose quantum neural network models – To perform quantum processes and extract information, choose the quantum data structure and quantum models for neural networks.
3. Sample or Average – Select the quantum state as samples from the classical distribution and give techniques for averaging through multiple runs with steps (1) and (2).
4. Analyze a model of traditional neural networks.

Some of the critical steps to enable quantum AI for further development are as follows:

- More effective and error-free quantum computing systems
- Use of open-source training and models
- A broad and competent development ecosystem
- Innovative quantum-based AI techniques that outperform traditional computing

Research communities in quantum computing and machine learning can study models of both natural and artificial quantum systems using the tools provided by this framework. It helps to discover new quantum algorithms to apply to real-time complex problems.

9.6 How quantum computing can benefit artificial intelligence

Some limitations have existed for solving complex problems in the past few decades. Researchers need a powerful machine, not just software, to meet their requirements. Due to some limitations in software optimization, researchers are trying to find a better way where the huge volume of data is processed and extracted. Below are some methods that quantum computing may impact how artificial intelligence develops in the future [11];

9.6.1 Handling large datasets

Even advance technologies, such as Machine Learning and AI, designed to interact and extract data consume more memory and take high computation time to process those data. It becomes difficult while evaluating massive datasets thorough traditional computers. QC addresses these issues by finding the optimized solution and solving the problem quickly by uncovering patterns. Some additional features of quantum computing are error correction code, managing quantum bits, and making better decisions in business problems.

9.6.2 Solve complex problem quickly

Traditional computers take more computation time to process the huge volume of datasets for computing and for business model strategy processes. It processes the data as data units of either zeros (0) or ones (1). In order to increases the speed of calculation and efficiency of the process, quantum computers use qubits, which represent the combination of zero and one at the same time, based on superposition. Developers can solve multiple problems simultaneously. For example, a calculation is performed on Google's quantum computer 100 million times more quickly than on a conventional computer.

9.6.3 Building better models

Companies are breaking their links with traditional computers due to the growing volume of data produced in many sectors, including banking, life sciences, and pharmaceuticals. These companies need some new and better data framework model for complex situations. Quantum technology helps to create better models, which help the medical field cure diseases more effectively, avoid financial collapse in banking, and improve the supply chain in the manufacturing sector.

9.6.4 Integration of multiple datasets

Data provider is the core challenge for data analysis in which data may be either too much or limited and not enough. Both situations are not advisable for processing data. One of the additional problems for data processing from a data provider is the variety of datasets. Data comes from different places that cannot be managed and integrated by normal computers. To make data analysis easier and to increase the speed of the process, quantum computing technologies are used to manage and integrate multiple datasets for solving business problems.

9.6.5 Combat fraud detection

Quantum computing helps to identify fraud detection in the banking and financial sectors. It trains the models using quantum computing technologies that can easily detect patterns and manage the volume of information. Banking and financial sectors need to create customer-centric business models through which they can provide good products and services with zero percentage of service loss and fraud detection. Quantum-based computing model enhances the system performance in the banking sector.

9.6.6 Technical obstacles

The stability of qubits is one of the major concerns with quantum computing. A lot of work is spent into trying to ensure that the bits on the computer chip do not interfere with one another because every bit in a computing operation must be in a state of either one or zero. Qubits, on the other hand, can represent any combination of 0 and 1, and they can communicate with one another. Controlling these interactions becomes very difficult and, as a result, qubit volatility makes it possible for inputs to be lost or changed, which has a negative impact on the accuracy of outputs.

Similarly, some of the additional difficulties are:
- Sensitivity
- Interaction with the environment
- Error correction
- State preparation constraints
- Wrapping up

9.7 Application of quantum computing

Quantum computing is the prominent technology that is used in various applications in the real world that are discussed below.

9.7.1 Artificial Intelligence and machine learning

The most dominant areas of today's word are Artificial intelligence (AI) and Machine Learning (ML). AI [17] enables computer programs to learn from their experiences and imitate human intelligence by using iterative processing and algorithmic training. Nowadays, all digital applications process the data – voice, audio, image, and other multimedia formats. Additionally, storage and managing the volume of data is a huge problem for researchers. At the same time, the number of applications are increasing, and along with increasing volume of data and data variety, it becomes a huge challenge for traditional computers to process tasks. In order to deal with these issues and provide good result faster and accurately, quantum computing-based advance algorithms are designed to study and implement in AI and ML. Some of AI-based applications in different places are the web search engine in Google, recommendation systems in YouTube and Amazon, understanding human speech in Siri and Alexa, self-driving cars in Tesla, etc. Quantum computers plays a major role in level 4 and level 5 self-driving cars [14, 15], that is Over the Air (OTA)-based electronic communications and on-board AI system models.

In Machine Learning, a model is generated and improved automatically through experience and the use of trained and existing datasets, called trained data. It processes huge volume of data and makes predictions or decisions within seconds. Unsupervised learning (clustering and dimensionality reduction) [16] and supervised learning (classification and regression), as shown in Figures 9.4 and 9.5, respectively, and reinforcement learning are three categories of machine learning (semisupervised learning).

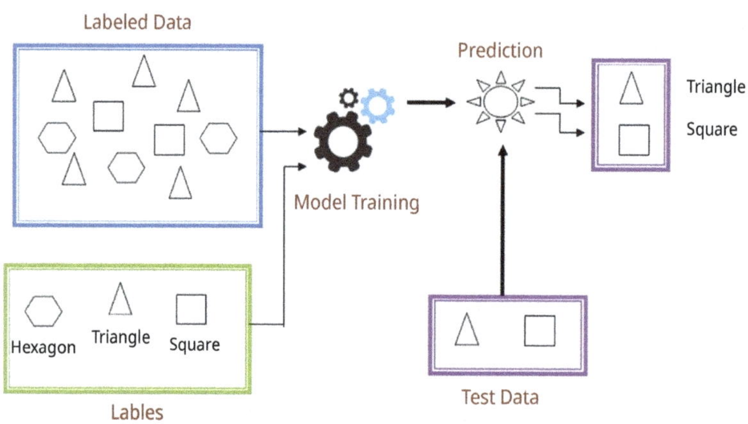

Figure 9.4: Supervised machine learning model using quantum computing.

9.7.2 Cybersecurity and cryptography

In general, the basic foundation of quantum physics is used to build the model for quantum cryptography [17]. To solve the many discrete logarithmic problems and send secured data transmission, the quantum future along with some cryptography methods such as Advanced Encryption Standard (AES), RSA (Rivest-Shamir-Adleman), and Data Encryption Standard (DES) provide feasible solutions. In actual, it is employed to create a shared, secret, and random sequence of bits for two systems, let's say Alice and Bob, to communicate with one another. The term "quantum key distribution" refers to this.

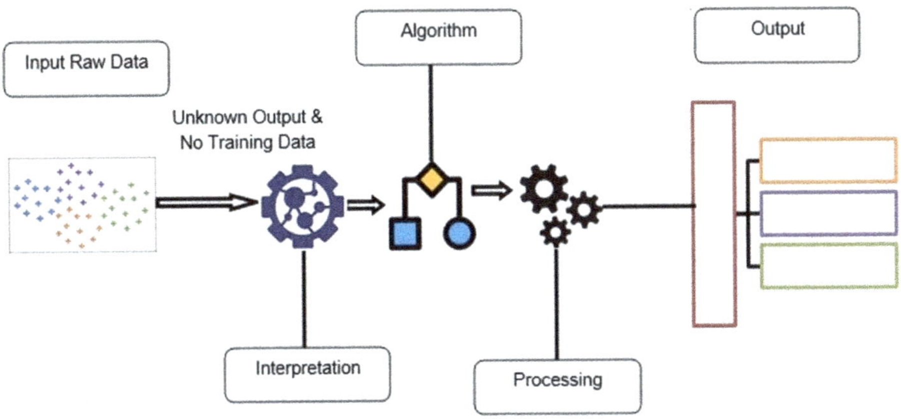

Figure 9.5: Unsupervised machine learning model using quantum computing.

Following the exchange of this key between Alice and Bob, additional information can be exchanged using well-known cryptographic techniques. Symmetric and asymmetric cryptography are the two main categories of cryptography. Symmetric cryptography encrypts and decrypts data using the same secret key and algorithm by both the sender and the recipient. AES and DES are the popular symmetric cryptography techniques. Asymmetric cryptography, widely known as public-key cryptography (PKC), uses pairs of keys, each of which has its own private and public key, to address the key distribution problem. This type of asymmetric cryptography is mainly used for digital signature. The online security business has recently been severely exposed due to the increasing frequency of cyberattacks that occur every day throughout the world.

Companies are setting up appropriate security architecture in their organization, but for traditional digital systems, the procedure is overwhelming and unworkable. Consequently, cybersecurity has remained a major issue for people all over the world. Finally, quantum computing comes to the aid to deal with all these issues and make the traditional model more secure and lossless in the reverse order of encryption.

9.7.3 Weather forecasting

Weather forecasting is one of the places where data is analyzed to predict the weather conditions minute by minute with help of quantum computers. Initially, traditional computers were used for this purpose. Due to the large amount of time traditional computers take while processing the huge volume of data and also due to the processing speed, quantum computing is included to deal with this process. It is essentially used for predicting the weather report for the current time due to climate changes in the world. This process, along with quantum technology, designs the advanced model to enhance the learning of weather patterns accurately and to increase the processing speed.

In order to predict the weather condition, several parameters are considered such as air pressure, temperature, and air density through advanced machine and computing techniques. Quantum machine model helps to improve the pattern recognition, which turns the model to predict the weather events accurately, analyze the climate models, and save thousands of lives a year.

9.7.4 Industry-based applications

Quantum computing makes use of the recognized laws of quantum mechanics in physics. This area of physics investigates the subatomic processes that govern the environment. Superposition and entanglement, two of its features, can be exploited to advance current computer techniques. The characteristic known as superposition enables two separate states to define a system. It can be both at once; it is not just one or the other. In traditional computing, bits with values of "1" or "0" are applied by computers. Qubits, the equivalent used in quantum computing, can have two values at once. When quantum components remain coupled, the phenomenon is known as quantum entanglement. Quantum particles keep a relationship with one another, no matter how far apart they are. One particle's impact can have an impact on another. The use of these quantum features to computing technology offers exciting potential. These are particularly helpful when investigating options or sorting through vast amounts of data. This is a completely different way of computing compared to the way that is currently done. Quantum computing has the potential to dramatically enhance innovation despite being a relatively new technology.

Designers may conclude that quantum computing has an equal or greater impact than Blockchain, based on the growth drivers and prospects. As a result, a variety of businesses are working on quantum computing to maximize its already known advantages. Messages that are encrypted can be sent more quickly utilizing quantum communication. The power that some characteristics like superposition and entanglement offer can be useful to certain businesses to a degree that its influence has not yet been quantified.

This article discusses how certain industries will be affected by quantum computing. Some of the sectors or industries that employ quantum computing are:
- Banks and Finance Segment
- Transportation and Logistics
- Manufacturing and Industrial Design
- Government and Academia
- Space and Defense
- Energy and Power
- Healthcare [29]
- Chemicals

9.7.4.1 Manufacturing and industrial design

To manufacture high-quality products [18, 19], efficient methods and designs are necessary. Designing can be a very difficult task. To create a functional product, industrial designers must take into account a variety of factors. This is crucial in the fields of electronics, transportation, and equipment. When creating a high-speed jet, for instance, designers may require numerous prototypes. This procedure guarantees that they have the best wing design for travelling at high speeds. It also applies to other significant components of the device. Design engineers can quickly sort through the various options with the aid of quantum computing. With the use of this technology, they could produce better designs more quickly. Additionally, it can improve troubleshooting for manufacturers. They can input data about machine failure into a quantum computer, and it can help identify the problem locations.

Quantum computing may also assist in reshaping the industry. Quantum computing has a huge potential to aid in the development of ground-breaking goods and services that will change and reinvent the industry. Chemical research, product development, and process improvement are three manufacturing sectors projected to experience significant advancements as a result of quantum computing. Early adopters have the chance to secure advantages that will be extremely difficult to challenge, which will have a significant impact. Figure 9.6 shows the complete picture of quantum computing use cases in the manufacturing sector.

The four categories of find, design, control, and supply are each the subject of prospective manufacturing use cases that are examined in this article.

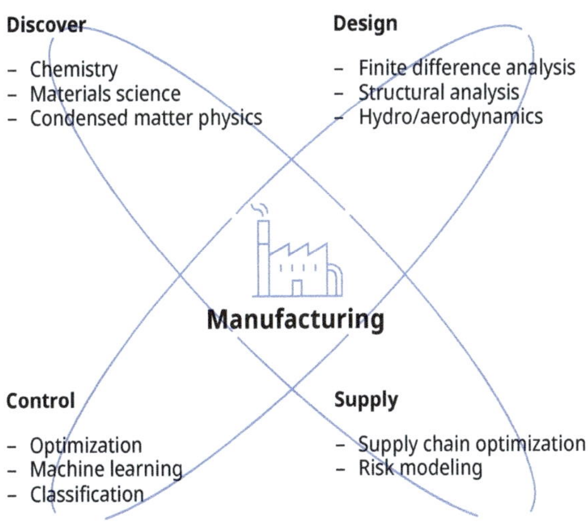

Figure 9.6: Categorizing the potential quantum computing use cases in manufacturing.

In the stage of discover [20], it is well known that even moderately large molecules are difficult for traditional computers to model accurately from a chemical standpoint. With its exponentially huge state space, quantum computing should be able to provide modelling of extremely complex molecules, which would greatly aid in the development of new materials and medications.

The automotive, aerospace, and electronics industries could gain from the expected capabilities of quantum computing in the manufacturing sector by:

- Better strength-to-weight ratios in materials
- Batteries with a vastly increased energy density
- More effective catalytic and synthetic methods that could aid in energy production and carbon capture

In the stage of designing, computer simulation is being used to design and test a variety of products. Hardware parts and subparts for the automotive and aerospace industries are 3D-modeled with particular engineering safety margins. These margins can add up, resulting in over-engineered, overweight, or expensive items that can hinder their potential to be commercially viable. However, it is predicted that in the future, quantum computers will be able to reproduce the interactions between the parts of intricate hardware systems more thoroughly and accurately, computing the system loads, load routes, noise, and vibration. This comprehensive analysis can reduce the total impact of many separate safety margins, improve cost without compromising system performance, and optimize the production of individual components within the context of the complete system.

In the stage of control, modern factory control methods test the limits of advanced analytics, particularly when using machine learning and analyzing numerous factors. Beyond the capabilities of classical computing, quantum computing may assist in uncovering novel correlations in data, improve pattern recognition, and advance categorization. The application of optimization to the combination of quantum computing and machine learning is anticipated to have a substantial impact on manufacturing in a number of domains.

9.7.4.2 Logistics

Logistics is often a time- and place-sensitive enterprise. Therefore, process optimization would greatly benefit it. There are several considerations to make while relocating anything from one place to another. Users can consider supply chains, vehicle accessibility, traffic, and customer expectations, among other things. Quantum computing can be used by businesses to identify the best pathways for each item. This method also considers genuine factors like the climate and traffic. Quantum technologies have the potential to change the logistics process and satisfy client demands. DHL and other logistics companies have already acknowledged this trend as having enormous potential.

9.7.4.3 Finance

In financial activities, a number of complex mathematical processes are regularly used. Analysts must take into account a wide range of factors in order to predict probable market outcomes. Major events sometimes need speedy responses, which traditional computers find difficult. Quantum computing allows for the creation of more accurate market simulations and predictions. They carry out Monte Carlo simulations far more successfully than do traditional methods. Financial analysts can investigate a wide range of probable outcomes from a number of variables using Monte Carlo simulation. These results help us comprehend probable outcomes and threats, especially when it comes to predictive modeling. Due to quantum technology, these jobs take less time and effort to complete. The biggest names in banking and finance are aware of the potential applications of this new technology. JP Morgan Chase and Wells Fargo have already expressed interest in quantum computing, which will power the future of banking.

In order to significantly speed up trade operations, transactions, and data processing, quantum computing has grown in favor in the banking and finance services industry. To efficiently calculate probability in a data-intensive environment, ever more-powerful computers are needed. In this context, several banks are utilizing new generation of computers that utilize quantum physics concepts to rapidly process enormous amounts of data. In August 2017, Telstra, the Federal Government, the New South Wales Government, and the University of New South Wales joined forces with the Commonwealth Bank (CBA) in a $83 million project to establish Australia's first quantum computing company.

Additionally, NatWest and Fujitsu worked together on a proof-of-concept research to enhance its mix of premium liquid assets, including bonds, money, and government securities.

In 2019, Google claimed that its Sycamore quantum processor completed a task in a little over 3 min, which would have taken a supercomputer thousands of years to perform. Given the experiment's limitations [21], it did illustrate the enormous potential of quantum computing. Today's leading financial and banking sectors profit from the use of quantum computing, which are shown in Figure 9.7.

Quantum computing will have a significant impact on the financial services industry over the next five years. Financial firms that implement quantum computing early might have considerable competitive advantages, including the capacity to surpass rivals and assume lead in the market. Quantum computing's combinatorial optimization skills may support investment managers in boosting portfolio diversity, fine-

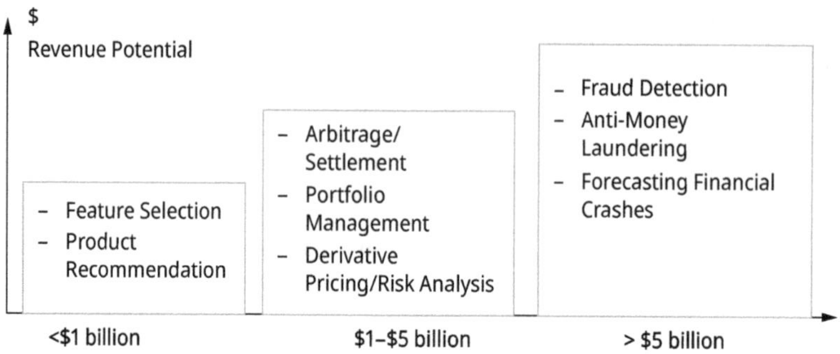

Figure 9.7: Leading financial and banking sectors using quantum computing.

tuning assets to better suit changing market conditions and investor objectives, and streamlining trade settlement processes more efficiently.

The following major situations are going to determine whether quantum computing has commercial value for financial institutions:
- Boosting investment returns
- Lowering capital requirements
- Developing new investment opportunities
- Improving compliance as well as risk detection and management

9.7.4.4 New product development and chemical engineering

Quantum computing [22, 23] could improve in the formation of mixtures by enabling a better understanding of the complicated molecular-level processes involved and by aiding the optimization of such mixtures to produce more effective products for the full spectrum of applications that chemicals support. Today, creating a new cleaning product composition relies on a combination of theoretical models, a technician's experience, and essentially trial-and-error experimentation. To better understand how, for instance, detergent molecules interact with a wine stain on a fiber and to determine the optimal active chemicals and formulas to remove it, quantum computing could help this process with optimization computations. The required calculation time might be reduced to seconds by a team employing a quantum computer and the right algorithm [27].

Additionally, quantum computing may benefit research and development efforts in the field of complex assemblies, which includes composites, effect pigments, and opto-electronic displays. To determine the optimal materials to utilize (in terms of thicknesses, concentrations, and structure), these need hybrid methodologies that integrate simulations, optimization, and artificial intelligence – all of which directly relate to the promise of quantum computing. Some chemical companies are already involved in this work, creating OLED designs for usage in TV and smartphone screens, utilizing current computing techniques. Here, using quantum computing has the potential to significantly speed up the research process [28].

Chemical engineering focuses on the control of atoms and molecules. Quantum mechanics is used by the field in research. It is a broad field, as well. Chemical engineering is used in manufacturing, healthcare, building, food processing, electronics, and in other fields. The variety of alternatives makes it possible that the perfect chemical configuration may be hard to find. With the use of quantum computing, these procedures can be hastened. This use will help advance the development of medications and vaccines. An early response is required in light of the COVID-19 epidemic.

9.7.4.5 Medical computing

One of the biggest causes of deaths worldwide is cancer. In fact, 1.7 million people died from respiratory cancer diseases alone in 2016 [24], according to a recent World Health Organization (WHO) survey. However, the chances of recovery through treatment are significantly higher if cancer is discovered at an early stage. There are numerous approaches for treating cancer. One is surgical removal and the other is radiation. In radiotherapy, beam optimization is crucial since it is essential to ensure that the radiation affects healthy cells and tissues close to the tumor site as little as possible. There have been numerous traditional computer-based optimization techniques for radiation in the past. A novel method for optimizing radiotherapy was developed in 2015 by researchers at the Roswell Park Cancer Institute using quantum annealing computers, such as those made by D-Wave. This method is three to four times faster than a conventional computer. A radiation beam [25] is guided by a quantum computer to precisely target and kill the cancer cells while sparing the nearby tissues. This quick identification of a small cluster of cancer cells and the subsequent elimination of those cells is only one of many potential medical advancements that might take place before quantum computing is ever realized.

9.7.4.6 Gaming computing

The gaming industry is one of the key applications of quantum computers. It is extremely fast and uses cutting-edge game algorithms. They can be used to build a great gaming setup that can run games at blazing high frame rates, which is what gamers may be interested about. Since existing hardware has not yet achieved "quantum supremacy," which is when quantum hardware can compute more quickly than the finest computers now available, the field of quantum computers is still in its beginnings. This is due to the fundamental differences in how classical and quantum computer algorithms operate. Quantum gaming yet appears to be feasible despite it. A few video games have been created that make use of quantum computers. One of them is based on the board game Battleships, and is titled Quantum Battleships [26]. Additionally, Microsoft has been developing a programming language called Q# that utilizes both conventional and quantum computing technology. Since it is quite similar to C#, it is very likely that games that utilize quantum hardware might be created using Q#.

Applications of quantum computing in gaming are interesting. Some of the characteristics used in gaming sector with quantum computers are listed below:
- Sharp, clear graphics
- True-to-life visuals and features
- Random environment generation
- Realistic simulations
- Increased storage space
- Machine learning integration – the game learns about the players and their playing style
- Video game creation and optimization to analyze the consumer data

9.8 Conclusion

In this chapter, we looked at how advanced designed algorithms of AI and ML-based models with quantum technology help to solve complex problems. QC enables the traditional computer. This chapter has discussed the key components of quantum computing as well as some applications in the areas of artificial intelligence, machine learning, cryptography, and cyber security. These applications have shown a significant contribution in both AI and ML models to process huge volume of data and give better results in terms of speed and accuracy. The review article provides merit to the argument of integrating quantum algorithms with artificial intelligence techniques. It can improve machines' capacity for learning as well as minimize the time required for different algorithms to train.

References

[1] J. Barzen, F. Leymann, Quantum humanities: A vision for quantum computing in digital humanities, SICS Software-Intensive Cyber-Phys. Syst., 35, 2019, 1–6.

[2] P. J. Coles, S. Eidenbenz, S. Pakin, A. Adedoyin, J. Ambrosiano, P. Anisimov, D. Gunter, Quantum algorithm implementations for beginners, 3, 2018, 1–92. arXiv preprint arXiv:1804.03719.

[3] V. Dunjko, T. Kapourniotis, E. Kashefi, Quantum-enhanced secure delegated classical computing, Quantum Inf. Comput., 16, 2014, 1–26.

[4] D. Aharonov, M. Ben-Or, E. Eban. Interactive proofs for quantum computations, ICS, 2, 2010, 453.

[5] http://jojo.us.com/elements-of-quantum-computing/.

[6] https://www.azoquantum.com/Article.aspx?ArticleID=133.

[7] V. R. Gotarane, S. S. Madhukar, Quantum Computing: Future Computing, Int. Res. J. Eng. Tech., 03(02), 2016, 1424–1427.

[8] https://www.insightsforprofessionals.com/it/leadership/what-is-quantum-computing.

[9] A. Pathare, B. Deshmukh, Review on cryptography using quantum computing, Int. J. Mod. Trends Sci. Technol., 18(1), 2022, 142–146.

[10] S. Kute, C. G. Desai, Quantum cryptography: A review, Indian J. Sci. Technol., 10(3), 2017, 1–5.

[11] https://analyticsindiamag.com/will-quantum-computing-define-the-future-of-ai/.

[12] https://quantumtech.blog/ 2022/02/22/at-the-intersection-of-quantum-computing-artificial-intelligence-and-machine-learning/.

[13] https://research.aimultiple.com/quantum-ai/.

[14] G. Biggi, J. Stilgoe, Artificial intelligence in self-driving cars research and innovation, Scientomet. Bibliomet. Anal., 4, 2021, 1–28.

[15] https://www.insidequantumtechnology.com/news-archive/true-self-driving-cars-will-be-impacted-by-quantum-computers/#:~:text=For%20Level%204%20and%20Level%205%20true%20self-driving,by%20the%20self-driving%20car%20up%20to%20the%20cloud.

[16] https://medium.com/mit-6-s089-intro-to-quantum-computing/supervised-and-unsupervised-machine-learning-using-quantum-computing-ac9d60a14a8f.

[17] https://analyticsindiamag.com/top-applications-of-quantum-computing-everyone-should-know-about/.

[18] https://www.computer.org/publications/tech-news/research/industry-applications-of-quantum-computing.

[19] A. Bayerstadler, et. al., Industry quantum computing applications, Quantum Technology and Application Consortium – QUTAC et al., EPJ Quantum Technol., 25, 2021, 1–17.

[20] https://www.ibm.com/thought-leadership/institute-business-value/report/quantum-manufacturing.

[21] https://www.greyb.com/impact-of-quantum-computing/.

[22] https://www.mckinsey.com/industries/chemicals/our-insights/the-next-big-thing-quantum-computings-potential-impact-on-chemicals.

[23] D. P. Nazareth, J. D. Spaans, First application of quantum annealing to IMRT beamlet intensity optimization, Phys. Med. Biol., 60(50), 2015, 4137–4148.

[24] https://listverse.com/ 2019/01/10/top-10-unexpected-future-applications-of-quantum-computers/.

[25] D. Solenov, J. Brieler, J. F. Scherrer, The potential of quantum computing and machine learning to advance clinical research and change the practice of medicine, J. Mo Med., 115(5), 2018, 463–467.

[26] https://decodoku.medium.com/quantum-battleships-the-first-multiplayer-game-for-a-quantum-computer-e4d600ccb3f3.

[27] G. Sasubilli, A. Kumar, “Machine Learning and Big Data Implementation on Health Care data,” 4th International Conference on Intelligent Computing and Control Systems (ICICCS), 2020, pp. 859–864, doi: 10.1109/ICICCS48265.2020.9120906.

[28] S. M. Sasubilli, A. Kumar, V. Dutt, “Machine Learning Implementation on Medical Domain to Identify Disease Insights using TMS,” 2020 International Conference on Advances in Computing and Communication Engineering (ICACCE), 2020, pp. 1–4, doi: 10.1109/ICACCE49060.2020.9154960.

[29] S. M. Sasubilli, A. Kumar, V. Dutt, “Improving Health Care by Help of Internet of Things and Big data Analytics and Cloud Computing,” 2020 International Conference on Advances in Computing and Communication Engineering (ICACCE), 2020, pp. 1–4.

M. Rohini, D. Surendran, Oswalt Manoj S

Chapter 10
A leap among quantum ML and DL models: a review

Abstract: Classical computing systems are based on electronic circuits represented at a macroscopic level by the electric properties of the circuits. They store information in the form of bits (classical bits) that take values of either 0 or 1, which can be physically realized in the form of electrical signals (for example, low voltage = 0 and high voltage = 1) on electronic circuits. Quantum computer systems differ from classical systems in that their bit level information is directly represented by the interaction of microscopic systems namely, photons, electrons, and atoms, which are monitored and distributed throughout the implementation by quantum theory. The information is stored in the form of qubits (quantum bits) represented as ground states $|0|$ and $|1|$. Thus, quantum theory represents the weightage of each outcome based on elements composing the physical systems performing each computation of smart systems. In quantum theory, we refer the readers to a timeline depicting the formative years of quantum modules. A growing number of computer scientists and physicists are coming forward, bringing in software packages and quantum programming languages based on classical programming languages. The realization of quantum technology has thus become an international and interdisciplinary effort.

10.1 Introduction

Artificial intelligence (AI) is perhaps going to help the human resources in improving their efficiency, reducing the repetitive tasks, tracking and removing the errors, maintaining consistency of results, and thus, supporting their professional growth. AI talent has become expensive as human resource has shortfall of skilled AI talent pool although AI enables human resource to create and deliver personalized services to the clients and customers, by understanding their needs and desires by offering the customized goods and services that are enabled solutions with the help of AI. Deep Learning (DL) is used in several areas like automobile Industry for self – driving cars, security enhancement in the defense sector, and medical field for disease identification. DL

M. Rohini, Department of Computer Science and Engineering, Sri Krishna College of Engineering and Technology, e-mail: rohinim@skcet.ac.in
D. Surendran, Department of Computer Science and Engineering, Karpagam College of Engineering, e-mail: d.surendran@gmail.com
Oswalt Manoj S, Department of Computer Science and Engineering, Sri Krishna College of Engineering and Technology, e-mail: oswaltmanojibm@gmail.com

https://doi.org/10.1515/9783110791402-010

algorithms are used in state-of-the-art problems like image recognition, object detection, and natural language processing that demands the quantum methodologies.

The DL field is becoming predominantly popular with improved Convolution Neural Network Architectures. Presently, the quantum computer systems available in the market (or the physical world) include near-term, intermediate-term, or small-scale devices ranging up to 100 qubits. These devices are now being used to test the power of quantum computing advantages [21, 22, 24]. Machine learning (ML) algorithms and the problem of optimization serve as some of the problems where the power of quantum computing systems can be evaluated. This gave rise to the field of "quantum machine learning," first coined [1] by Schuld et al. [2] and discussed as the most exploring work joining the two disciplines also elaborated in their book.

Among the several designing quantum process methodologies and algorithms, the first aims at harvesting algorithmic implementations by conversion of traditional models into quantum process computations. The second strategy is an exploratory approach, which aims to provide several hybrid technics that employ smart intelligence in quantum space, in addition to achieve runtime ML algorithms. This is achieved by seeking machine learning models that suit the physical characteristics and constraints of a quantum computer. This approach uses any system obeying the laws of quantum mechanics for training a model that learns from data.

The four approaches of integrating quantum computing with ML were introduced by [3]. They include the classical-classical (CC), classical-quantum (CQ), quantum-classical (QC), and quantum-quantum (QQ) implementation depending on which system the data-generation and the information-processing are done. It is found that quantum algorithms can provide exponential speedups over classical algorithms to solve certain tasks, for example, solving a system of linear equations of the form $Ax = b$. Another such notable example, where the evidence of exponential speedup was proved, is the Shor's factoring algorithm, which serves as a basis for cryptosystems like RSA. There is a notable difference in the behavior of the laws of probability in quantum mechanics, as compared to classical physics.

10.2 Related works

Feynman was the first to question the effect of this behavior of quantum mechanics on computation, which could be realized in a process that follows quantum principles. He claimed that by computational quantum mechanics that works with smart intelligence, one could compute more efficiently than by using the legacy software. More tasks aligned to quantum mechanics and computing showcased the modeling of a Turing machine and quantum circuits [4] further explained that computers are bound by time and space complexity when solving larger problems. In theoretical computer science, algorithms exhibiting a polynomial growth in the order of time and space are classified as efficient. Several other computational complexity classes namely, PSPACE[1], NP[2], BPP[3], $P^{\#P}$ [4] are related as $P \subseteq BPP$, $NP \subseteq P^{\#P} \subseteq PSPACE$.

The purpose of using quantum computation is to process the task of image and object segmentation through each filter framed by algorithms, with increase in the computer's computational power. But this work did not show how any problem could be solved in a specific quantum time, because all these problems were easily addressed by [5], who laid the foundations of quantum computation theory of computational complexity (check their paper). Their work demonstrated the construction of the Oracle problem (problem involving a "black box" subroutine), which was further improved by [6] depicting its resolution in polynomial time on a quantum Turing machine (quantum computer) and requiring exponential time on a classical computer. This was indicative of the potential of quantum computers to solve complex problems in polynomial time, which would, otherwise, be very difficult to solve in decision making, recruitment, and selection processes, thus enabling a smooth flow of operations.

10.3 Quantum error correction

The results of a quantum computation are likely to deviate from the ideal results under the following circumstances:

- If the gate operations are not perfect.
- If the quantum mechanical systems are not well isolated from the environment, they are affected by spurious interactions with the environment, resulting in unwanted transitions and decoherence.
- If the quantum system considered in the design of the quantum computer differs from the idealized model system.

The above points play a crucial role in preserving the quantum information (in effect, quantum coherence), which is vulnerable compared to classical information, thus impacting the error-control mechanism in quantum systems [7]. Currently, the error correction of quantum systems involves the use of "ancilla" bit for storing the earlier qubit state and verifying the changes. This is unlike the error correction mechanism in classical systems [8].

10.4 Near-term quantum devices

The current available quantum computer systems are noisy intermediate-term devices [9], which are not universal in nature and can only implement a subset of gates to reliably solve a specific problem at hand. These systems are limited by the number of qubits and, hence, also called small-scale devices. Finally, near-term devices, as the name suggests, are noisy in nature because they are still evolving in error-detection and correction mechanisms to preserve the coherence of qubits. Thus, it is suitable to design smaller circuits with lesser number of gates than can be applied on the qubits to obtain useful and correct results. Recent work proposed how higher-dimensional quantum systems particularly used in classification tasks are vulnerable to adversarial perturbations compared to lower-dimensional systems. Such a phenomenon is independent of the decoherence effects or the classification algorithm itself and erase any exponential advantages sought by the algorithm. Gokhale et al. [16] extended this study to practically demonstrate vulnerability of quantum machine learning systems to adversarial attacks in real-life scenarios and suggest practical defense strategies to fight a number of attacks. Our proposed implementation adapted from [17] is an example of a hybrid classical-quantum smaller circuit design.

10.5 Variational circuits

For near-term quantum devices at hand, a class of hybrid classical-quantum (CQ) algorithms has shown growing popularity for testing various applications on the quantum device, while processing the data on the classical systems. Here, the aim of the quantum device is to compute terms of an objective function and use the traditional algorithm-based analysis for $C(\theta)$ with an objective function that calculates the cost function of each quantum module performing the computations. The final states of the qubit (or expectation values) are then measured on the classical system, which depend on the weightage of attribute θ. Since the training is performed by a combination of quantum and classical systems, it is termed as a hybrid scheme [18]. The advantages of such a scheme are: (a) suitability to execute on near-term quantum devices since only a fraction of the overall algorithm runs coherently on the quantum circuits (smaller circuits); (b) can be a structure-based process of quantum elements due to the variety of input parameters from circuit; (c) learns to deal with systematic errors about the circuit; (d) can deal with noise since the optimization effort happens iteratively.

10.6 Quantum computing gates

Two basic operations with the demand of achieving process elements in smart technologies include the quantum logic gates and the computational basis measurements to measure the outcome of the qubits. In order to have a quantum computer learn from classical data (CQ case), it becomes necessary to prepare the data in an appropriate manner referred to as "state preparation." The different ways in which classical data can be encoded are Hamiltonian encoding, Basis encoding, Hybrid Amplitude encoding, and Qsample Encodeing. Data encoding is a crucial step for preparing the classical data to be processed on the quantum system. For more details, refer to [10, 23].

10.7 Evaluating the performance of quantum systems and optimality of measurement

For estimating the quality of the classical-quantum model as well, the most important metrics include accuracy and loss. The loss term measures the model's predictions to the target labels. Accuracy increases with minimum loss. The error term measures the number of misclassified labels and depends on a number of factors that can affect this value. The laws of quantum mechanics only permit unitary transformations of the state. A unitary matrix is one whose conjugate transpose is equal to its inverse (i.e., X is unitary iff, $XX^{\dagger} = XX^{-1} = I$, or $X^{\dagger} = X^{-1}$).

10.8 Quantum machine learning for pattern classification

We discuss some of the related work in the field of classical-hybrid algorithms applied to pattern classification. A quantum Fourier transform classifier is used for identifying and localizing patterns in images. His proposed work exhibited an exponential speed-up compared to its classical counterpart. Marciano and Pagels [20] demonstrated that classifiers based on adiabatic quantum computing successfully outperformed the classical ML classifier AdaBoost. Kavitha Rani [19] introduced a quantum machine learning (QML) approach for binary classification of qubit states that is free from the use of quantum memory and performs with the minimum allowed error rate for any size of the training dataset. Earlier work on classifying qubit states (also represented as positive-operator valued measurement and not the actual quantum states) was also demonstrated by [20].

Deutsch [4] proposed a novel version of the quantum adiabatic algorithm to classify M vectors into k clusters. This method could be performed in logarithmic time providing exponential speed-ups both in the number of vectors and their dimension. Additionally, they demonstrated that QML also helped in enhancing privacy by making fewer calls to the quantum database during the cluster assignment. Similarly, [5] also presented a quantum version of the popular support vector machine classifier [6], demonstrating its performance in logarithmic time with exponential speed-ups both in the size and number of the training vectors. A quantum pattern classification algorithm applies the distance-based classification principle similar to the K-nearest neighbor (kNN) approach.

The implementation was based on creating a superposition of the training dataset and computing the Hamming distance [7] between the input state and the amplitude of each vector in the superposition. The quantum circuit was realized using three quantum registers; first for the input state; second for the superposition; and third for the ancilla qubit. Preskill [8] demonstrated the power of their quantum nearest-neighbor algorithms designed using proven faster quantum methods of computing distance metrics on several real-world classification tasks involving the popular MNIST handwritten digits database, the half-moon dataset, and datasets involving diseases. They obtained exponential or even super-exponential speed-ups over the classical counterparts. Schuld and Petruccione [9] proposed the least squares approach applied to the given data in training plane. The algorithm performed in polylogarithmic time and could also check if the data set under investigation qualifies for linear regression.

Very recently, we see the advancement in applying supervised quantum classification techniques providing higher accuracy on real-world datasets. Christensen et al. [11] introduced a supervised learning quantum neural network (QNN) model capable of representing classical or quantum data. Their work was an exploratory study providing a classical simulation of a small quantum system designed with near-term quantum processors in mind. The proposed QNN demonstrated binary classification of handwritten digits belonging to class 3 vs. class 6 from the popular [10], which proposed a quantum version of slow feature analysis (QSFA) approach by combining the dimensionality reduction technique and a classification procedure based on quantum Frobenius distance (QFD). They demonstrated their work on the MNIST dataset, providing a 98.5% classification accuracy in polylogarithmic time. They inserted the perceptron's response in a neural network and proved that such a QNN has at least the same power as classical neural networks.

Thus, the quantum calculation methods have transformed the classical machine learning algorithms into quantum algorithms to be implemented on quantum computers. The classical-classical approach (CC) studied by Christensen et al. [11] involves the development and execution of quantum-like algorithms on classical computers. Many works attempted to present a good comparative table of various QML techniques for solving classification and regression problems based on hybrid classical-quantum, quantum and quantum-like implementations. In Table 10.1, we also summarize some literature studies of engineering principles in quantum design.

10.9 Classical transfer learning for image segregation detection and analysis

A fully convolutional network (FCN) based on the *VGG-16* initially learns the surface label and transfers this learning to detect edges of the spliced region. The *ResNet50* used pretrained models in conjunction with their proposed convolutional neural network (CNN) to extract features from the spliced and authentic images, which are later classified using the support vector machine (SVM) classifier. An implementation of transferring the learning from pretrained models namely, *VGG-16* and *ResNet50*, to classify spliced vs. authentic images from the IEEE IFS-TC Image Forensics Challenge dataset is done. Reed et al. [12] *VGG-16, VGG-19, GoogLeNet* and *Inception-v3* [13] pretrained models to extract features from the illuminant maps of the DSO-1 spliced images, which were further classified using SVM. Such systems are computationally expensive and require the use of GPU systems to obtain the desired output.

10.10 Quantum computing for image splicing detection

- Quantum mechanics helps us achieve a transition from a state of high uncertainty to low uncertainty, illustrated using an interesting coin toss example.
- Quantum computing systems can solve computationally expensive problems involving large input sizes in polynomial time compared to classical computing systems, which can, at best, solve them in exponential time.
- Optimization problems in machine learning represent the nature of quantum computing problems (or can be viewed as quantum computing problems), which require repetitive computations to obtain the best result (or the optimum result). Thus, quantum computing can be applied in solving optimization tasks in machine learning.
- Application areas of machine learning involving image pattern classification also involve large input sizes and repetitive computations. Quantum machine learning can, therefore, help in solving these problems effectively.

These reasons motivated us to test the application of quantum transfer learning (QTL) to the problem of image splicing classification. The problem is modeled in a similar fashion as the ants vs. bees classification. The proposed implementation of quantum machine learning i.e., transfer learning is a classical-quantum (CQ) supervised machine learning, where the class labels of the input data under consideration are available. It is observed from the above analysis that there is no significant relationship between the gender of the respondents and AI enablement in healthcare management systems. Therefore, all the predictor variables like gender are not related with dependent variable AI enablement in healthcare management systems of Digital Healthcare Service Providers.

Here, when the gender of the respondents from Digital Healthcare Service organizations varies between male and female, the AI enablement in healthcare management systems does not vary. Although AI-enabled tools are essential at the workplace for both the male and female employees, the AI enablement in healthcare management systems does not vary with the gender. Hence, the incorporation of AI technology within healthcare management systems, such as AI-powered electronic medical records, electronic health records, and progress notes, along with AI-driven clinical decision support system software that assists healthcare professionals in recalling procedures and tests, as well as AI-enhanced clinical decision support systems (DSS), demonstrates the significant impact of AI in healthcare management.

Software to remind healthcare professionals [21] on reduction of medication errors, AI-enabled device alerts on drug reactions (dosage errors and drug interactions) using Alternative Dispute Resolution (ADR), AI-enabled bar coding medications and automatic dispensing, and other AI enablement helps healthcare professionals to view medical imaging and test results on digital gadgets at their convenience; these enable AI initiatives in the Digital Healthcare Service Space, and it does not influence on the gender of the employees as a result of which AI systems developed in healthcare [22, 23] are neutral based and not biased. Therefore, AI enablement in healthcare management systems [24] does not influence the organizations with respect to gender friction, which always holds well. Hence, the null hypothesis gender and AI enablement in healthcare management systems is accepted.

10.11 Proposed model design

Classical transfer learning has proved beneficial for image splicing detection. Figure 10.1 represents the block diagram of the QTL implementation applied to image splicing. In this, previously experimented resources trained in the image quantification model with large dataset to solve a particular task are used in conjunction with a smaller network to be trained on a different but related dataset for solving another task, and a hybrid classical-quantum model with learning transferred from the popular deep learning model, ResNet18 with 18 layers of residual connections pretrained on the ImageNet dataset to address the problem of ants vs. bees classification was implemented. The results obtained by executing the model on the quantum simulator (*PennyLane* = 96.7% accuracy) and quantum processors (*ibmqx4* = 95% accuracy, *Aspen 4–4Q-A* = 80% accuracy) showed a promising outcome. In our proposed design, we adopt this approach to consider the advantages of image segregation problems in training plane. Here, a classical pretrained model, ResNet50 that is used in image based dataset for classifying images belonging to 1,000 classes is deployed. The last but the final layer of this model with 2,048 output features is passed onto the classical preprocessing layer of the hybrid classical-quantum network, referred as the "novel quantum circuit." The learning from the classical pretrained model is transferred to the smaller dressed quantum network to classify spliced vs. authentic images. Observe that both the pretrained model and the target classical-quantum model use different but related datasets i.e., images. Figure 10.2 explains the proposed design.

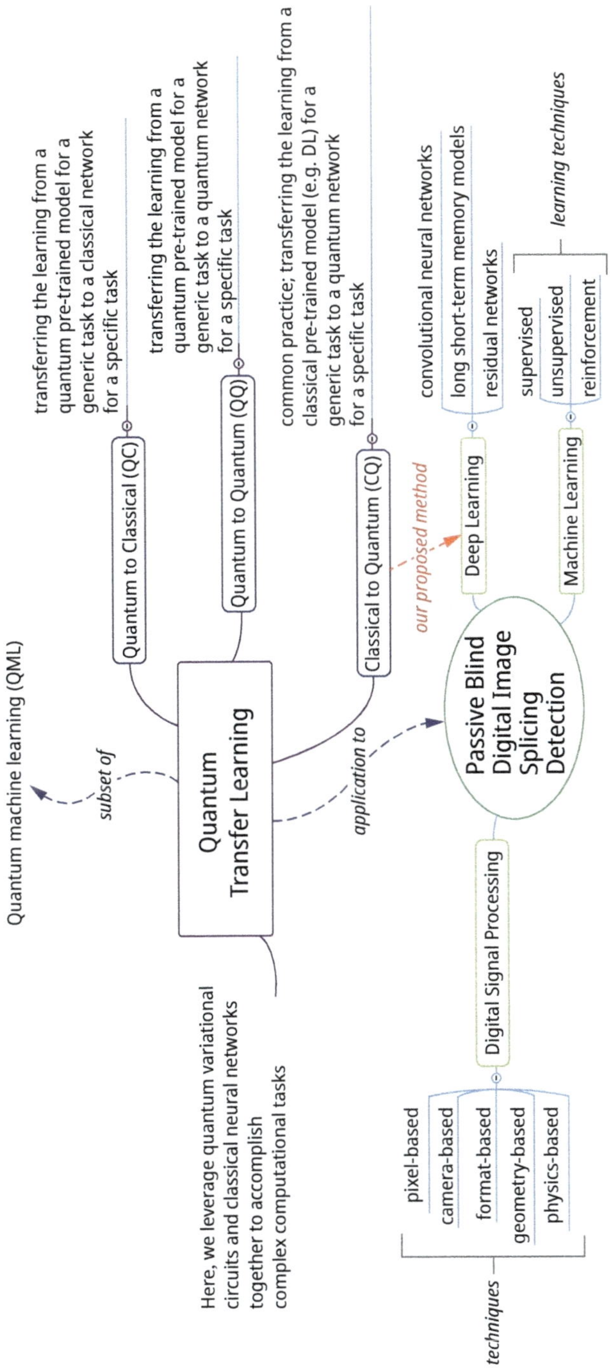

Figure 10.1: Block diagram of the QTL implementation applied to image splicing.

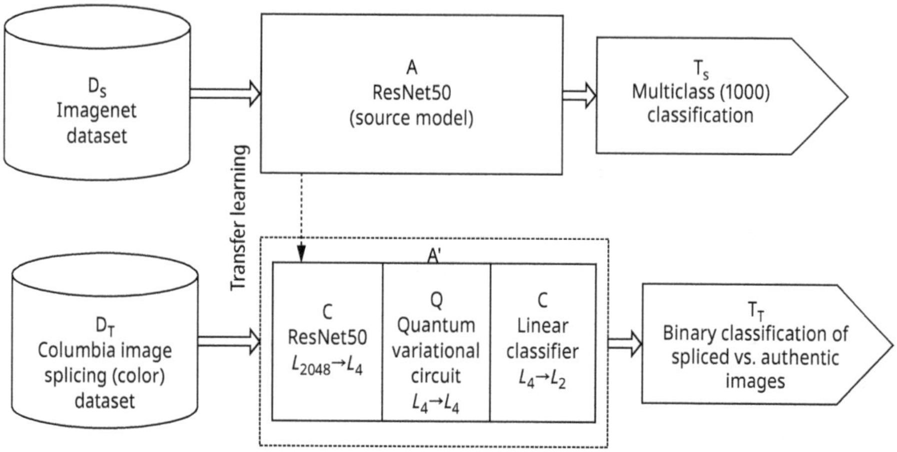

Figure 10.2: Block diagram of image splicing.

10.12 Explaining the QTL approach equations

The study found that a greater number of patterns in delivering digital services should meet the customer expectations and perceptions by the companies in terms of delivering services, and the challenges they faced were mechanism of delivery, responsiveness of staff, lack of empathy factor, lack to build promises and assurance, and also poor accountability towards their mistakes. This study will give a theoretical phase but lacks the concrete evidences, but as focus on delivering services these factors can be considered for the future research. A Variational Quantum Circuit of depth q can be defined as a concatenation of many quantum layers (L). Each quantum layer can be defined as a unitary operation parameterized by different weights acting on an input state $|x\rangle$ and producing the output state $|y\rangle$. This unitary operation could be a sequence of single-qubit rotations followed by a fixed sequence of entangling states. The Hilbert space dimension of the input states in a quantum layer is preserved. A quantum layer (eq. 10.1) and a variational quantum circuit Q can be represented as(eq. 10.2):

$$L:|x\rangle \rightarrow |y\rangle = U(w)|x\rangle \tag{10.1}$$

where U is a unitary operation and w is an array of classical variational parameters.

$$Q = L_q \circ L_{q-1} \circ \ldots L_2 \circ L_1 \tag{10.2}$$

The QNN is realized by three layers namely, the embedding layer (E), the variational layer (Q), and the measurement layer (M).

This variational embedding layer can be realized through single-qubit rotations say, (R_y, R_x) or single-mode displacements parameterized by x mapping traditional vector elements in 2D spaces (x) to Hilbert quantum element space ($|x\rangle$) (eq. 10.3). After embedding the classical vector into the quantum state ($|x\rangle$), a concatenation of variational quantum layers (VQC) is applied to the input state $|x\rangle$ to produce the output state $|y\rangle$ (refer to eqs. (10.1) and (10.2)). Finally, we extract the classical output vector y from the VQC by measuring the expectation values of the output state $|y\rangle$. For n_q input states $|x\rangle$, there are n_q local observables $|y\rangle$ denoted by $\hat{y} = [\hat{y}_1, \hat{y}_2, \ldots \hat{y}_n q]$. This process is defined as the measurement layer (M), which maps a quantum output state $|y\rangle$ to a classical output vector y (eq. 10.4).

$$E : x \rightarrow |x\rangle = E(x)|0\rangle \tag{10.3}$$

where E is a unitary operation used for embedding the classical input vector $|x\rangle$ with reference to the ground state $|0\rangle$.

$$\mathrm{M} : |y\rangle \rightarrow \mathrm{y} = \langle y|\hat{y}|y\rangle \tag{10.4}$$

The full QNN can be then written as given in eq. (10.5) (read from right to left).

$$\begin{aligned} Q &= M \circ Q \circ E \\ &= |y\rangle \rightarrow y \circ |x\rangle \rightarrow |y\rangle \circ x \rightarrow |x\rangle \qquad (10.5) \\ &= \langle y|\hat{y}|y\rangle \circ U(w)|x\rangle \circ E(x)|0\rangle \qquad (10.6) \end{aligned}$$

We can simplify the above understanding as follows (eq. 10.6) (read from left to right):

$$Q = x \rightarrow |x\rangle \rightarrow |y\rangle \rightarrow y \tag{10.7}$$

For the hybrid classical-quantum neural network, the authors introduce the concept of a *dressed quantum circuit*, which receives input from the last but final layer (L) of the pretrained classical network, processes in the quantum network just defined and outputs the results back to the classical network for the final outcome. This involves pre- and post processing by the classical network. Equation (10.7) (read from right to left) depicts the dressed quantum circuit with n_{in} classical input features, n_q embedded quantum states, and n_{out} final classical output states:

$$\tilde{Q} = L_{nq} \rightarrow n_{\mathrm{out}} \circ Q \circ L_{n\mathrm{in}} \rightarrow n_q \tag{10.8}$$

10.13 Design of image-based approach

The hybrid classical-quantum adapted transfer learning [19] for the ants vs. bees classification and applied it to the problem of image splicing classification. The dataset used for the image-based approach is the Columbia image splicing dataset. The code implementation for the QTL approaches is for the classical model. The choice of our pretrained classical DL model remains ResNet50. The initial experimental trials on the Columbia image splicing dataset with a 80–20% train-test split provided improvements in the detection accuracy by implementing the QTL approach.

This study considers the role of neural pattern by uploading data through a software version on human intelligence, considering the factors like memory, emotions, instincts, and thoughts are being trained through a memory-driven algorithm that does prediction and produces intelligence in making decisions and considers these two capabilities as the structure of reality that surrounds us. As stated in the study, according to Hawkins, the memory system also adds a set of sense to extract the patterns of sensitiveness like light, sound, touch, and feel characteristics, as mentioned in machine terms as sonar, radar, and infrared. The difficulties in forecasting and evaluating how precisely the enterprises will adapt to AI have resulted in a clear general principal stating that most companies and executives will struggle to adapt to the change, but the organizations can forecast the future in a rapid sense, foresee how the advantage of AI powered in chemoinformatic and drug discovery will be grasped, and react to opportunities. The analysis also revealed that "rapid experimentation and learning is the most efficient technique."

If professionals in the field of ML do not scale up experiments and do not complete their jobs during the course of the following ten years, AI will not replace individuals, but individuals who use AI will replace them. We split these 363 images into 80% (290) training set and 20% (73) testing (validation) set. (Later, for testing the image-based approach on realistic image splicing images, we replace the Columbia dataset with sample images from the DSO and AbhaS datasets with a 70–30% train-test split.) Refer to Table 10.1. For the patch-based approach, the images from the Columbia dataset are replaced by sample patches from the Columbia dataset.

$A^j = \tilde{Q} \cong L_{4\to 2} \circ Q \circ L_{2048\to 4}$, our predefined and designed quantum elements with classical preprocessing linear layer with 2,048 input and 4 output features; a quantum network with 4 input and 4 output qubits; and a post-processing classical linear layer with 4 input and 2 output features.

T_T = Test dataset for unused data obtained from the previously designed quantum circuit that builds several classes of slitted variable used in this study.

10.14 Patch-based approach

In order to further test the efficacy of quantum computation-based learning (QTL) to image segregation and detection, we provided a higher number of training and testing samples by following a patch-based approach. Here, we extracted the spliced regions of size 64 × 64 from the fake images by overlaying the ground-truth masks. We also randomly extracted patches of size 64 × 64 from authentic images (refer to Section 4.3.3) for details of the patch-based extraction process and Table 10.2 for training and validating patches for the Columbia dataset. The results obtained for the patch-based approach showed a rise in the level of average test accuracy and recall, throughout the classical and hybrid classical-QTL implementations.

10.15 Input qubit values (x) and the variational layer weights (wk)

The values of the variational layer weights, w_k in the quantum network (Q) denoted as self.q params in the code implementation are explained with the help of Python code as follows:

$$\text{self.q params} = \text{nn.Parameter}(0.01^{*}\text{torch.randn}(6^{*}4))$$

$$\text{self.qparams} = \text{nn.Parameter}(0.01^{*}\text{torch.randn}(24))$$

10.16 Results

The experiments executed in the Kaggle environment summarize the results obtained from the earlier image-based approach and the proposed patch-based approach. In the image-based approach, the quantum simulator and processor outperform the classical model in terms of the average test accuracy. The results obtained on the quantum simulator for the patch-based approach are very close to those obtained for the classical transfer learning (CTL) approach. But the execution on the quantum processor seems to provide a 100% test accuracy and recall.

Overall, we note that CTL performs slightly better than the QTL for the patch-based approach but for the image-based approach, the performance falls down. This verifies an important claim on factoring numbers and explains the improvement observed in the performance of quantum computing systems with an increase in the input size. The same holds true in this case; when the input size is 64 × 64 (patch-based), we observe a similar performance in the CTL and QTL approaches, while with an input size ranging from 757 × 568 to 768 × 1,152, there is a significant improvement in the performance of QTL.

Transfer-learning approach on the classical-classical system and the hybrid classical-quantum (CQ) simulator. The trained CQ model is then executed on the *ibmqx2* hardware for evaluating the model on the test dataset alone. It is important to note, here, that for each test image that is being validated, the actual quantum hardware runs 1,024 shots to give the final predictions (of 4 outputs), which are passed onto the classical post-processing layer to output the authentic versus spliced classes. In the table, we only highlight the results for the average test accuracy and recall of the best performing models. We mention the best results obtained only for adding more clarity in the results.

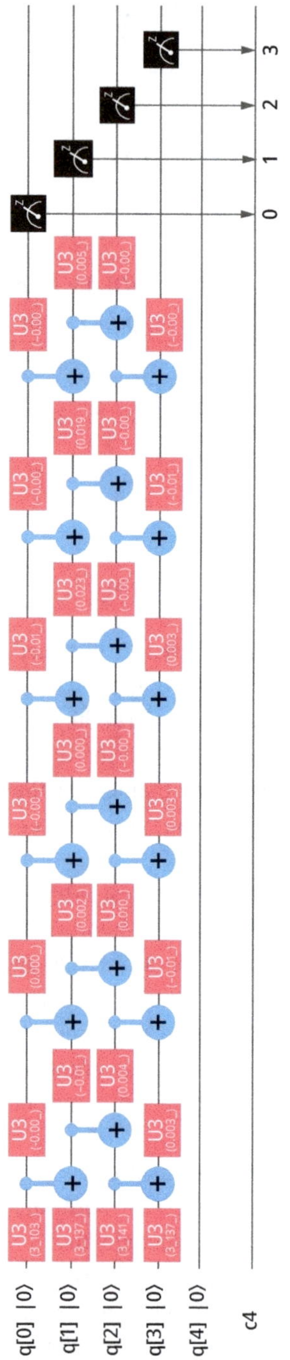

Figure 10.3: Circuit diagram of the quantum network (*Q*) obtained from the IBM circuit composer.

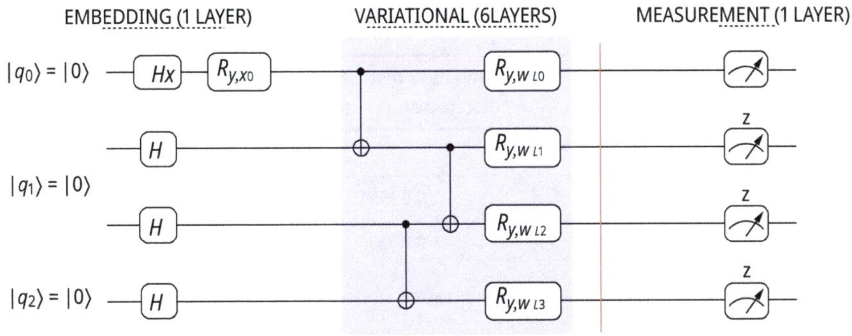

Figure 10.4: Circuit diagram explaining the first element of the input.
Vector x of dimension (1, 1, 4) *scaled π times)*
Note. The value of *L* ranges from 0 to 6 (no. of variational layers, *Circuit diagram explaining eqs. (10.1) to (10.3) for the second element of the input vector x of dimension* (1, 1, 4) *scaled π times), where Figures 10.3 and 10.4 represent the circuit diagram of the quantum network (Q) design and implementation.* Figure 10.5 represents the value of L ranges from 0 to 6.

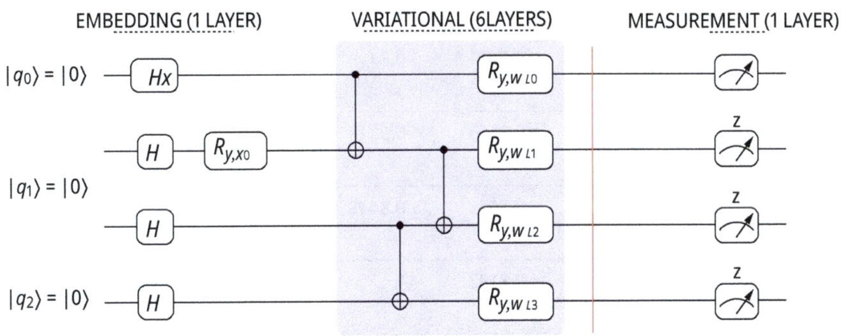

Figure 10.5: The value of *L* ranges from 0 to 6 (no. of variational layers).

Table 10.1: Average test accuracy and recall on the Columbia (color) dataset.

Trial	QTL/ CTL	Execution details	Average test accuracy	Average test recall
Image-based approach				
1	CTL	Kaggle.com CPU and GPU	0.8057	0.8911
2	QTL	Kaggle.com CPU and *PennyLane* (Xanadu) simulator	0.8194	**0.9167**
3	QTL	Kaggle.com CPU and *ibmqx2* (IBM quantum processor)	**0.8500**	0.8718

Table 10.1 (continued)

Trial	QTL/ CTL	Execution details	Average test accuracy	Average test recall
Patch-based approach				
1	CTL	Kaggle.com CPU and GPU	0.8986	1
2	QTL	Kaggle.com CPU and *PennyLane* (Xanadu) simulator	0.8900	1
3	QTL	Kaggle.com CPU and *ibmqx2* (IBM quantum processor)	**1**	**1**

Note. The values in bold indicate the highest recorded average test accuracy and average test recall.

Table 10.2: Execution details.

Trial	QTL/ CTL	Execution details	Average test accuracy	Average test recall	Average test loss
1	CTL	Kaggle.com CPU and	**0.793**	0.84	0.4514
2	CTL (best)	Kaggle.com CPU and	0.8455	0.9818	0.4061
3	QTL	Kaggle.com CPU and	0.7645	**0.8448**	0.5218
4	QTL (best)	Kaggle.com CPU and	0.8182	1	0.4741
5	QTL	Kaggle.com CPU and	0.7818	**0.9364**	0.5684

Average test accuracy and recall on the Columbia dataset for the 70–30% train-test split

In the case of the Columbia dataset we observe that the average test accuracy obtained on the CC model (CTL implementation) is higher compared to the CQ model implementation. From Table (earlier table for Columbia), we note that a larger training set (80–20%) helps the model to learn better and provide a higher accuracy (mention it here). The recall, on the other hand, is higher in the CQ model implementation. For the DSO-1 dataset, we see that the accuracy improves with CQ implementation as compared to the CC implementation. However, the recall considerably falls down with the run on the *ibmqx2* machine. We observe a similar pattern where the CQ implementation increases the test accuracy and recall. The run on the *ibmqx2* quantum machine gives a high recall, which is closer to the results obtained in the best runs, but the accuracy drops lower than the average test accuracy of the CC implementation discussion. Table 10.3 shows the average test performance results of DSO data repository.

Table 10.3: Average test performance results of DSO data repository for the 70–30% train-test split.

Trial	QTL/ CTL	Execution details	Test correctness of system	Test evaluation of systems	Average failure case
1	CTL	Kaggle.com CPU and	0.5556	0.5911	0.67
1	CTL (best)	Kaggle.com CPU and	0.6167	1	0.6493
2	QTL	Kaggle.com CPU and	**0.5822**	**0.7389**	0.6775
2	QTL (best)	Kaggle.com CPU and	0.6333	1	0.6706
3	QTL	Kaggle.com CPU and	**0.5833**	0.3333	0.6853

Note. The values in bold indicate the highest recorded average test accuracy and average test recall.

Table 10.4: Average test accuracy and recall on the proposed AbhAS dataset for the 70–30% train-test split.

Trial	QTL/ CTL	Execution details	Average test accuracy	Average test recall	Average test loss
1	CTL	Kaggle.com CPU and	0.5598	0.6467	0.7224
2	CTL (best)	Kaggle.com CPU and	0.6207	1	0.6975
3	QTL	Kaggle.com CPU and	**0.5931**	**0.66**	0.6669
4	QTL (best)	Kaggle.com CPU and	0.6552	1	0.6316
3	QTL	Kaggle.com CPU and	0.5517	**0.9333**	0.7017

10.17 Conclusion

Thus, the present works aim at the evaluation of the strength and applicability of quantum computing for solving a real-life use case in image analysis algorithm compared to the benchmark developed results obtained from classification framework. Additionally, for the Columbia dataset, we also followed a patch-based approach to evaluate and compare the strength of the QTL approach. For each of the above tests, we also evaluated the performance of CTL. The quantum implementations were tested on the PennyLane, which is the quantum environment management IDE that simulates and the IBM quantum processor ibmqx2. This explains the fact that the performance of a quantum processor also depends on the physical properties of the superconducting devices used to build the quantum machine, which may possibly affect the coherence of the quantum states. It was observed that the "dressed quantum circuit" (QTL) significantly improved the classification performance and benchmark results of the image- based approach. For the patch-based approach, the results for QTL (on the simulator) and CTL were very close.

The test on the quantum processor for the patch-based approach surprisingly provided a 100% test accuracy and recall. The overall results verified an important claim made by Shor in his work on factoring numbers, which suggests that the performance of a quantum computing system increases with an increase in the input size. The image-based implementation uses entire images as against smaller patches, and hence, we observe an improvement in the performance of QTL over CTL. The time required for training the model in the PennyLane simulator is longer than that on the classical system coupled aGPU device. But we see an improvement in the test accuracy and test recall of the QTL implementation.

These results provide evidence for the potential of testing image splicing detection on hybrid classical-quantum systems. In the future, we encourage researchers to test these claims on the other variants of the QTL design. The goal of this exploratory approach was to identify innovative ways in which quantum theory could contribute to pattern recognition in passive-blind image content manipulation. Intermediate-term quantum technologies are paving the way for QML research. In future, ML may, in turn, be used to contribute methods to the toolbox of quantum research. The current aim of this research is to add to the existing knowledge base than to find commercial applications.

10.18 Limitations of the proposed work

The proposed work focused on implementing the hybrid classical-quantum (CQ) transfer learning setting to image splicing classification. This work highlights the potential of a QML approach to blind image splicing forensics. The model implementation follows a supervised learning approach. The patch-based approach could only be tested for the Columbia dataset, since the number of training and testing patch samples were comparatively lower than the other datasets (DSO-1) [14] proposed in this work. The current restrictions with the infrastructure at hand, in terms of the network bandwidth, the amount of free RAM allocated, and the hours of GPU/ TPU provided by platforms limit the scope of the experiments. We avoided the use of the Worldwide Web dataset for the image-based approach [12] because of the imbalanced nature of the dataset.

The implementation code set the seed value to 1 in order to reproduce the results. We agree that choice of different seed values or evaluating the model several times would eventually lead to subtle differences in results within a particular range. However, for the initial implementation, the results presented provide a glimpse of the positive benefits of applying [15] a QTL approach. Data encoding depends on the input size of the data and can cost linear time. Basis encoding has several spatial limitations, and hence, ways to reduce the number of input quantum states by adequate pre-processing are being researched.

Though the computation time required by a quantum machine (simulator and processor) is relatively higher compared to a classical machine, we argue that this problem will be dealt with, in the near future, as more quantum machines get realized. This is the first work to test the potential of QTL for image splicing detection. For our future work, we wish to fine-tune the existing model and test different elements in five qubit circuits to improve the performance of the model. We also want to extend our work to detecting copy-move image forgeries and apply quantum image processing approaches in combination with QML to address image content manipulation detection and localization.

References

[1] S. Lloyd, M. Mohseni, P. Rebentrost, Quantum algorithms for supervised and unsupervised machine learning, arXiv preprint arXiv:1307.0411, 2013 Jul 1.

[2] M. Schuld, I. Sinayskiy, F. Petruccione, An introduction to quantum machine learning, Contemp. Phys., 56(2), 2015 Apr 3, 172–185.

[3] E. Aïmeur, G. Brassard, S. Gambs, Machine Learning in a Quantum World, in: Conference of the Canadian Society for Computational Studies of Intelligence, Springer, Berlin, Heidelberg, 2006 Jun 7, pp. 431–442.

[4] D. E. Deutsch, Quantum computational networks, Proc. R. Soc. Lond. A Math. Phys. Sci., 425(1868), 1989 Sep 8, 73–90.

[5] C. H. Bennett, E. Bernstein, G. Brassard, U. Vazirani, Strengths and weaknesses of quantum computing, SIAM J. Comput., 26(5), 1997 Oct, 1510–1523.

[6] D. R. Simon, On the power of quantum computation, SIAM J. Comput., 26(5), 1997 Oct, 1474–1483.

[7] J. Stolze, D. Suter, Quantum Computing: A Short Course from Theory to Experiment, John Wiley & Sons, Weinheim, 2008 Sep 26.

[8] J. Preskill, Quantum computing in the NISQ era and beyond, Quantum, 6(2), 2018 Aug, 79.

[9] M. Schuld, F. Petruccione, Supervised Learning with Quantum Computers, Springer, Berlin, 2018 Aug 30.

[10] E. Farhi, H. Neven, Classification with quantum neural networks on near term processors, arXiv preprint arXiv:1802.06002, 2018 Feb 16.

[11] J. H. Christensen, K. K. Kanikicharla, E. Aldrian, S. I. An, I. F. A. Cavalcanti, M. De Castro, W. Dong, P. Goswami, A. Hall, J. K. Kanyanga, A. Kitoh, Climate Phenomena and Their Relevance for Future Regional Climate Change, in: Climate Change 2013 the Physical Science Basis: Working Group I Contribution to the Fifth Assessment Report of the Intergovernmental Panel on Climate Change, Cambridge University Press, 2013, pp. 1217–1308.

[12] G. M. Reed, M. B. First, C. S. Kogan, S. E. Hyman, O. Gureje, W. Gaebel, M. Maj, D. J. Stein, A. Maercker, P. Tyrer, A. Claudino, Innovations and changes in the ICD-11 classification of mental, behavioral and neurodevelopmental disorders, World Psychiatry, 18(1), 2019 Feb, 3–19.

[13] M. Zampoglou, S. Papadopoulos, Y. Kompatsiaris, "Detecting image splicing in the wild (web)," 2015 IEEE International Conference on Multimedia & Expo Workshops (ICMEW), IEEE, 2015 Jun 29, pp. 1–6.

[14] M. Schuld, F. Petruccione, Supervised Learning with Quantum Computers, Springer, Berlin, 2018 Aug 30.

[15] T. D. Carvalho, A. A. Giaretta, Bioacoustics reveals two new syntopic species of Adenomera Steindachner (Anura: Leptodactylidae: Leptodactylinae) in the Cerrado of central Brazil, Zootaxa, 3731(3), 2013 Nov 1, 533–551.

[16] A. L. Gokhale, D. Pramod, S. D. Thepade, R. Kulkarni, MissMarple: A novel socio-inspired feature-transfer learning deep network for image splicing detection, arXiv preprint arXiv:2112.08018, 2021 Dec 15.

[17] D. Surendran, V. Karthiga, K. B. Monica, R. Pooja, "Alzheimer's Disease Diagnosis Based on Ensemble of Multi-model Convolutional Neural Networks," 2021 5th International Conference on Trends in Electronics and Informatics (ICOEI), IEEE, 2021 Jun 3, pp. 1451–1457.

[18] M. Rohini, D. Surendran, Toward Alzheimer's disease classification through machine learning, Soft. Comput., 25(4), 2021 Feb, 2589–2597.

[19] P. Kavitha Rani, M. Rohini, S. Sureshkumar, S. Sathiya, Automated Data Visualization System to Reinforce Mid-Day Meal Tracking Scheme, in: Innovations in Computer Science and Engineering: Proceedings of the Ninth ICICSE Springer, Singapore, 2022, pp. 193–205.

[20] W. Marciano, H. Pagels, Quantum chromodynamics, Phys. Rep., 36(3), 1978 Feb 1, 137–276.

[21] M. Rohini, D. Surendran, S. O. Manoj, Prognosis of Alzheimer's disease progression from mild cognitive impairment using apolipoprotein-E genotype, J. Elect. Eng. Technol., 17(2), 2022 Mar, 1445–1457.

[22] S. M. Sasubilli, A. Kumar, V. Dutt, "Machine Learning Implementation on Medical Domain to Identify Disease Insights using TMS," 2020 International Conference on Advances in Computing and Communication Engineering (ICACCE), 2020, pp. 1–4, doi: 10.1109/ICACCE49060.2020.9154960.

[23] S. M. Sasubilli, A. Kumar, V. Dutt, "Improving Health Care by Help of Internet of Things and Bigdata Analytics and Cloud Computing," 2020 International Conference on Advances in Computing and Communication Engineering (ICACCE), 2020, pp. 1–4, doi: 10.1109/ICACCE49060.2020.9155042.

[24] P. K. Kotturu, A. Kumar, "Comparative Study on Machine Learning models for Early Diagnose of Alzheimer's Disease: Multi Correlation Method," 2020 5th International Conference on Communication and Electronics Systems (ICCES), 2020, pp. 778–783, doi: 10.1109/ICCES48766.2020.9137872.

B. Kiruba, M. Manicka Raja, I. Jasmine Selvakumari Jeya, C. Angel

Chapter 11 A perspective study on quantum machine learning models for the areas of medicine, materials, sensing, and communication

Abstract: Quantum machine learning incorporates quantum algorithms into machine learning programs. Medical researchers now routinely use machine learning to predict disease in a variety of ways, such as by generating models to predict drug response, identifying disease subtypes, and developing innovative therapies. A computing model has the ability to enhance medical pictures make analysis by using techniques such as edge detection, and verify whether medical pictures are matching or not. Quantum computers, which are exponentially faster and more capable than conventional computers, have the potential to transform the entire industry, with proponents claiming they could save the healthcare system hundreds of billions of dollars while improving patient care. Quantum models were initiated as a multidisciplinary model, with effort encompassing sectors like digital systems and quantum physics. Now is the period to expand the discipline by embracing novel associations and firms with materials science. This chapter examines a project that use quantum computing to solve a significant issue with medical datasets: small size, which is a result of their limited samples, materials, sensing, and communication.

11.1 Introduction

Richard Feynman initially suggested the concept of quantum modelling and computing (QC) in 1982 after observing the rapid effort needed for cutting-edge performance. In contrast to conventional systems, each processing unit state is predetermined to be a nil or a one, but in quantum computing, the condition might be neither nil or any value between nil and one. Due to this exclusive property, QC is able to explore numerous equivalent trails parallelly inside a solitary computation component, which

B. Kiruba, Department of Artificial Intelligence and Data Science, Sri Krishna College of Engineering and Technology, Coimbatore, Tamil Nadu, India, e-mail: kirubamugesh@gmail.com
M. Manicka Raja, Department of Computer Science and Business Systems, Sri Krishna College of Engineering and Technology, Coimbatore, Tamil Nadu, India, e-mail: manickaraja89@gmail.com
I. Jasmine Selvakumari Jeya, Senior Associate Professor, School of Computing Science and Engineering, VIT Bhopal University, Madhya Pradesh, India, e-mail: wjasminejeya@gmail.com
C. Angel, Department of Computer Science and Engineering, DMI Engineering College, Kanyakumari, Tamil Nadu, India, e-mail: angel.dhas19@gmail.com

https://doi.org/10.1515/9783110791402-011

is substantially impossible for conventional computers; it needs multiple passes, consequently increasing the layers of complexity. Figure 11.1 depicts the correlation between quantum computing and machine learning [1].

Figure 11.1: Correlation between quantum computing and machine learning.

Quantum machine learning (QML), which combines machine learning with quantum computing, is a growing field. The main goal of QML is to improve the act of conventional machine learning procedures by using the characteristics of quantum computing, such as principle of superposition. Two of QC's most significant issues are the restricted amount of available qubits as well as the high level of background noise that occurs whenever it is used [2].

Despite employing quantum physics concepts such as entanglement and superposition, the bulk of quantum machine learning routines are based on traditional routines and procedures. To circumvent the limitations of NISQ-era quantum computers, hybrid systems that integrate classical and quantum computing are occasionally employed.

In order to circumvent some limitations of the quantum computers of the NISQ period, hybrid systems that combine classical and quantum computers are frequently employed. The quantum neural network best exemplifies a guided quantum algorithm (QNN). Dimensionality reduction and clustering are the two most prominent unsupervised strategies for quantum machine learning. Meanwhile, due to the record covering the vectors to be gathered that should be searched less and register higher volumes due to the reduced number of calls required by the quantum methods, privacy improvement is one area where quantum clustering techniques can be utilized. Optimization algorithms, quantum-enhanced Reinforcement Learning (RL), and generative models are examples of semisupervised quantum algorithms.

Three important advantages of quantum machine learning are as follows:
- Run-time advantages include speedier results;
- Greater associative or content-addressable memory are part of the gains to learning ability;
- Enhancements to learning efficiency include simpler models that utilize less training data to produce the same results or the capacity to learn more intricate associations from the same data.

We present an example for each of the three groups.

11.1.1 Quantum hybrid Helmholtz machine for enhancing performance

Using distinct benefits such as efficient amplification, hybrid algorithms run sections of the program conventionally and some portions on a quantum processor. These hybrid algorithms are the solution to generative modeling challenges caused by traditional computers. For instance, these propagative replicas are utilized to acquire possibility supplies across data sets. It is feasible to uncover more abstract representations of the information as depth of a propagative model rises, but this comes at the cost of inference as training is impossible. Variation approximations and Markov Chain Monte Carlo sampling are calculated and utilized for both inference and training [3].

11.1.2 Quantum Hopfield neural network improvement of volume

Neural networks consist of nodes that may be linked in a variety of ways and interrelate with one another via biased edges. Algorithms for machine learning fall under this category. This rule is not applicable to Hopfield neural networks (HNN). They make connection with one another by symmetric edges, and lack own connections. Due to their association, HNNs can serve as associative memory, allowing them to recall a collection of outlines and link noisy inputs to the nearest recorded design. Using training techniques like Hebbian learning enables the network to acquire memory patterns. Hence, weights will be calculated rapidly from all design patterns, demanding minimum computer labor.

Using unitaries learned on a training set, a quantum feed-forward model of an HNN is demonstrated. The routine is related to a conventional HNN, which could be using triple error rates that get progressively more stringent. If there is a no possibility of finding one bit of any memory pattern, the strict error rate only reflects the patterns that the HNN must learn. A less rigorous phrase, "message error rate," is equal to the proportion of probe vectors for which the memory cannot be retrieved accurately. While employing probe vectors as well, the "bit error rate" accounts for each bit that cannot be recovered precisely independently. Minute patterns can be examined by using a quantum computer [4].

11.1.3 Enhancing cognition: variational quantum circuit for machine learning

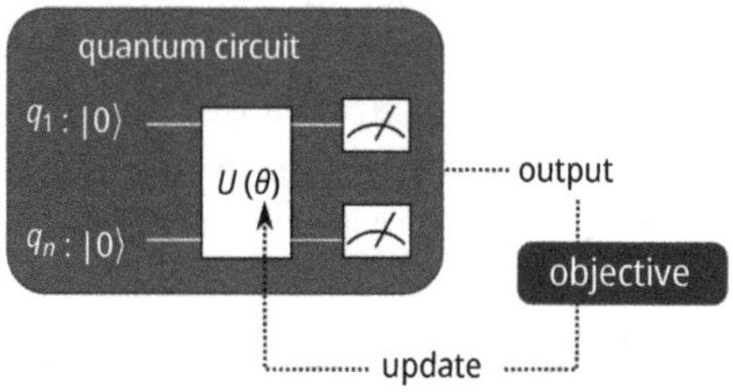

Figure 11.2: Variational quantum circuit.

Variational Quantum circuits, as represented in Figure 11.2, are a sort of hybrid quantum-classical method that functions identical to a classical dense layer. The approach is based on a quantum circuit U that is reliant on a set of parameters, Qubits q1, q2, and an objective function that specifies a scalar to be optimized. The circuits can be seen as a black box, and optimization is commonly performed using an iterative approach, such as gradient-based optimization. Variational quantum circuits (VQC), which involve optimized quantum gates, gets attention as a technique aimed at discovering and implementing novel hybrid QML algorithms. Using these quantum circuits, it is feasible to evaluate certain cost functions. The cost function may be optimized using a number of conventional approaches, which may once more require a quantum circuit.

11.2 Quantum machine learning in medicine and healthcare

For real-time machine learning to guide clinical practice, expanding databases connected to several data sources that include continually updated medical record information are required (e.g. wearable technology). In order to handle this level of complexity, future machine learning systems will require access to the instantaneous processing power of quantum computing. With more data feeding predictive models and quantum computing power, a physician may be able to pick a medication that is exactly suited to a patient, based on continually running and updated models that anticipate treatment response, while taking into account the various patient characteristics.

Quantum-enhanced machine learning techniques may work for early, precise treatment, in addition to granular risk forecasting. Precision medicine is the finest medical use of quantum machine learning. The objective of precision medicine is to individualize preventive and therapeutic approaches. In certainty, health sectors contribute only 10–20% to results; the additional 80–90% is attributable to health-related behaviors, socioeconomic characteristics, and environmental factors [5]. The interdependencies and correlations between these numerous elements provide challenging computational challenges for optimizing therapeutic effectiveness. As a result, due to individual variation, many existing medications fail to achieve the expected outcomes.

Approximately one-third of individuals treated with drug-based cancer therapy are successful. In other instances, the effects of pharmacological therapy can be catastrophic; each year, about 200,000 people die in Europe due to adverse drug responses. Customizing medical methods requires significant proactive effort. As was previously said, early diagnosis, treatments, and preventative actions typically result in significantly improved results and cost-effectiveness. Classical machine learning has shown promise for inferring future illness risk for a series of patient categories built on EHRs. However, tests endure owing to the qualities of EHRs and extra health-related data, such as quantity of noise, the extent of the appropriate feature space, and the difficulty of the feature connections.

Consequently, it is feasible that supervised and unsupervised quantum-enhanced machine learning algorithms might provide exact, timely, and comprehensive risk forecasting. Eventually, medical experts may be able to understand how a patient's risk for a certain sickness changes over time, if they employ simulated diagnostics built on continuous fact streams from entities. However, this information is insufficient to assess an individual's disease risk. Understanding how to perform an effective medical intervention on every given patient is thus essential. The learning of drug compassion at the cellular level is one method aimed at achieving this goal.

Since the discovery of chloral hydrate, the first synthetic medication, in 1869, pharmaceutical companies have competed ferociously to develop the next revolutionary drug. However, the limitations of conventional computer technologies usually impose constraints on contemporary drug development strategies that employ advanced computing, artificial intelligence (AI), and machine learning (ML). For this reason, innovative firms and researchers are increasingly turning to quantum computing (QC) and quantum machine learning (QML) to identify novel medications. Quantum machine learning (QML) uses quantum algorithms to complete hard machine learning tasks.

Using classical machine learning, drug discovery has already demonstrated tremendous potential. By turning them quantum circuits using QML, scientists may run conventional ML algorithms effectively on incredibly powerful quantum computers. It is possible that the capabilities of these quantum computers to assist researchers in comparing larger-scale molecules will usher in a golden age of faster, more cost-effective, and more accurate drug discovery as they become more prevalent and as quantum technology becomes less prone to errors.

These innovations could pave the way for the next phase of R&D, which will be incredibly quick, precise, and effective, but will also involve vast amounts of data that are too large for conventional approaches to effectively manage. Google's announcement of a successful quantum computing (QC) experiment that attained so-called "quantum supremacy," a computation that was too difficult to complete using conventional computing (CC), was hailed as a turning point for QC. However, it was also a significant advancement for medication discovery, as QC has the computer power to enable the analysis of larger-scale compounds and combinations than ever before.

In the case of colon cancer, early detection increases survival rates by a factor of 9 and reduces action costs by a factor of 4. However, present diagnostics are difficult and exclusive for an extensive variety of diseases.

Scanning techniques such as CT, MRI, and X-ray scans have been essential diagnostic tools for practitioners over the past century. Methods for medical picture recognition and diagnosis with computer assistance have been advancing rapidly. The noise, low resolution, and limited reproducibility of many of these photographs have a negative impact.

Without a concurrent growth in the amount of therapeutically relevant information available after multiple bases, quantum computing is useless to medicine. When information can be combined from many sources, such as current patient data systems, bio-banks, wearable health devices, etc., it will be possible to overcome obstacles to achieving a healthcare system in which quantum computing leverages eras of information from billions of experimental encounters a year.

Quantum computing may improve medical image analysis, including edge detection and image matching. However, quantum computing is useless to medicine without an increase in therapeutically relevant data from multiple sources.

Categorization and single-cell diagnostic strategies could benefit from the use of quantum-enhanced machine learning approaches like quantum support vector machines [6–12].

Furthermore, the identification and characterization of biomarkers may require the analysis of large datasets, such as those from the fields of genomics, transcriptomics, proteomics, and metabolomics.

They may be able to regularly assess and evaluate people's health. Health plans and benefactors may help from lower conduct costs due to earlier diagnoses. They may also be able to conduct meta-analyses for more complex diagnostic procedures to determine when to perform them.

The powerful use cases of quantum-enhanced machine learning in current research scenarios, the expansion of new clinical trials, the administration of patient data and records, and the acquisition of enduring disease detection make it essential in the field of healthcare. As a result, the healthcare industry is employing quantum computing to maintain patient-centered care for healthcare clients.

Quantum-enhanced machine learning algorithms, including quantum random forest (QRF), quantum K nearest neighbor (QKNN), quantum decision tree (QDT), and quantum Gaussian Naive Bayes (QGNB), outperform conventional machine learning algorithms.

In addition, the rates of execution of conventional machine learning algorithms and quantum-enhanced machine learning algorithms were related, with the quantum random forest having the lowest implementation duration – a margin of 150 microseconds.

Quantum computers, which can be cast-off in CT or magnetic resonance imaging, can be used to intelligently analyze and manage information significantly and more effectively than predictable computers. QML can be used to develop innovative methods for using magnetic flux and other more reliable methods to calculate the heart rate. Enhancement is the procedure of removing a portion of a likelihood distribution from quantum models.

Even though quantum-enhanced machine learning (QEML) increases computing speed and can achieve information storage using a variety of methods, its practical limitations are readily apparent. Investigation is limited to a model with a minimally viable solution. The current model may need to be transformed into a readily deployable services mode to enable widespread use in clinical trials. This may necessitate additional research in heart disease diagnosis.

11.3 Quantum machine learning in materials

Consider a mechanically stable and well-isolated object, such as an atom, molecule, or rock. Each item has a distinct energy, or E. E is the total sum of all the parts' energy if the item is composed of many weakly connected parts.

The discovery and characterization of novel materials have traditionally relied heavily on experiments. Due to the high resource and equipment requirements of new investigation, it must be organized in a long time for a very small amount of resources. As a result, serendipity or human intuition were often used to make significant discoveries.

Combining computer models and experimentation has significantly reduced the time and cost associated with material design. Large-scale material groups could be studied computationally in high-throughput studies to find the best experimental candidates as a result of the ongoing improvement of computer power and the development of more effective programs.

Quantum computers have a bright future in the realm of solving problems involving electronic structure. For quantum chemistry problems involving up to hundreds of electrons, much effort has been put into developing and optimizing quantum algorithms. However, polynomials increase the runtime, and the number of constituents can still hinder quantum simulations of large-scale systems.

Due to their ability to compress high-dimensional data into low-dimensional representations, ML methods have traditionally been used primarily in data science. However, interest in the potential applications of ML methods in the physical sciences, particularly quantum physics, has increased in recent years.

To handle the multiple Schrodinger issue (using reinforcement learning), to learn quantum states from measurements (unsupervised learning), and to learn materials or chemical characteristics from datasets (supervised learning), for example, a quantity of machine learning (ML) approaches are yet to be developed.

Typically, these approaches entail searching through vast databases for nontrivial connections between crystal or molecule structures, such as atomic positions and nuclear charges [13].

QML, which are machine learning extensions to quantum regimes, are supposed to be extra effective for maximizing the potential of quantum qualities. Quantum machine learning techniques help to solve quantum many-body systems for the correct electronic structure calculations of lattice models, molecular systems, and, more recently, periodic systems.

Classical simulation refers to the use of a classical computer to simulate the gates. After confirming that the algorithm works when applied conventionally, we apply it using Qiskit [14].

Quantum physics' increased performance benefits applications ranging from quantum-enhanced sensing to secure communication and quantum information processing.

11.4 Quantum machine learning in sensing

Quantum machine learning techniques expedite the arrangement of information gathered from a sparse network of quantum sensors.

The study of Zhang, Zhuang, and colleagues investigates a way for quantum-enhanced data processing, resulting from the combination of quantum machine learning with the most recognized quantum technologies: quantum metrology, then sensing [15].

By combining and analyzing the quantum outputs of many sensors using quantum information processing techniques, a quantum advantage can be realized. The possible benefit results from an important aspect of quantum metrology: The benefit of processing radar data coherently scales as the square root of the dimension of the so-called Hilbert space, which is a representation of the quantum condition observed by the system.

In an experimental setup, they investigate how a traditional indication can induce a quantum condition in a pair of quantum sensors – comparison made between the ideal signal-discriminating measurements for two scenarios.

Reinforcement learning algorithms produce effective results in meteorology. Control theory is used to describe a wide range of physics problems. Reinforcement learning (RL) provides a logical way to formulate the pursuit of optimum control.

The problem of optimal single-photon adaptive phase-estimation was studied in the context of global parameter estimation, i.e., when the parameter is a priori unknown. RL has also been applied to control problems in quantum metrology.

Two ongoing study topics in the field of intelligent interpretation of remotely sensed pictures are automatic target identification (or recognition) and high-resolution remotely sensed picture categorization. In order to do these tasks, initially, the low-level characteristics included in the raw photos must be computed.

The use of remote sensing (RS) is necessary for a wide variety of earth observation tasks, including the detection of ground items, the monitoring of climatic and environmental change, the monitoring of urban growth, and the evaluation of damage caused by natural disasters. RS pictures give exact global observation and insights into the sustainability and health of the ecosystem for both naturally produced and anthropogenically caused activities. These observations and insights may be applied to both natural and anthropogenically induced processes [16–19].

There are two types of RS sensors: passive and active. Passive sensors are typically used to measure and record the light reflection of the features of Earthly things because they do not provide energy to the objects being monitored. Aerial photography is an important method of passive sensor-based remote sensing that can collect data in the visible, near-infrared, and even longer wavelengths of solar radiation.

Artificial neural networks were the source of inspiration for deep learning. However, vanishing gradient concerns may lead neural network learning algorithms to get trapped in poor local optimums as the complexity of a neural network model increases [20].

11.5 Quantum mechanics in communication

Meanwhile, the world is waiting for the development of powerful quantum computers. Players in the optical telecommunications business need to familiarize themselves with the implications of quantum machine learning that future quantum computing brings with it. This is vital for them to be able to plan for the future of the industry. It is important to begin making preparations as soon as possible in order to begin performing prototype testing. It would be good to start making preparations to harness the potential power.

One of the potential advantages offered by quantum technology is the ability to locate answers to problems in a shorter amount of time than is currently possible. The development of quantum computing around the world has received a considerable boost from the increase in financing that has been provided by both the public sector and the private sector. The whole ecosystem of quantum computing has been the focus of extensive study on a global scale. This research has been conducted on every facet of the ecosystem. This encompasses the compiler modules, the implementations of middleware, the programming languages, the software and hardware, the algorithms, and a wide range of physical implementations (such as superconducting, trapped ions, optics, cold atoms, and artificial atoms). It is currently more necessary than it has ever been for academics working in the field of telecommunications to understand quantum computing and investigate ways in which it might be used to solve problems in the field of telecommunications. This is the case because quantum computing has the potential to solve problems in a number of different ways.

In contemporary settings for telecommunication, a problem's classical data may be broken down into two categories: the classical input data and the classical output data. As a consequence of this, the classical input data needs to be converted into quantum states before being fed into a quantum circuit, and the classical readouts of the problem's output quantum states need to be converted into classical readouts through the use of quantum measurements. Both of these steps are necessary. Quantum algorithms are an excellent match for classical issues that do not need exact responses since quantum observations are inherently uncertain. The approaches and procedures that are used in the field of machine learning are also extremely well suited to the kind of work that is being done [21]. Despite the fact that conventional and quantum machine learning algorithms are equal in this respect, it is not simple to create quantum machine learning algorithms that operate well or even approach, if not exceed, the performance of their classical versions.

Quantum computations are naturally dependent on chance, and as a result, they are perfectly suited for use in more traditional settings that do not demand the creation of accurate answers. This is because these contexts do not necessitate the production of exact answers. The discipline of computer science – known as machine learning – is responsible for the development of a variety of approaches that are well suited to accomplishing these objectives. In spite of the fact that conventional and quantum machine learning algorithms are on level with one another in this regard, developing quantum machine learning algorithms that perform as well as, or even better than, their conventional counterparts is not an easy undertaking.

Because of recent developments in fast digital signal processing, optical coherent communication has become the method of choice for high-speed, long-distance communication. These developments have enabled optical coherent communication to achieve its current status. After the presentation of digital carrier-phase estimation in 2005, there was a noticeable increase in people's interest in optically coherent communication.

A digital signal processor takes the entering digital data and converts them into a waveform that is suitable for transmission by both the quadrature-phase and the in-phase components of an optical light field. This waveform is then output by the digital signal processor. After being generated by a laser, the output optical signal is then subjected to modulation by a modulator before being sent to the receiver via an optical fiber. This article provides a demonstration of the four-quadrature amplitude modulation technique (QAM). The quality of optical signals suffers significantly as they are transmitted over optical fibers because of the nature of the medium. Inaccuracies, such as polarization rotation, chromatic dispersion, polarization mode dispersion, and nonlinear distortion brought on by the Kerr effect, an electro-optical phenomenon, are examples of these problems. The phenomenon known as polarization mode dispersion takes place when the two distinct polarizations of light move at different speeds. Polarization rotation, chromatic dispersion, and polarization rotation are some of the other problems with the image.

Some examples of these flaws include polarization rotation, chromatic dispersion, polarization mode dispersion, and nonlinear distortion brought on by the Kerr effect. Chromatic dispersion is the phenomenon that takes place as a result of the different spectral components of a signal moving at different speeds (which is an electro-optic effect that leads to a dependence of the fiber refractive index on the transmit signal power).

Signals with a high order of modulation frequently suffer from nonlinear distortion when they are sent over long distances and at high speeds. In order for the receiver to successfully recover the broadcast signal with the highest possible accuracy and the lowest possible error probability, it is necessary for the receiver to take into consideration the nonlinear effects. In most cases, it is difficult to assign a symbol or decision area to an incoming received signal and divide the optical phase space into decision areas for all 16 symbols. This is because there are 16 possible symbols. This is important in order to account for the fact that it may be difficult to decide which symbol should be used. Because of this, it is vital to take this into consideration.

Examples of machine learning algorithms that have been suggested as possible answers to the issue include artificial neural networks (ANN), support vector machines (SVM), and *k*-means clustering, to name just a few. One of the benefits of using the *K*-means clustering approach is that it does not call for any training before using.

The centroids that may be located in the phase space make up the fundamental building blocks of the *k*-means clustering algorithm. There is a centroid that can be located inside the phase space that correlates to each signal point. This centroid may be discovered by using the phase space. It was placed there on purpose in order to be the centroid that was the most closest. When a new signal point is analyzed, a designation is assigned to it, and it is placed in the centroid that is situated in the region that is geographically closest to it. This results in the addition of the new point to the collection of points that are identified by the centroid. Following this step, the location of the centroid is brought up to date, which ultimately leads to the establishment of a new center for the point cluster. The computation of the (Euclidean) distance is replaced by a statistical estimation procedure in the quantum *k*-means algorithm. This allows for a more accurate representation of the data. As a consequence, it is possible to generate an estimate of the distance by performing just one iteration of an estimation unit. When many different estimations are combined into one, the accuracy of the final result is improved.

However, the potential speedup of the quantum *k*-means design is contingent on deftly encoding the initial classical information onto the quantum state. This is necessary in order to realize the speedup. It is possible that any speedup that would have been offered by the fundamental quantum approaches would be lost due to the difficulties associated with loading the initial data. This is a possibility that cannot be ruled out.

In order to determine whether or not the quantum algorithms are advantageous and outperform their conventional counterparts in a whole system, including the data loading component, additional research is still required at this time. This will be ascertained by comparing the efficiency of the quantum algorithms with that of their more traditional analogues. In addition to this, it is very possible that quantum algorithms would be beneficial in certain parameter ranges (such as the feature dimension). Quantum algorithms are not somethings that should be used on a whim in order to guarantee speedup; rather, one should thoroughly research the application issue at hand, together with the real parameters, before using them. This is essential in order to ensure that there is no slowdown.

Concerning artificial neural networks (ANN), a section of ANN research has been focused on the subject of nonlinearity compensation in optical communication in an attempt to discover solutions. This has been done in an effort to find solutions. An ANN is capable of carrying out the role of a classifier and has the ability to mimic the nonlinear input-output connection of a function that is either unknown or difficult to carry out in its entirety. It is possible to produce the required training data for the ANN's parameters by routinely sending a predetermined sequence of signals from the transmitter to the receiver. This will allow the data to be trained. It is possible to carry out these steps in order to generate the training data. The artificial neural network that has been trained is subsequently used to create predictions about the outputs (for example, information symbols) that are the direct outcome of the incoming signals (the distorted information symbol points in the optical phase space, for example)

A significant body of research has formulated a variety of implementation techniques for the neural network that make use of quantum circuits. These strategies have been developed as a result of the study. However, the primary focus of many of these studies is on the neuroscience-based Hopfield networks, which are able to successfully include associative memory. To the best of our knowledge, there has not been any suggestion made for a quantum version of neural networks that is both completely functional and effective [22–30].

Even while the introduction of 5G wireless networks has already begun in certain areas of the globe, the ultimate objective of creating a totally intelligent network that offers everything as a service and provides users with a genuinely immersive experience is not yet a reality. As 5G approaches its restrictions in the next decade or so, research on the design objectives for its successor has already started to be undertaken in the literature. This study is being done in preparation for when 5G reaches its limitations. The concept of 6G communication networks is now the subject of discussion within the scientific community. This idea is known by a number of distinct labels, one of which is the B5G vision.

Along with the rising need for communications that are speedy, trustworthy, secure, intelligent, and friendly to the environment, there has been a corresponding growth in the necessity that the systems have a high computing capability. The inherent parallelism that is supplied by the underlying notions of quantum physics and the potentials that are exhibited by the discoveries of contemporary QC technology strongly imply that there is the potential to surpass the capabilities of traditional computer systems.

It is possible to trace the enormous power of quantum computing all the way back to basic notions such as quantum superposition, quantum entanglement, and the no-cloning theorem. Utilizing quality control in very large tensor product spaces might make concurrent processing of multidimensional data sets of enormous sizes much easier. These collections of data can be of any size. QC-assisted communications is a relatively new area of research that is expected to prove fruitful in the future in terms of achieving very high data rates and maintaining the integrity of connections in the field of communications. This is due to the fact that it is anticipated that QC-assisted communications will help maintain the integrity of connections. In order to do this, noiseless classical communication was combined with quantum communication by using quantum channels that maintained a constant transmission rate. This was done so that the two types of communication could work together.

Conventionally, it is believed that the use of ML is justifiable when there is no precise mathematical model of the system, there is a significant amount of training data, the system or model being studied is stationary (slowly varying over time), and the numerical analysis is acceptable. These four conditions must all be present for the use of ML to be justifiable. The use of ML can only be considered defensible if each of these four prerequisites is satisfied.

In recent times, there has been a lot of interest in ML techniques because of their capacity to deliver data-driven solutions to a number of difficult problems that are present in communication systems. This interest can be attributed to the fact that ML techniques are capable of delivering solutions to these difficult problems. The use of machine learning (ML) in communications is quickly gaining popularity, particularly for the purpose of creating self-sustaining, adaptive networks that can satisfy the requirements for dynamic configurability of future devices and services. This particular use of ML in communications is gaining popularity particularly for the purpose of creating self-sustaining, adaptive networks. This is particularly true for the aim of developing self-sustaining, adaptable networks that can fulfil the criteria for generating self-sustaining, adaptive networks. This is a prerequisite for creating self-sustaining, adaptive networks. When there is a scarcity of training data and/or time, it is probable that supervised learning, which is used in vehicular communications for channel correction, may present difficulties. This is a possibility. This is due to the fact that the nodes themselves may move around. This occurs as a consequence of the fact that higher mobility leads in a bigger Doppler spread, which in turn produces a decrease in coherence time, which in the end results in fast fluctuation in the channel statistics.

Supervised learning may have potential uses at the physical layer for downlink communications, including efficient power management and interference cancellation. Supervised learning is utilized for a broad range of successful applications in the network, application, and transport levels, as well as in other layers, in addition to being employed for the physical layer. Even while satellite communications have the advantage of providing global network coverage, the fact that they have extraordinarily high connection latency reduces their appeal. This is the case despite the fact that they do have the benefit of providing international network coverage.

Learning through unsupervised methods entails gathering input data samples that are currently accessible in order to train a system without having any prior knowledge of the answer that is expected from the system. This type of learning can be accomplished by collecting input data samples that are currently accessible. At the physical layer, for example, the symbols from the incoming noisy data may be transferred to the constellation maps by clustering the sample points in the decision space. This is done in order to reduce the amount of noise in the data. This results in the creation of effective nonlinear decision boundaries, which are required for the process of mapping.

The process of calculating a certain output in response to a particular observation and then recognizing a feedback performance indicator, also known as a reward, from the environment is how reinforcement learning is accomplished. Reinforcement learning is performed by progressively adapting to more desirable patterns of behavior in order to accumulate the maximum amount of positive reinforcement (performance).

Because there are no direct training data that is associated with the intended output, this learning strategy may be thought of as a compromise between supervised learning and unsupervised learning. This is because there is no direct training data that is linked with the intended output. The reason for this is because there is no direct training data that is matched with the output that is intended. A kind of indirect supervision may be provided by having a previous idea of how the system should operate at its best. Control and classification issues are typical applications of reinforcement learning; Markov decision processes and Q-Learning (QL) are two well-known examples of systems that fall under this category.

The Quality-of-Service (QoS) of a cellular network may be improved by attaching an agent to each serving station in the network. This allows the network to learn the optimal scheduling parameters that will improve the network's overall QoS. The agent will have a better understanding of how to enhance the network as a result of this. Reinforcement learning might one day be used for the goal of power control and optimization on the physical layer of communication networks. In order to find a solution to this problem, a model-free distributed reinforcement learning approach for power distribution has been developed. [Citation needed] Changing the transmit power is accomplished with the use of Channel State Information (CSI) and Quality of Service (QoS) indicators using this technique [31, 32].

Genetic programming, which takes its inspirations from biological evolution, optimizes the problem at hand by evolutionary assessment of the fitness targets while taking into consideration restrictions and limitations. Genetic programming derives its inspiration from the process of biological evolution. The method of biological evolution is where the concept of genetic programming was first conceptualized. The use of genetic algorithms is one of the methods that has received the most attention from researchers as a potential solution to a variety of issues involving optimization and estimation at different tiers of communication systems. This attention has been focused on the use of genetic algorithms as a method that could potentially solve these issues.

At the physical layer of communication systems, optimal antenna selection in MIMO systems, power management, and symbol identification in MIMO systems have all been accomplished by using Genetic Algorithms. This has been done at the physical layer of communication systems (GA). An extensive investigation of the range of potential uses and applications of evolutionary algorithms in wireless communication systems is provided in this article [33, 34].

It is possible to pick a model for an algorithm that is used for machine learning in accordance with the amount and kind of data that is being acquired. The use of batch learning methods is best suited for software programs that already have enormous volumes of training data accessible to them. Learning strategies that make use of batches operate under the assumption that there is an infinite supply of processing time available and look over the whole space of all conceivable data and knowledge structures. Such offline approaches, which require data to be manually gathered, categorized, and then processed in batches, generally run into the restriction that there is very little data available in applications that are based on the real world. This limitation can be overcome by using online approaches, which do not require manual data gathering, categorizing, or processing.

Because of this, the batch-learning algorithms that we have been discussing do not lend themselves very well to applications that demand real-time data processing. As a result of this, we have been discussing them. The use of training that is carried out online as a potential solution for these applications that stream data is recommended. Nevertheless, there is a limit placed on the amount of time that may be spent analyzing each data sample while participating in the online course. Intelligent caching and channel tracking are two examples of frequent communication system applications that employ off-line (batch) and online (real-time) learning, respectively. Both of these types of learning may be performed in real time.

Model-based learning will commonly make use of readily accessible objective functions and good computing efficiency when attempting to optimize performance indices. This is because of the nature of the problem being solved. Learning that is only based on data samples, on the other hand, makes use of all of the data samples that are currently accessible in order to extrapolate and/or interpolate the samples, which necessitates a significant amount of both memory and time. This type of learning is not as efficient as learning that is based on a larger amount of data. Both model-based learning and sample-based learning have a variety of applications that are comparable to one another. Two examples of each application are symbol decoding and content demand prediction.

The term “artificial neural network” (ANN) refers to a kind of data processing structure that has its roots in biology and is meant to learn new abilities by seeing and analyzing data. Its beginnings may be traced back to the field of artificial intelligence research. ANNs are often used in order to discover any patterns that could be present in the input data. This is one of its primary purposes. Processing the input data via many layers of simulated neural connections allows for the achievement of this goal. In an ANN, the activities of combining and/or limiting are carried out by each node (neuron), while the activities of scaling are carried out by each link. Combining and/or limiting may also be thought of as limiting. An ANN is composed of input, hidden, and output layers of neurons, and all of these layers are interconnected with one another.

It is possible for the connections that exist between the various layers of neurons to be either complete or imperfect, pooling or recurrent, feed-forward or recurrent, or any combination of these four types of connections. The topologies of the connections that are present between the various layers of neurons in an artificial neural network are going through a period of rapid evolution as more and more applications for artificial neural networks are created (ANNs). Among the most notable structures are the multilayer perceptron (MLP), the convolutional neural network (CNN), the recurrent neural network (RNN), the Hopfield network (HN), the general annealing network (GAN), the echo state network (ESN), and the neural Turing machine (NTM). Other essential structures are also included [35].

It is not completely out of the question for quality control and machine learning to work closely together for the benefit of communication systems as a whole. New subfields, such as "Quantum Information Science" and "Quantum Computer Science," have emerged as a direct consequence of the high degree of parallelism that can be accomplished with quantum computing (QC).

Within the area of quantum physics, the notion of parallelism may be traced back to the origins of qubits, entanglement, and superposition as its primary sources of inspiration. As a result of the fact that a qubit is capable of simultaneously holding both the binary states "0" and "1", the end result is that any n interacting qubits can simultaneously represent $2n$ different binary patterns, whereas classical computers are only capable of simultaneously representing a single binary pattern at-once – a qubit is able to concurrently retain both of the binary states. It is commonly known that the concepts from quantum physics may produce statistical data patterns that defy logic and that are challenging or even impossible for ordinary computers to process. These patterns can also create anomalies in the data.

Machine learning often involves processing and labeling high-dimensional data vectors, for which the needed time polynomial grows exponentially with the data dimension.

Quantum communications are designed to make use of the quantum nature of information, which presents both new challenges and new opportunities for the design of communication protocols for 6G and the subsequent generations of networks. The purpose of quantum communications is to utilize the quantum nature of information. Quantum communications have the enormous potential to provide complete randomness and security, transport a great deal more information, and significantly improve the quality of transmission, when compared to traditional binary-based communication systems. In addition, quantum communications have the ability to transport a great deal more information.

Quantum ideas are adaptable and may be implemented in a number of different communication settings, including satellite networks, terrestrial wireless networks, and underwater communications.

Optical fiber communications is one of the application areas of quantum communications that has garnered considerable attention. The conventional method of doing this task is founded on the classical electromagnetic fields, and it can be affected by fluctuations that are not intended.

In addition, the performance of photo detectors may be hindered by noise that has its roots in quantum mechanics. These challenges can be conquered by developing optical communication systems inside a quantum mechanical framework.

In addition, satellite communications is an important sector that presents a number of opportunities for the implementation of quantum technologies in a range of contexts (SatCom). In addition, another fascinating field of application is the quantum Internet, which allows the transfer of information. Additional significant use of quantum communications is quantum broadcast, which implements classical bits as opposed to quantum bits in order to move a particular quantum state from one location to another using quantum devices. Quantum teleportation is accomplished by utilizing the quantum entanglement principle.

11.6 Conclusion

This chapter delivers a comprehensive examination of the application of quantum machine learning algorithms in important fields such as medicine, materials, remote sensing, and communication. We are able to execute traditional algorithms noticeably more quickly than before, thanks to a variety of machine learning strategies as well as the fusion of these strategies with quantum computers. It is possible to specialize in a inclusive diversity of fields within this sector, such as the production of memory, the creation of algorithms for clinical data analytics, the development of routers, networking, space communication, and so on. In the coming decades, quantum computing and communication will completely supplant their conventional analogues in the fields of communication and computing. This transition will take place on a global scale. Quantum machine learning refers to the field of study that emphasizes the development and application of quantum software to enable machine learning at a rate that is significantly faster than that of traditional computers. In spite of the fact that current research has led to the development of quantum algorithms, which have the potential to serve as the basis for machine learning algorithms, major hardware and software barriers still remain.

References

[1] S. B. Ramezani, A. Sommers, "Machine Learning Algorithms in Quantum Computing: A Survey," 978-1-7281-6926-2/20/$31.00 ©2020 IEE.

[2] J. Schmidt, M. R. G. Marques, S. Botti, M. A. L. Marques, Recent advances and applications of machine learning in solid-state materials science, NPJ Comput. Mater., 5, Article number: 83 2019.

[3] F. Phillipson,"Quantum Machine Learning: Benefits and Practical Examples," International Workshop on Quantum Software Engineering & Programming, March 2020.

[4] N. Meinhardt, N. M. P. Neumann, F. Phillipson, Quantum Hopfield neural networks: A new approach and its storage capacity, TNO Report, 2020.

[5] C. M. Hood, K. P. Gennuso, G. R. Swain, B. B. Catlin, County Health Rankings: Relationships Between Determinant Factors and Health Outcomes, Am. J. Prevent. Med., February 1 2016. https://www.ajpmonline.org/article/S0749-3797(15)00514-0.

[6] M. Amin-Naji, H. Mahdavinataj, A. Aghagolzadeh, "Alzheimers disease diagnosis from structural MRI using Siamese convolutional neural network," 2019 4th International Conference on Pattern Recognition and Image Analysis (IPRIA), Tehran, Iran, 2019, pp. 75–79.

[7] J. Luyapan, X. Ji, D. Zhu, T. Mackenzie, I. Amos, J. Gui, "An efficient survival multifactor dimensionality reduction method for detecting gene-gene interactions of lung cancer onset age," 2018 IEEE International Conference on Bioinformatics and Biomedicine (BIBM), Madrid, Spain, 2018, pp. 2779–2781.

[8] P. Rani, S. V. Kavita, S. Verma, G. N. Nguyen, Mitigation of black hole and gray hole attack using swarm inspired algorithm with artificial neural network, Access, 8, 2020, 121755–121764.

[9] S. Ramisetty, S. Varma, S. Varma, The amalgamative sharp wireless sensor networks routing and with enhanced machine learning, J. Comput. Theor. Nanosci., 16(9), 2019, 3766–3769.

[10] G. E. Hinton, R. R. Salakhutdinov, Reducing the dimensionality of data with neural networks, Science, 313, 2006, 504–507.

[11] Y. LeCun, Y. Bengio, G. Hinton, Deep learning, Nature, 521, 2015, 436–444.

[12] F. A. Faber, A. Lindmaa, O. A. von Lilienfeld, R. Armiento, Machine learning energies of 2 million elpasolite (abC2D6) crystals, Phys. Rev. Lett., 117, 2016, 135502.

[13] V. von Burg, G. H. Low, T. Häner, D. S. Steiger, M. Reiher, M. Roetteler, M. Troyer, Quantum computing enhanced computational catalysis, Physical Review Letters, Volume: 117, Page number: 1–6. arXiv Preprint arXiv:2007.14460, 2020.

[14] S. H. Sureshbabu, M. Sajjan, O. Sangchul, S. Kais, Implementation of quantum machine learning for electronic structure calculations of periodic systems on quantum computing devices, Journal of Chemical Information and Modelling, 8 May 2021.

[15] S. Lloydm, Quantum machine learning for data classification, Physics, 14, 79, 1 June 2021, Massachusetts Institute of Technology, Cambridge, MA, USA.

[16] K. Burke, Perspective on density functional theory, J. Chem. Phys., 136, 2012, 150901.

[17] W. M. C. Foulkes, L. Mitas, R. J. Needs, G. Rajagopal, Quantum Monte Carlo simulations of solids, Rev. Mod. Phys., 73, 2001, 33–83.

[18] K. Nakano, C. Attaccalite, M. Barborini, L. Capriotti, M. Casula, E. Coccia, M. Dagrada, C. Genovese, Y. Luo, G. Mazzola, et al., Turborvb: A many-body toolkit for ab initio electronic simulations by quantum monte carlo, J. Chem. Phys., 152, 2020, 204121.

[19] M. Rupp, A. Tkatchenko, K.-R. Müller, O. A. Von Lilienfeld, Fast and accurate modeling of molecular atomization energies with machine learning, Phys. Rev. Lett., 108, 2012, 058301.

[20] R. Ramakrishnan, P. O. Dral, M. Rupp, O. A. von Lilienfeld, Big data meets quantum chemistry approximations: The δ-machine learning approach," J. Chem. Theory Comput., 11, 2015, 2087–2096, pMID: 26574412. http://dx.doi.org/10.1021/acs.jctc.5b00099.

[21] J. Schuff, L. J. Fiderer, D. Braun, Improving the dynamics of quantum sensors with reinforcement learning, New J. Phys., 22, 2020, 035001.

[22] C. Padwick, et al. "WorldView-2 pan-sharpening," Proceedings of the ASPRS 2010 Annual Conference, San Diego, CA, USA. 2010.

[23] M. Drusch, et al., Sentinel-2: ESA's optical high-resolution mission for GMES operational services, Remote Sens. Environ., 120, 2012, 25–36.

[24] B. Thai, G. Healey, Invariant subpixel target identification in hyperspectral imagery, Algorithms for Multispectral and Hyperspectral Imagery V, International Society for Optics and Photonics, 1999.

[25] D. B. Heras, et al. "Towards real-time hyperspectral image processing, a GP-GPU implementation of target identification," Proceedings of the 6th IEEE International Conference on Intelligent Data Acquisition and Advanced Computing Systems, IEEE, 2011.

[26] N. M. Nasrabadi, Hyperspectral target detection: An overview of current and future challenges, IEEE Signal Process. Mag., 31(1), 2013, 34–44.

[27] H. V. Nguyen, et al., EXIT-chart aided quantum code design improves the normalised throughput of realistic quantum devices, IEEE Access [Internet], 4, Jul 2016, 10194–10209.

[28] D. Alanis, et al., Quantum-aided multi-objective routing optimization using back-tracing-aided dynamic programming, IEEE Trans. Veh. Technol., 67(8), Aug 2018, 7856–7860.

[29] M. A. M. Izhar, et al., Quantum turbo decoding for quantum channels exhibiting memory, IEEE Access [Internet], 6, Feb 2018, 12369–12381.

[30] P. Botsinis, et al., Quantum-assisted indoor localization for uplink mmWave and downlink visible light communication systems, IEEE Access [Internet], 5, Jul 2017, 23327–23351.

[31] G. Sasubilli, A. Kumar, "Machine Learning and Big Data Implementation on Health Care data," 4th International Conference on Intelligent Computing and Control Systems (ICICCS), 2020, pp. 859–864, doi: 10.1109/ICICCS48265.2020.9120906.

[32] S. M. Sasubilli, A. Kumar, V. Dutt, "Machine Learning Implementation on Medical Domain to Identify Disease Insights using TMS," 2020 International Conference on Advances in Computing and Communication Engineering (ICACCE), 2020, pp. 1–4, doi: 10.1109/ICACCE49060.2020.9154960.

[33] S. M. Sasubilli, A. Kumar, V. Dutt, "Improving Health Care by Help of Internet of Things and Bigdata Analytics and Cloud Computing," 2020 International Conference on Advances in Computing and Communication Engineering (ICACCE), 2020, pp. 1–4, doi: 10.1109/ICACCE49060.2020.9155042.

[34] P. K. Kotturu, A. Kumar, "Comparative Study on Machine Learning models for Early Diagnose of Alzheimer's Disease: Multi Correlation Method," 2020 5th International Conference on Communication and Electronics Systems (ICCES), 2020, pp. 778–783, doi: 10.1109/ICCES48766.2020.9137872.

[35] S. R. Swarna, A. Kumar, P. Dixit, T. V. M. Sairam, "Parkinson's Disease Prediction using Adaptive Quantum Computing," 2021 Third International Conference on Intelligent Communication Technologies and Virtual Mobile Networks (ICICV), 2021, pp. 1396–1401, doi: 10.1109/ICICV50876.2021.9388628.

R. Vedhapriyavadhana, G. Ignisha Rajathi, R. Niranjana, N. Pooranam

Chapter 12 Quantum computing: application-specific need of the hour

Digital transformation technology and tools: shaping the future of primary healthcare

Abstract: In this tech-savvy era, as new technologies are developed every day by researchers and scientists, this also paves way to discover replacements to classical computers. One such advancement is the uncovering of quantum computing. They are brought up to light for the purpose of solving problems and scenarios solved by classical computers but in a more effective way and at least a span of time. The study of quantum computing focuses on finding new ways to compute by utilizing quantum physics phenomena. A qubit, in contrast to a typical computer bit, can be either 0 or 1, or it can be a superposition of both 0 and 1. Quantum theory serving as the base to operate supercomputers is based on two aspects of quantum physics: superposition and entanglement. This quantum computing plays extraordinary roles in different zones of applications such as weather forecasting, logistics, finance, healthcare, and cybersecurity. They are discussed with clarity in this chapter by throwing bright light on these important inevitable topics.

12.1 Introduction

The field of computing is devoted to creating computer technology based on the ideas of quantum technology which describes how energy and matter behave on an atomic–subatomic scale. Quantum computers use some of the virtually ethereal quantum mechanical phenomena to make significant improvements in processing speed. Even the most advanced supercomputers of the present and the future are expected to be outperformed by quantum machines [1].

R. Vedhapriyavadhana, School of Computer Science Engineering, Vellore Institute of Technology, Chennai, e-mail: vedhapriya.vadhana@gmail.com
G. Ignisha Rajathi, Computer Science and Business Systems, Sri Krishna College of Engineering and Technology, Coimbatore, e-mail: ignishaelton@gmail.com
R. Niranjana, Electronics and Communication Engineering, Sri Krishna College of Engineering and Technology, Coimbatore, e-mail: niranbme@gmail.com
N. Pooranam, Computer Science and Engineering, Sri Krishna College of Engineering and Technology, Coimbatore, e-mail: pooranam.n@gmail.com

https://doi.org/10.1515/9783110791402-012

The basic unit of quantum computing is quantum bits or qubits. The capacity of bits to exist in several states simultaneously gives the quantum computer a large portion of its processing capability. They are capable of carrying out tasks utilizing a combination of 1s, 0s, and even a 1 and a 0 at the same time. Bits are a sequence of electrical or optical pulses that represent 1s or 0s and are used by modern computers. The tweets, emails, iTunes music, and YouTube movies are all basically just lengthy lines of these binary digits [2].

The study of quantum computing focuses on finding new ways to compute by utilizing quantum physics phenomena. A qubit can be either 0 or 1, or it can be a superposition of both 0 and 1. With additional qubits, quantum computers' power increases exponentially. Adding extra transistors to a classical computer only increases power linearly; this is not the case here. A connected set of qubits can offer significantly more processing power than the same number of binary bits due to some peculiar quantum features. These characteristics include entanglement and superposition, respectively.

To do its magic, quantum computers need entangled qubits connected in a quantum daisy chain. The reason there is so much talk about the potential of machines is because they can accelerate calculations using specially created quantum algorithms.

12.1.1 Superposition and entanglement

Quantum theory that serves as the base to operate supercomputers is based on two aspects of quantum physics: superposition and entanglement. This provides quantum computers the ability to perform processes at speeds exponentially faster than ordinary computers while consuming very less amount of energy. Quantum computing is of much use in various domains such as finance, military affairs, intelligence, drug research, aircraft design, utilities, polymer design, machine learning, artificial intelligence (AI), big data search, and digital manufacturing which could all reap enormous benefits.

There are numerous applications of quantum computing, such as facilitating safe transfer of data. Fighting cancer and other health issues, such as generating new medications, are further strategies. Quantum computers can also aid in enhancing radar's capacity to identify objects like missiles and aircraft. Other fields include the environment and utilizing chemical sensors along with quantum computing to keep the water pure.

When compared to supercomputers and conventional computers, quantum computers are many times quicker. Sycamore, Google's quantum computer, is alleged to have completed a computation in 200 s that would have taken IBM's Summit, one of the fastest computers in the world, 10,000 years to complete. The time required, according to IBM, would be 2.5 days, which is still more than 1,000 times slower than Google's quantum computer [3].

In terms of how it works and what it is used for [4], quantum computing is distinct from classical computing. While conventional computers utilize transistors, which can only be either 1 or 0, quantum computing uses qubits, which can be both 1 and 0. Because of this, quantum computing is far more potent and can be utilized for simulations or massive data processing. However, no commercially viable quantum computer has yet been created.

12.1.2 Need of the hour

Quantum computing offers promising phenomena to improve the performance of supercomputers to a wider extent. As a consequence, some kinds of issues will probably experience significant speedups, thanks to them. Researchers are striving to create these algorithms which tend to demonstrate quantum speedups as well as identify the issues which would serve as the best fit for them. In general, it is anticipated that quantum computers will be of great assistance with problems involving optimization, which are crucial to everything from financial trading to defense.

There are various other qubit applications that are not directly connected to computation or simulation that also exist and are the topic of active research in recent days, but they are outside the purview of the study. The two most significant domains include quantum networks and communications, which may result in completely new ways to share information, and quantum sensing beyond the classical short noise limit.

There are numerous advantages that quantum computing can offer. For instance, considering our big data era, quantum computing devices will speed up the analysis and pattern recognition of enormous datasets. Additionally, this technology might be used to improve communications, create incredibly precise medical sensors, and train AI systems.

This quantum computing also finds its use in modern cybersecurity issues, stock profile management, aeronautical issues, molecular research, and other issues are a few instances. Protein modeling is a different illustration. The scientific community attempted to find a computational tool that could simulate and deactivate a single protein in a shorter amount of time during the COVID-19 epidemic. The world could have been spared from the worldwide health calamity if such a tool had been accessible.

Quantum computing also comes into play while handling huge amounts of data. Modern computers and supercomputers are prone to mistakes, which negatively impacts the performance. Furthermore, traditional computers cannot handle computational tasks like analyzing the chemical impacts of medications. Instead, because they can process large amounts of data more quickly, quantum computers are better suited for such tasks.

12.1.3 Statistics on quantum computing

In 2016, the market for quantum computing was estimated to be worth USD 89 million. By 2025, it is expected to have grown to USD 949 million, with a CAGR (compound annual growth rate) of 30% (MarketWatch). According to McKinsey, the automobile industry will be one of the key value pools for quantum computing, with a significant influence beginning around 2025. By 2030, they predict that associated technologies will have an economic impact on the automotive sector of $2–3 billion (McKinsey). By 2030, there may be 2,000–5,000 quantum computers operating globally. The bulk of those involved in and developing quantum computing are startups (40%) and institutions (33%) (PwC). From 2 qubits in 1998 to 128 qubits in 2019, qubit counts of quantum computers have been rising rapidly (Statista). The three businesses where quantum computing is most appropriate are finance (28%), global energy and materials (16%), and high-tech (source: McKenzie, Consultancy.co). The global market for quantum computing was worth $89 million in 2016. By 2025, this amount is projected to increase by more than ten times, reaching $949 million. This projection was created using a 30% CAGR between 2017 and 2025.

Quantum computer statistics indicate that some businesses own multiple machines. The year 2019 saw the development of the 53-qubit IBM quantum computer and the 128-qubit Rigetti quantum computer. As a consequence of the rapid development of quantum computing, there has been a continuous rise in the number of announcements in the field, indicating a significant push for the sector. Microsoft and Amazon already provide quantum computing services through Azure Quantum and Braket, respectively. Even Google wants to build 1 million qubit quantum computers by 2029. Having mentioned the above facts, it is expected that the market size of quantum computing will advance in the coming decade, as mentioned in the recent statistics using data from the Statista Digital Economy Compass. By 2030, the global quantum computing business may generate $9 billion in sales, up from $260 million in 2020. Between 2020 and 2030, the market's annual average growth might exceed 40%, with growth accelerating after 2025. Data simulations will be considerably complicated and faster because compared to the huge computational capacity provided by quantum computers, this may result in technical advances in a variety of fields. Quantum computers may display both states in a single qubit, unlike binary computers, whose bits can only be seen as 1 or 0.

12.2 General applications of quantum computing

Quantum computing accomplishes computations faster than supercomputer which any mechanical system would perform using advanced problem-solving techniques and is expected to rule in the next decade [5]. It finds its applications in many areas

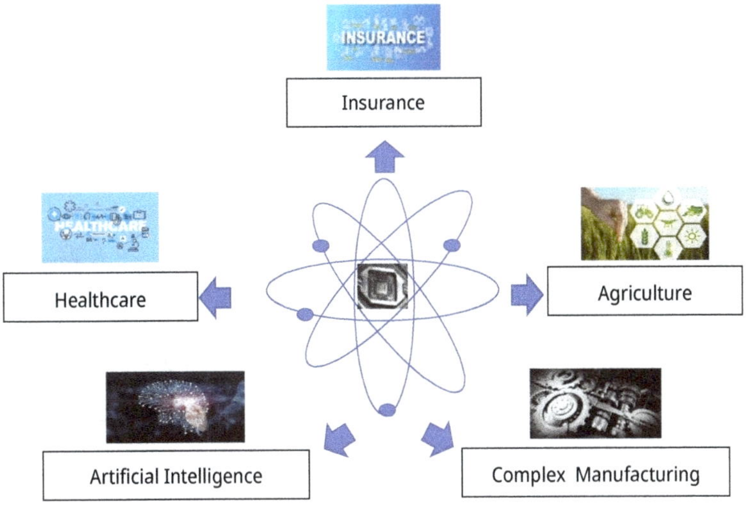

Figure 12.1: Applications of quantum computing.

[6] not restricted to AI, financial services, complex manufacturing, agriculture, healthcare, and so on, as shown in Figure 12.1.

12.2.1 Insurance

Finance is another domain that consists of huge amounts of data or in other words big data which also includes combinatorics calculations. When it comes to financial markets, the computing speed plays a huge role as every millisecond matters when it comes to obtaining the value of a share or the price information in hedge funds which is made easy with the help of quantum algorithms. The complex data analysis is done with the help of AI to predict the future of economic status, especially with the very high volume of flexible data which is pushed to cloud for storage, and the insurers can rely on it totally as storing complex variables are little challenging in the normal computer and where classic computers take the role in saving them.

12.2.2 Healthcare

The healthcare data maintained in hospitals in the form of electronic health records need to be analyzed using the complex system as right from the diagnosis till therapy procedures and many complex decisions are taken and it is dynamic too with respect to individual cases. The data sources pertaining to healthcare are quantum enriched using machine learning algorithms. This classical modeling saves many lives and economically efficient not only in personalizing medicine but also in price optimization. Mishra and Vedhapriyavadhana [7] also extended this to specific applications for physically challenged people, where assistance to the visually challenged elderly people through machine learning concepts is briefly discussed.

12.2.3 Agriculture

The numerous applications start from spreading the fertilizers over field, quality of weeds sown, climatic advances, soil strength evaluation, protecting crops from pests and insects, smart irrigation, and planting systems too. The farmers were supported to continuously monitor the well-being of the crop from AI.

12.2.4 Artificial intelligence

To make faster and better predictions and decisions, AI requires combinatorics processing of very large quantities of data that can be made possible using the growing research field in quantum machine learning which opens a lot of opportunities. Even though it is more of a remote possibility with our current limitations, it, for sure, does make thinking machines a reality than science fiction.

12.2.5 Complex manufacturing

The complex products are simulated and optimized with the help of quantum computing which performs optimization in the fastest way. As every product developed is software controlled right from the design till production, the software validation and quality check are done dynamically at all levels. These software systems are thoroughly analyzed and can support complex designs than the classical computers.

12.3 Role of quantum computing

12.3.1 Weather forecasting

We know that weather, in general, is never constant so it is always hard to be perfectly accurate when it comes to forecasting the weather. The current models can predict snowstorms and hurricanes which are in regional scale, whereas localized events like thunderstorms cannot be predicted which is where we would require a huge amount of computing power that will allow us to keep an eye on the world so that we can identify to prevent and reduce the property damage and loss of life. Therefore, the prediction of weather should be extreme and further in advance for that we need to have an increased accuracy value that could be able to prevent the targeted region [8], and they can be prepared in advance as even a simple storm can become a life-threatening one. But this is currently not possible as we still need even more computational brute force.

Being able to attain this improvement, we can reduce further damage caused by disasters like hurricanes, tornadoes, and other extreme weather events which result in deaths and huge damages in the region. Lots of work have been undertaken to reduce this problem and try to rectify the current demerits and improve it furthermore to get the higher accuracy over the year by developing the sophisticated computational models. But still it needs lot more development to get the right warning at the right time because with the help of current model it will be able to cover only on the regional scale and that to not more localized events and to cover the damages, to prevent the unexpected weather changes it needs more accuracy on both scales [9]. This is where we need the emergence of quantum computing to bring out the massive distribution and to create the enormous value.

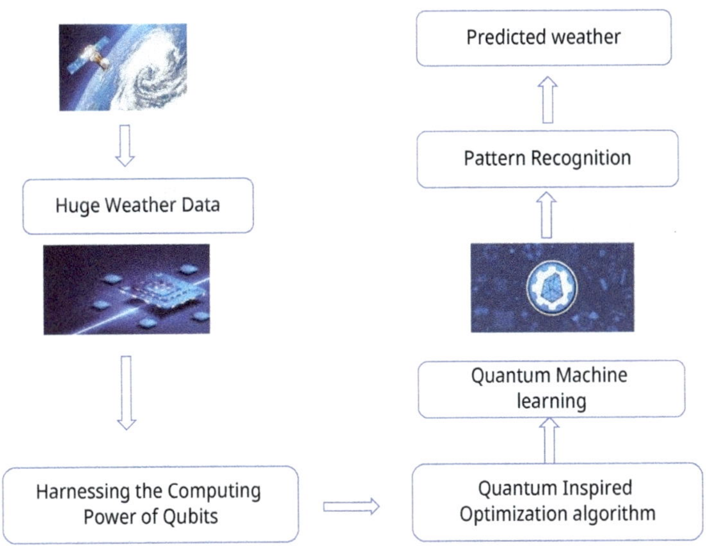

Figure 12.2: Quantum computing in weather forecasting.

Basically, for analyzing the weather forecasting it requires a large amount of data and also that data should contain all the possible dynamic variables and values such as the pressure, density, and temperature as shown in Figure 12.2. Even though there has been a lot of development when it comes to computational models, there has been lot of limitations too when it comes to classical computers and even supercomputers just cannot simply achieve those result as there are many factors and dynamic variables, which are ever-changing.

To overcome this situation and to get higher accuracy value, quantum computing has the potential and through using this we can improve and boost the conventional numerical method for tracking and predicting the current meteorological condition by handling over the large data which carries the different variables [10]. Considering the supercomputers we might be able to get the phenomenal value for micrometeorological with this quantum computing can become feasible it will be like the creation of cloud or a wind eddy.

Here, for maintaining the data and to handle it in a quicker and efficient manner, two things need to be done: harnessing the computing power of qubits and optimization of algorithms (quantum-inspired). Furthermore, by using the quantum machine learning, we can find the pattern recognition and other needed information. The PASQAL's quantum is basically an idealized method for simplifying the complex computation of BASF and to influence the utmost advantage of PASQAL for this weather forecasting. The model needs to validate its ability to achieve its classical high performance.

PASQAL's focus is crucially on to the problem which is basically implicit in the complex nonlinear differential equations, and for this they implemented the quantum neural network on the processor which is neutral-atom quantum [11]; here, the basic approach used is physics-informed neural network. And the main advantage by using this quantum computing that this can be well served on weather forecasting on both the large and local scale levels and it can provide the advanced and accurate value for the extreme weather events through this we can give the prior warning to the people and decrease the damage impact on the society.

12.3.2 Logistics

When it comes to logistics, quantum supercomputers can analyze all available routes, and taking the most efficient one will make route planning much easier [12]. It can also maximize the simultaneous packing of parcels, support reallocation of assets, and rapidly test designs and materials for logistic use.

The requirements of the customers are met then and there, and that is the main objective of any supply chain organization. But the chain trade-off across departments keeps changing dynamically based on the strange routine from the regular one and to challenge the situation created because of the unpredictable scenario. The permutation for the same is really high and the volume of data cannot be dealt with the classic computers, and there was the usage of quantum computers that rule the world. The classical computer stuck at the variety of weight changes and possibilities.

The optimal plant location is chosen and production is carried out. Through various promising hubs, the products are distributed, and further for local/global shipping, the best optimization technique is chosen and delivered at lower cost. With the help of

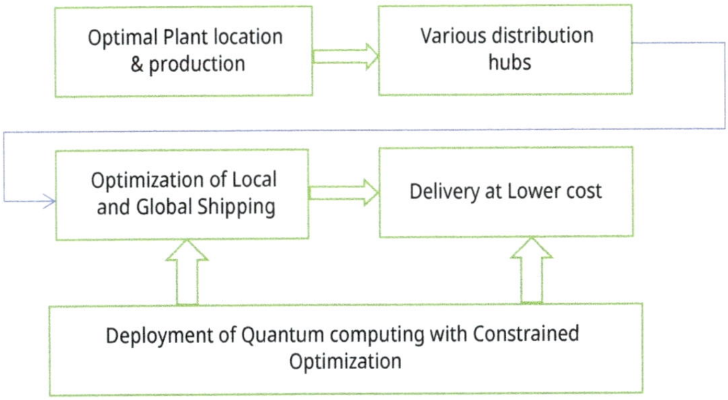

Figure 12.3: Deployment diagram of quantum computing with optimization.

constrained optimization, the loads under local and global shipping are optimized, and the efficiency can be increased with fastest delivery at lowest cost as shown in Figure 12.3. The shipment [13] scheduling becomes faster when lakhs and lakhs of products are transported at the best cost by achieving the maximal production level. For supply chain, quantum computing is a boon with latest technological innovations, and the future of the supply chain still holds in it.

12.3.3 Finance

In the digital transforming world to solve the optimization problems pertaining to finance, there are primarily three classes where quantum computing can be put to use, especially when it deals with the uncertainty in the financial domain [14]:

i. targeting and prediction,
ii. trading optimization, and
iii. risk profiling.

The classical computers provide solutions to the problems which are not optimal in the financial sector. These are then covered by the quantum computers by providing quantum solutions as shown in Figure 12.4 through optimization techniques based on the defined use case development, and commercialization is done based on the best use case identified. The investors who invest have the high legal risk when floated in the market comes to forecasting and prediction with the help of the quantum hardware implemented, and thereby the solutions are arrived without any delay in order to accelerate only the advantageous effects of quantum solutions arrived.

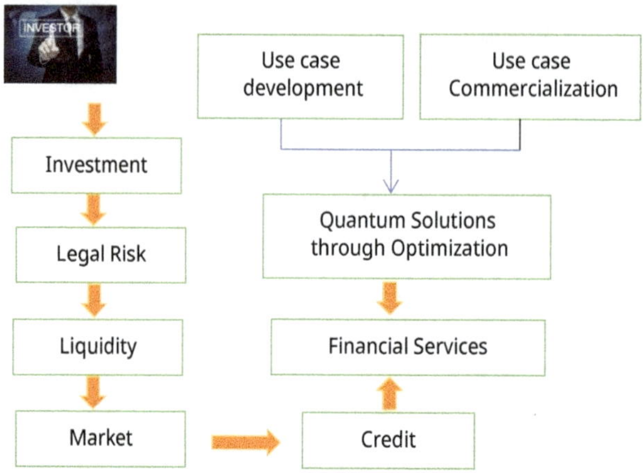

Figure 12.4: Quantum computing in finance.

12.3.3.1 Targeting and prediction

Nowadays, customers require financial services that provide personalized products and services that adapt rapidly to their evolving needs and behaviors. The speed and accuracy elevates steeply when quantum computing is used to solve the complex models, especially the calculations that are not possible with the traditional one. Scrutinizing through behavioral data and getting more accurate results is made possible by quantum computing. Financial institutions are also losing around USD 10 billion to 40 billion due to bad data management and fraud.

12.3.3.2 Trading optimization

In the current trading landscape, real-life constraints like market volatility cannot be incorporated into portfolio optimization by investment managers. Due to computational limits and transaction costs, keeping up with market movements has become hugely constrained. Monte Carlo simulations are used to represent short-term impact using partial differential equations, whereas optimization algorithms are used for long-term evaluation with high dimension.

All its aims are the maximum profit and minimal risks. Classical data needs to be converted to quantum data, and the bottleneck is that the advantages of quantum computing may be neutralized because of the conversion process and also memory shortage is a great threat. The best quantum application can be developed on the best quantum hardware platform for the problem defined by the user. So start with the use cases for the problem to be solved.

12.3.3.3 Risk profiling

Due to the estimation error limitations, Monte Carlo simulations which are required to analyze the impact of risk and uncertainty is limited. As time goes on, like the Basel III, we can expect more overlapping amendments to regulations, directives, and standards. So, quantum computers can be useful in providing faster risk scenario simulations with more precision. The financial elements need to be studied at microlevel in a whole network of institutions. The market value of the institution should be predicted based on several criteria and one such is the ownership of the assets that are distributed among these institutions. The prices of the assets change with respect to many other factors too.

To evaluate the business value of any institution/industry, a roadmap needs to be developed with the business strategy. The open source available for quantum computing simulations can be best utilized by the new developers to arrive with the solutions to all the real-world problems. As the future growth of any institution relies on this prediction, the best use model should be commercialized with the optimization algorithm using quantum computing techniques.

By 2030, we can anticipate to have computing systems with fault tolerance and a quantum advantage for lots of applications. We will also be able to see an increase in powerful NISQ (noisy intermediate-scale quantum) machines being introduced in the next 5 years, and even though it might not be powerful for most applications, few of them can still take advantage of these. We can expect to see the rise of error-corrected fault-tolerant quantum processors in the end of the 5-year period after which large-scale quantum computing can be merged along with real-world applications.

A lot of industries do require quantum computing but the current technologies make the possibilities to be limited; hence, in the upcoming years, we can see improvements in quantum computing which in turn can open up lot more possibilities.

12.3.4 Healthcare

This segment deals with how quantum computing works and few ways in which the healthcare industry is able to benefit many fold by applying its concepts at the fundamental level.

Quantum computing is able to fundamentally change healthcare [15, 16]. The tech giant Google has recently announced that it has achieved quantum supremacy. They claim that their quantum processor was able to effortlessly perform a complex mathematical calculation in 3 min and 20 s, for which the latest supercomputer would take 10,000 years to complete. This proved that quantum speedup is achievable in a real-world system, and that one day, these machines will be able to outperform all classical computers. Quantum computers are hundred million times faster than today's computers when compared with AI.

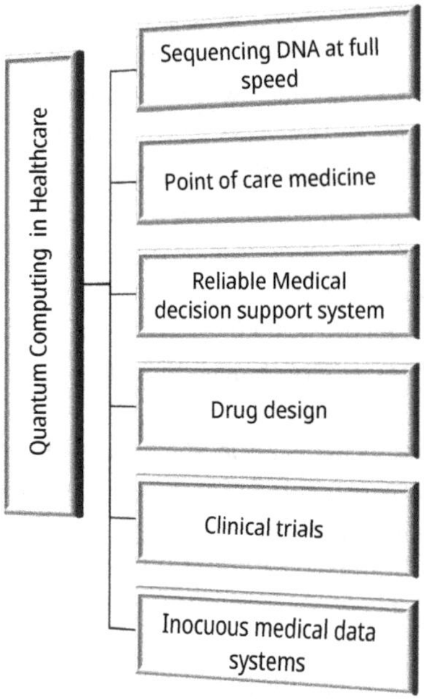

Figure 12.5: Applications of quantum computing in healthcare.

We are able to discover oncology drugs, using a simulation platform that helps lower the astonishing 96% failure rate in clinical trials as shown in Figure 12.5. The major scientific leap enabled by quantum computing in this particular application would be the accurate simulation of protein folding, and how these little nanomachines behave is crucially important to understand how our cells work. And it is now believed by researchers and scientists in the field of quantum mechanics that a lot more that is unknown in biology could be unlocked by the processing ability proposed by quantum mechanics.

12.3.5 Sequencing DNA at full speed

Even though we are capable of full-genome sequencing, and its costs are down to an affordable level, future genome sequencing will be like a regular blood or urine test, and quantum computing will be able to analyze the vast amounts of data. Quantum computing can ensure faster sequencing, more comprehensive analysis, more efficient comparative analysis with millions of other patient data and with similar lifestyle and health parameters. It will be able to make reliable predictions about what could go wrong with a patient in future, and also what can be done now in the present to prevent it. Quantum computing can finally help crack the human genome, its impact on health and disease, and discard the guesswork that is normally associated with personalized medicine.

12.3.6 Point-of-care medicine

Quantum computing promises much higher level of digital health a reality. Raihan Ur Rasool et al. [17] say that conventional machine learning approaches and algorithms have proved to be quite effective in predicting the occurrence of disabilities or infections that a person may concur in future. Nevertheless, there are still limitations in these algorithms caused due to noise, quality of information obtained from imaging or signal acquisition modalities, or charting up the correlation between the features obtained.

We already have smartphone apps, wearables, health sensors, and portable diagnostic devices that are keeping us informed about our health and disease management. But quantum computing will be able to make sense of the astronomical datasets and track patients' health on the go. Connected sensor systems might render physical hospitals useless and truly make patients the point of care. They would receive diagnosis and treatments wherever they are.

12.3.7 Reliable medical decision support system

It is practically impossible to read, learn, remember, and apply the knowledge of every possible clinical trial and study done all over the world while making a decision for a particular ailment. There will be millions of medical statistics and data already done, and 1 or 2 million on the verge of coming up. Liu et al. [18] proposed that quantum computing can take this to a whole new level. They can take the entire lexical and empirical knowledge in real time, and could provide the best decision support for the doctors [18]. They could skim through all the studies, clinical trials, textbooks, and publications, all at once. Quantum computing could find correlations and root causes that the humans can never find, and provide the diagnosis and treatment options that humans could have never figured out.

12.3.8 Drug design

This is where the quantum computing's capabilities are best used to their maximum. Drug design does not only leverage the speed and processing of quantum computing, but the very nature of it. The way the electrons of the molecules of the drug interact with the human body works by the same principles as quantum mechanics. Simulating quantum phenomena on quantum computers can lead to much faster and much more precise results.

12.3.9 Clinical trials

Quantum computing can lead to an era of in silico trials as well. In silico trials are where no human beings, animals, or even a single cell is required for testing a particular therapy or drug. It is a human way and theoretically could be fast and prove to be a more effective path than in any other viable trials adopted to this day. Today completely simulated trials are still not possible, but with quantum computing, virtual humans can undergo trials and can have new drugs taking the plunge into the markets, within weeks of their discovery.

12.3.10 Innocuous medical data systems

One of the biggest factors of concern in quantum computing is safety. Given the processing power of quantum computing, it is mostly understood that even the lowest level quantum computer could crack the encryption code on a classical computer, as qubits could overcome any wall of security. But on the contrary, quantum computing can also be used to make use of quantum uncertainty to create encryption, and quantum security is literally unbreakable. Hackers, if they try, will have to break the laws of quantum physics in order to gain access to health records encrypted this way.

12.3.11 Cybersecurity

For as long as there have been encryption systems to make data secure, there have been attempts by hackers, working on breaking these codes. Recently, new advances in the control of quantum systems are paving the way for a paradigm change in cybersecurity.

If this issue is not addressed within time, before the quantum computing takes over every sector of online communications, it could lead to the exposure of data across the globe, at a level which we cannot fathom. The privacy of emails, credit card numbers, military communications, health records, genetic information, juvenile/criminal records, executive communications, drug trial data, intellectual property, information transmitted between cloud data centers, and so on could all be compromised.

The search for a solution for this issue was redirected toward quantum computing again. This led to the emergence of quantum cryptography, leveraging the laws of quantum physics to encrypt data. Data that is encrypted using the nature of quantum mechanics turned out to be almost unbreakable.

Striking a trade-off with its potential risks, quantum computing makes use of physics rather than mathematics. Quantum cryptography has also evolved over the ages. It started out with quantum key distribution (QKD). QKD could barely survive

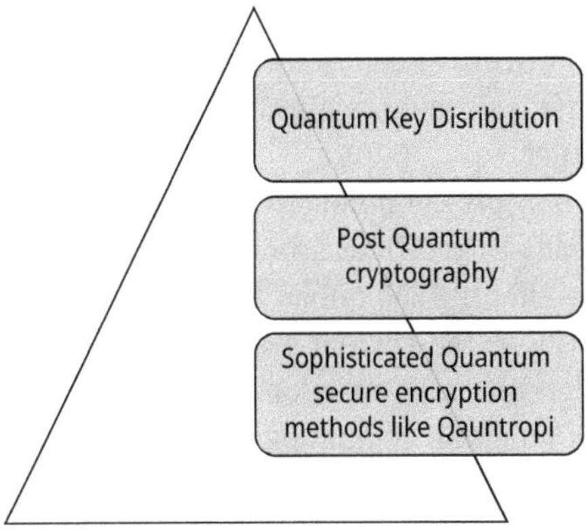

Figure 12.6: Evolution of quantum cryptography methods.

outside the lab. It uniquely requires new dark fiber infrastructures. It is expensive, scalable, and underperforming.

The next evolution that came into the picture was post-quantum cryptography (PQC) in Figure 12.6. Various algorithms have been developed in order to implement this like isogeny-based cryptography, SIKE algorithm, Shor's algorithm, and Grover's algorithm. Some of them possess an inherent probabilistic nature, which is believed to help pave the way for higher levels of cybersecurity. But now with advancements springing up, these approaches are starting to get unfamiliar and mostly untested. Their long-term security possibilities could not be completely relied on.

The next evolution is a platform like Quantropi. Their unique cryptographic solutions are founded in quantum mechanics, expressed as linear algebra, an established mathematics that is thoroughly understood and universally accepted. Their patented technologies are readily deployed on any existing network, over unlimited distances at full network speeds. This provides us with quantum secure key generation and distribution over today's network. The software platform involved in this environment requires minimal investment in new tech infrastructure. Customers can immediately dump it onto their hardware and process to ensure quantum resilience, before seamlessly transitioning to 100% permanent quantum security. It has proposed to provide the three requisites for quantum cybersecurity, namely, trust, uncertainty, and entropy. Trust can be established via a quantum-secured asymmetric encryption. Uncertainty should be ensured for attackers, rendering the data uninterruptable forever by a quantum symmetric encryption. Entropy can be serviced, by ensuring string random key generation and quantum-secure distribution. By this way, governments, enterprises, and developers can have confidence that their sensitive data and communications are always secure and private.

Takagi [19] proposed that mathematical calculations required for ensuring security of key crypto systems could be solved in polynomial space using the quantum physics mechanism. They propose the design and safety of cryptosystems built on principles of polynomial space, especially the lattice-based cryptosystems which are built on multiple variable design. Baldi [20] proposed that out of all the existing algorithms to implement PQC, Shor's algorithm and Grover's algorithm contribute an additional quickening of the processing speed, and is even more effectual as it works in polynomial space with prime factors of an integer.

As far as cryptography is concerned, it is not just the number of qubits. It is also the error rates and the accuracy that one needs to get. Hence, breaking something like 2,048-bit RSA cryptography would require almost 6,200 qubits and 2.7 billion operations. If the error rates are going to be 3×10^{-9}, the current error rates are 10^{-2}, which involves seven orders of magnitude difference in comparison.

To be more precise about the influence of quantum computing on cryptography, many cryptosystems are broken down into one of two different types. One is public key or asymmetric key cryptography and the other is called symmetric cryptography. The impact of quantum computing on public key cryptography is quite large scale.

12.4 Future of quantum computing

Quantum computers are not just yet another faster version of the computers we use today. They will operate in a fundamentally different level, as they work based on the laws of quantum physics. So, we can do computing tasks that are outside the range of even the best of best supercomputers today. That clearly points to the fact that we can do calculations, predictions, searches, and so on at a more unimaginable level.

We want to make sure that the privacy of all of the data used for communication is well secured and encrypted. Quantum offers the best way to build encryption that can never be hacked. No matter how good the hackers are. This is yet another transformative example for how society can change.

Quantum Internet would be very different from the current Internet we use. But that does not mean it is going to completely replace our current Internet. Normal day-to-day usage of Internet including mails and social media does not require quantum processing capacity. Instead, there might be a quantum communication in the background with a quantum encryption network, working in unison with current Internet, to be able to improve privacy and security of all our communications. It may allow us to do certain things that are not possible with current Internet, like teleporting of information.

The Internet of things (IoT) includes devices and technologies over a very large spectrum. Therefore, it cannot be assured that quantum computing could influence IoT in just one way. But one aspect common in all IoT devices is that they are all communicating in one network. That network could be made top secure by quantum cryptography in ways that we cannot do right today. That could be considered the greatest impact of quantum computing on IoT.

Quantum computers can potentially help us address some of our biggest challenges, like climate change, how we use energy more efficiently, and also provide a more reliable decision support system for doctors in the medical field. It has all the potential to revolutionize the medical field further in various aspects like prevention of drug diversion, drug design, sequencing of DNA, and virtual clinical trials.

Quantum being a technology is basically going to impact us, whether as individuals, or as part of the government or corporate world. But the corporate world will have to think hard about their social responsibility, in terms of security, privacy, healthcare, impact on the environment, and so on, if they are going to take up the reins of this powerful technology for their applications. The governments too will have to work out certain clauses and restrictions to be included, and to safely use this extremely powerful technology. These factors need to be addressed before this technology is fully rolled out.

12.5 Conclusion

The spark of quantum computing with all its revolutionary realms has been glimpsed in this chapter. This entire chapter has briefed the evolution, statistics, need of the hour with quantum computing, different zones of applications such as weather forecasting, logistics, finance, healthcare and cybersecurity. Whenever technology offers us enormous power, there is always a reason to worry. In the end, how we use the technology is going to be in our hands. There is a huge amount of potential good that the quantum technology can bring to our society, and is bound to have some unintentional consequences. But if the risks involved with this technology are addressed ahead of time, quantum can create a paradigm shift in the society in a way that we would all be involved.

References

[1] https://www.ibm.com/topics/quantum-computing.

[2] https://www.investopedia.com/terms/q/quantum-computing.asp.

[3] https://www.sciencealert.com/quantum-computers.

[4] https://azure.microsoft.com/en-in/resources/cloud-computing-dictionary/what-is-quantum-computing/?cdn=disable.

[5] https://hbr.org/2021/07/quantum-computing-is-coming-what-can-it-do#:~:text=Quantum%20computers%20can%20be%20used,to%20incidents%20of%20product%20failure.

[6] https://builtin.com/hardware/quantum-computing-applications.

[7] A. Mishra, R. Vedhapriyavadhana, "Medicine assistance application for visually impaired people," Proceedings of the 2022 3rd International Conference on Intelligent Computing, Instrumentation

and Control Technologies: Computational Intelligence for Smart Systems, ICICICT 2022, 2022, pp. 850–855, doi: 10.1109/ICICICT54557.2022.9917621.

[8] https://1qbit.com/blog/quantum-computing/forecasting-the-weather-using-quantumcomputers/#:~:text=Quantum%20computing%20will%20serve%20to,and%20reducing%20property%20damage%20annually.

[9] https://www.analyticsinsight.net/quantum-predictions-weather-forecasting-with-quantum-computers/.

[10] https://www.passionateinmarketing.com/quantum-computer-for-weather-forecasting/.

[11] https://thequantuminsider.com/2022/07/20/basf-pasqal-to-use-quantum-computers-for-weather-prediction/.

[12] https://www.interlakemecalux.com/blog/quantum-computing-logistics#:~:text=In%20logistics%2C%20route%20planning%20would,taking%20into%20account%20all%20variables.

[13] https://www.supplychainbrain.com/blogs/1-think-tank/post/33547-how-quantum-computing-will-power-the-future-of-logistics.

[14] https://www.ibm.com/thought-leadership/institute-business-value/report/exploring-quantum-financial.

[15] S. M. Sasubilli, A. Kumar, V. Dutt, "Machine Learning Implementation on Medical Domain to Identify Disease Insights using TMS," 2020 International Conference on Advances in Computing and Communication Engineering (ICACCE), 2020, pp. 1–4, doi: 10.1109/ICACCE49060.2020.9154960.

[16] S. M. Sasubilli, A. Kumar, V. Dutt, "Improving Health Care by Help of Internet of Things and Bigdata Analytics and Cloud Computing," 2020 International Conference on Advances in Computing and Communication Engineering (ICACCE), 2020, pp. 1–4, doi: 10.1109/ICACCE49060.2020.9155042.

[17] R. Rasool, H. Ahmad, W. Rafiq, A. Qayyum, J. Qadir, Quantum computing for healthcare: A review, TechRxiv, 2021, 10.36227/techrxiv.17198702.v2.

[18] Z. Liu, X. Liang, M. Huang, "Design of logistic regression health assessment model using novel quantum PSO," 2018 IEEE 3rd International Conference on Cloud Computing and Internet of Things (CCIOT), IEEE, 2018, pp. 39–42.

[19] T. Takagi, Recent developments in post-quantum cryptography, IEICE Transactions on Fundamentals of Electronics, Communications and Computer Sciences, E101A(1), 2018, 3–11. https://doi.org/10.1587/transfun.E101.A.3.

[20] M. Baldi, "Post-Quantum Cryptographic Schemes Based on Codes," 2017 International Conference on High Performance Computing & Simulation (HPCS), 2017, pp. 908–910, doi: 10.1109/HPCS.2017.151.

Nisha Soms, S. Kalyani, S. Nagarani, M. Rohini, Oswalt Manoj S

Chapter 13
Industrial Internet of things and Industry 4.0: a learner's perspectives toward quantum technologies

Abstract: Industry 4.0, often known as the Industrial Internet of things (IIoT), is considered to be one of the most prominent business concepts related to industry in recent years. The IIoT and Industry 4.0 are commonly represented at a higher level by the consultants who speak to executive clients from a commercial perspective, notwithstanding the underlying technological complexity. When novel business models and financial gains are easily understood by their clients, consultants place an emphasis on the models related to business and also on the operational efficiency that are very enticing. Sadly, these presentations frequently inspire and enchant the officials who view the economic welfares, but frequently flop to convey the client about the technological abstraction of the industrial Internet's complexity in the lower layer. In this section, we thus seek to discourse this issue. We begin with an overview of the high-level possible benefits of IIoT business incentives and also the models, as well as its successful use cases. We then examine the technical hurdles associated with establishing an IIoT network. We also address the information needed to build and deliver an IIoT network to business and technology participants.

13.1 Introduction

Although it has been referred to by others, including Cisco, as the Internet of everything and Internet 4.0, General Electric (GE) was the first to coin the term "industrial Internet." Because the Internet of things (IoT) as a whole has a broader horizontal concept, it is crucial to distinguish between the vertical IoT strategies (see Figure 13.1) and the bigger conception, such as the consumer, commercial, and industrial forms of the Internet. Smart homes, personal connectedness through fitness monitors, entertainment-integrated technology, and personal in-car monitors are just a few examples of consumer-facing markets with high visibility. Services in the commercial sector

Nisha Soms, CSE, KPRIET, e-mail: nishasoms@gmail.com
S. Kalyani, S&H, SRIT, e-mail: kalyani.sh@srit.org
S. Nagarani, S&H, SRIT, e-mail: sureshnagarani@gmail.com
M. Rohini, CSE, SKCET, e-mail: rohinibtech@gmail.com
S. Oswalt Manoj, CSE, SKCET, e-mail: oswaltmanojibm@gmail.com

https://doi.org/10.1515/9783110791402-013

have a high marketability because they cover a wide range of financial and investment products, including banking, insurance, financial services, and e-commerce; all of which place a high value on knowing their customers and offering them goods and services that improves their lives. On the other hand, the enterprise IoT is a subset of IoT that includes businesses of all sizes. This chapter focuses on a specific subset of IoT called the industrial Internet of things (IIoT), which has applications in numerous fields, including but not limited to energy generation, manufacturing, agriculture, healthcare, retail, transportation, logistics, aviation, and space exploration [1].

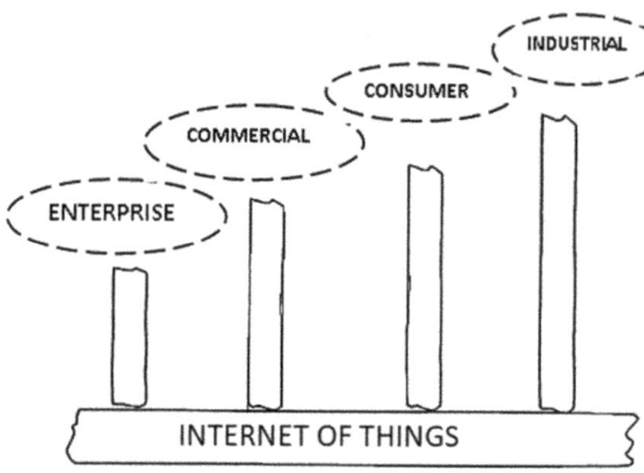

Figure 13.1: Internet of things: vertical and horizontal aspects.

We shall utilize the term "industrial Internet of things" as a general word in this work, following GE's example. However, when referring to philosophically and strategically distinct paradigms, such as Industry 4.0, we will use the specific word. Some of the most influential people in business believe that the industrial Internet will spur extraordinary expansion and productivity over the next decade. Leaders in business, government, academia, and the technology industry are all working around the clock to find ways to tap into and fully exploit this enormous opportunity. According to one market estimate, the US market is expected to reach $151.01 billion by 2020, growing at a compound annual growth rate of 8.03% from 2015 to 2020. For their part, businesses understand that the Internet usage can lead to economic growth. Manufacturers and governments alike are seizing the moment to reindustrialize by relocating previous offshore production back to the United States. Governments promote reindustrialization in the hopes of raising GDP by increasing added value from industry. There is precedent for the establishment of the industrial Internet, such as the stratospheric growth of B2C e-commerce in media, retail, and finance during the previous 15 years as a result of the entrance of the Internet. The popularity of web-scale, Internet-born businesses like Netflix, PayPal, eBay, and Amazon attests to the success of business-to-consumer sales methods.

The manufacturing, agricultural, energy, aviation, transportation, and logistics sectors are included in this optimistic outlook for the next decade. Since industry accounts for two-thirds of global GDP, the stakes could not be higher. In contrast, the industrial Internet is only getting started. However, business titans have been slow to embrace the Internet despite its widespread availability over the past 15 years. Their reluctance stems from the fact that they are unaware of how it might influence the status quo in terms of established markets, supply chains, business models, existing workforces, and final outputs such as productivity and goods. In addition, 87% of company leaders in the sector claimed to still not having a complete understanding of the technology or business models in January 2015. Given the widespread approach of defining the industrial Internet in terms of abstract concepts, it makes the characteristics of the underlying technologies meaningless. Sensors and gadgets that generate data for controlling operations have been common place in industrial organizations, for instance, for decades. The essential technologies of IIoT, such as machine-to-machine (M2M) connectivity and collaboration, have been around for at least a decade [2]. For instance, businesses have not been sluggish in amassing large amounts of data for future use, analysis, and forecasting. Many C-level executives in the industrial sector wonder why they would benefit from integrating their M2M architecture with the Internet.

This chapter is organized in such a way that the first section focuses on the novel and exciting business models that emerge from the IIoT, which are of great interest to business executives. The enabling technologies for IIoT will be discussed in the next section. Therefore, we discuss how to construct IIoT networks in practice utilizing various technologies and protocols. Technology and protocol convergence is crucial, but occasionally a platform or mediation service is required to bring everything together. The middle of the chapter discusses protocols, software patterns, and middleware IIoT platforms, which act as the binding agent or the lens through which we can visualize or connect our IIoT network. Lastly, we progress from broad IIoT principles and ideas to the more specific Industry 4.0, where the emphasis is on production. Since "Industry 4.0" refers to manufacturing specifically, this work may be interpreted as a means by which we might modernize production and reindustrialize our countries [2].

13.1.1 Industrial Internet: definition

Before discussing the benefits of IIoT adoption for businesses, it is important to define what the IIoT actually is. The industrial Internet enhances visibility and insight into a company's operations and assets by integrating machine's sensors, middleware, software, and backend cloud compute and storage systems [2]. As a result, it offers a means of modifying corporate operational processes by feeding back insights gleaned from analyzing massive amounts of data with cutting-edge analytics tools. Businesses benefit from increases in operational efficiency and productivity, which cut down on unscheduled downtime and boost overall efficiency and, in turn, earnings.

Some M2M technologies utilized in modern factories may come close to mimicking the IIoT in terms of the technologies and processes at play, but the IIoT operates on a far larger scale. The rapid online analysis of massive data streams using cloud-hosted advanced analytics is one application of big data in IIoT systems. Cloud-based storage systems allow for the widespread archiving of massive amounts of data in preparation for batch-based analytics to be done at a later date. Information and statistics can be gleaned from data using these huge batch job analytics in ways that were previously impossible due to either smaller sampling pools or just more potent or improved algorithms. Once the analytics have been run, the process engineers may use the results to optimize operations, giving the executives the data they need to make informed decisions that will increase output and efficiency while decreasing overhead.

13.1.2 Partially powerful two-thirds

Nonetheless, the concept of the power of 1% is intriguing in light of the industrial Internet. This refers to the fact that in most industries, savings of just 1% in operational costs or inefficiencies through the use of the industrial Internet can have a substantial impact. One percent annual savings on aviation fuel translates to $30 billion in savings. Similarly, a 1% reduction in fuel consumption for a power plant's gas-fired generators results in $66 billion in operating savings. In addition, a yearly decrease of 1% in capital investment on equipment in the oil and gas business would yield over $90 billion in savings [7]. The same may be said for the sectors of agriculture, transportation, and healthcare [11]. It follows that even a small increase of 1% in the majority of businesses would significantly boost the return on investment (ROI) of the capital and operating costs related to deploying the industrial Internet. If you are looking to execute an IIoT strategy, you may be wondering what kind of financial and technological investments will be needed.

13.1.3 Significant technologies for industrial IoT

Numerous cutting-edge technologies are combined to create the industrial Internet, which is a superior system. For instance, new sensor technologies create data which is more insightful than just precise, in addition to more of it. Sensors can be self-aware in addition to being able to estimate how long they have until they stop functioning. As a result, the sensor might offer data that is both precise and forward-looking. Machine sensors can acquire these same abilities through their controllers, allowing them to think for themselves, making predictions, and making comparisons. They can, for instance, assess how their current setup stacks up against a database of optimized parameters and thresholds. This allows for individuals to make their own diagnoses. The size and cost of sensors have shrunk considerably in recent years. This promoted the economic and technical viability of instrumenting machines, processes, and even humans.

A key enabler and driver for the IIoT is found due to the availability of historical, predictive, and prescriptive analysis made available by big data and advanced analytics. This makes it possible to comprehend a machine's or process' inner workings better. Analytics, when integrated with the next generation of self-aware and self-predicting components, can generate reliable predictive upkeep lists for equipment and assets, extending their useable lives while reducing wasteful repairs. This process has been accelerated by the advent of cloud computing over the past decade, as providers like Amazon Web Services now provide the low-cost, scalable, and elastic computing, storage, processing, and networking resources essential to effective big data analytics on a pay-as-you-go basis. However, other companies may feel more comfortable leaving their data in a private cloud or on premises due to security concerns.

13.1.4 Why is the time right for industrial Internet?

Looking at the capabilities and inefficiencies of legacy systems can help us understand why the industrial Internet is happening now, despite the fact that its technologies have been present for a while. One theory holds that human operators' capacity to spot inefficiencies in more complex industrial systems has waned, making it more difficult to effect changes in the conventional sense. This can lead to machines that function at a fraction of their potential, which in turn is creating operational incentives to implement new solutions. Due to declining prices in computation, bandwidth, storage, and sensors, IT infrastructures can now back widespread instrumentation, monitoring, and analytics.

This paves the way for widespread monitoring of factory machinery. The cost and storage space needed for large data sets are two problems that cloud computing aims to solve. Cloud service providers are also releasing analytic tools that can handle enormous data sets. This seems to be an important aspect, as these technologies are developing and becoming more generally available. Cloud adoption and software as a service are two examples of how IT has embraced these technologies. Only recently have industrial business leaders begun to feel more at ease about the steadiness and maturity of tools, solutions, and applications across multiple IT industries. Similarly, low-power radio wireless wide area network (WWAN) solutions have developed over time, making it possible to remotely monitor and control assets that were either too expensive or too unreliable to use before. It is now possible to deploy these wireless radio networks in an industrial context since they are stable and reasonably priced.

These changes could lead to the creation of brand-new exciting opportunities when they are functional to industrial businesses, machinery, fleets, and networks.

Thanks to the reduction in the cost of computation, storage, and networks made possible by the cloud computing model, businesses can now collect and analyze significantly greater volumes of data than ever before. Because of this, the industrial Internet is a more compelling option than the current M2M-only paradigm. But the industrial Internet also has its own problems that could be major roadblocks to the widespread use. These are the foundational elements and driving forces that must be in place before an industrial Internet deployment can be considered a success.

13.1.5 A well-equipped and trained team

Important analytics projects always necessitate the involvement of knowledgeable data scientists, process engineers, and electromechanical engineers; thus, this is a must. Colleges and institutions appear to be falling behind the times, still encouraging high school graduates to pursue programming careers rather than data science roles, making it difficult to recruit qualified candidates. There are no signs of this changing in the near future. Data scientists and electromechanical engineers are expected to be in great demand over the next 10 years, yet neither profession is currently creating enough of either.

Companies can more reliably extract useful information from their massive data lakes if they have employees with strong data analysis abilities, which is a crucial factor in the company's bottom line. This is not just any data, but data that actually helps the organization succeed and is in line with its long-term objectives. For this, you will need data scientists who are also business experts with a firm grasp on the company's overall strategy and near-term, intermediate-term, and long-term objectives. For this reason, the executive branch has developed a brand-new position known as the chief data officer.

13.1.6 The dedication to innovation

The adoption of IIoT requires a corporation to be innovative and to think about the ROI of the project over the long run. The initial investment in sensors, gadgets, equipment, and systems will need to be financed. The data collecting process and setting up the analytics' parameters and algorithms might not provide results right once; success might take some time to materialize. Additional resources and persistence might be required. After all, you might not get the answers you are hoping for from a statistical analysis. The right questions need to be asked. Data scientists might examine the business strategy and adjust their investigation, or the queries they ask of data warehouses so that their findings support the company's goals.

13.1.7 A robust security group experienced in reducing cybersecurity risks in industries and its environments

This is crucial since the IIoT is a melting pot of technology, which might leave it vulnerable to attacks if the interfaces and protocols used are not fully understood. The most valuable assets, as well as those at the greatest risk, can be uncovered through risk assessments, and strategic measures can be established to lessen their impact. In a traditional factory, for example, the machinery that actually produces the product, such as lathes that operate off of pre-programmed templates, is embedded with all the intellectual and design expertise required to manufacture the product. Security teams should also ensure that all supply-chain activities adhere to established policies and procedures.

13.1.8 IIoT and creative development in the industry

By those who support it, the industrial Internet is frequently referred to as the "third wave" of innovation. In light of the fact that the Internet revolution is the second wave of invention after the industrial revolution, this is relevant. It is commonly accepted that the next wave of innovation, known as the industrial Internet revolution, has already begun. The full potential of digital Internet technology has not yet been realized, and even if it has, it is still in its infancy, across the vast majority of industrial technological domains. Intelligent devices and systems are beginning to interface with industrial machinery, processes, and the cloud, but this is still in its infancy. Lack of standardization in the protocols, interfaces, and applications is needed to build an IIoT value chain. Examples include how diverse user requirements have helped spread modern communication and radio protocols and technology. In a nutshell, there is no silver bullet technology or protocol that can solve every problem. System integration is difficult within a company due to the availability of different protocols and technologies, but it might be problematic when working with external business partners due to the even greater complexity involved. Never forget that not even the largest corporations can manage their own value chains. Therefore, connecting to partners may be time-consuming, expensive, and unsafe until interfaces, protocols, and applications are standardized.

13.1.9 Sensible hardware

Intelligent gadgets are being created, and these can be brand-new items or modified pieces of technology. The focus of current innovation is on facilitating intelligent gadgets. Instrumentation can be connected to a variety of things, including actuators, components, sensors, machines, and also human bodies. It is true because it is can be made easy and affordable to add instruments to any object for the purpose of data collection. Considering the data flow cycle, in which information moves back and forth from smart devices to tools for data collecting and analysis and control feedback, the data that data scientists gather throughout this stage of the cycle will be most useful to them.

13.1.10 Possibilities and advantages

Not surprisingly, when questioned about the most important benefits they expect to receive from the IIoT, most adopters listed improved profits, increased revenue flows, and reduced operating expenditures. Using big data to improve operations through analytics appears to be like grabbing the low hanging fruit, it is easy to obtain. Predictive maintenance is typically the first strategy used by industrial organizations because of the speed with which it generates outcomes and ROI. Predictive maintenance and remote asset management are two uses for the IIoT. By utilizing a strategy of remote connectivity, sensors, and big data analytics, Thames Water can forestall equipment failures and react more swiftly to any urgent scenario that may emerge as a result of inclement weather. When it comes to the tactical implementation of IIoT, however, other industries, such as healthcare, have different objectives. Projects like inspecting oil and gas lines with drones and autonomous vehicles in inhospitable regions and deploying autonomous mining equipment are examples of the cutting-edge work being done in this area [7].

At the moment, Schlumberger uses a remotely operated vehicle to conduct underwater inspections. The unmanned vehicle is self-sufficient and autonomous, demanding no local human support while it glides over the ocean floor and monitors conditions for up to a year powered only by wave motion, so it may be installed in remote ocean locales. Formerly, submersible ROVs (remote operational vehicles) needed an umbilical line from a mother ship above to supply power and control signals and keep them afloat while they operated. In contrast, self-powered ROVs eliminate the need for nearby support vessels. There is also no umbilical cord to get caught on bottom features. The benefits of IIoT extend beyond the realm of "conventional" industries. When it comes to healthcare, we also see that different fields have different goals and approaches.

A better focus on patients and their needs is a priority in the healthcare industry [9]. Improvements in patient care are prioritized since patient survival rates are the most reliable indicator of a healthcare provider's performance. This is important since medical mistakes are still a primary source of avoidable mortality in hospitals. The joint project between Google and Dexcoms in order to advance disposable, tiny glucose monitors which can be read via a wrist band connected to the cloud is a nice illustration of how miniaturized sensors are used in hospitals [9, 10]. Hospitals can gain from gathering and analyzing big data and utilizing intelligent technologies to improve care for their patients without disturbing the patients in any way. By making it feasible to have more efficient, integrated, and wearable monitoring and analytic tools, medical equipment manufacturers' efforts to integrate and miniaturize their products are improving healthcare [9].

Because of the adaptability, downsizing, and use of medical technology, smart drug delivery systems are more precise and dependable than conventional methods of pharmaceutical administration. When spread over a network, smart devices can share data with one another in a manner similar to how information can be transferred between computers. This enables more intelligent analysis of sensor data from the patient, as well as monitoring and processing of this data in a shorter period of time, to ensure that alarms are only triggered when considering the collective data from other monitoring sensors indicating that the patient's health is in danger [10]. It is evident that the early adopters of the industrial Internet have all benefited in different ways by applying innovation and analytics to the exclusive difficulties faced in the business done.

13.1.11 The reason for the purchase

As a result of the IIoT, a new approach has emerged in business, particularly in manufacturing, and it centers on the idea that the producer should care more about the needs of the client than about the specifications of the product they purchase. One good illustration of this is the demand for commercial jetliners [1]. Is it because he/she fancies having one, or does he/she require it to ferry across the world hundreds of his/her customers? Manufacturers used to concentrate on creating products that could be sold on the open market for the lowest possible price. As a result, businesses had to devise ways to set their products apart from those of other producers. The product's quality, price, quantity, or the degree to which the client values it can all influence the value added. Competition with a low barrier to entry often quickly adapts to successful differentiation measures, rendering these efforts ineffective. In order to compete, rivals can, for instance, increase their lot sizes to the same or greater extent. To make matters worse, if the cost was the main selling point, the rival business might simply undercut them, triggering a downward spiral of ever-lower costs.

13.1.12 We are not selling light bulbs, but light

A light bulb's primary purpose is to provide light, as opposed to the preceding example where the consumer wanted the goods to provide a service (air transportation). As a result, businesses started to reconsider the problem from a different perspective: what if you sold light itself instead of light bulbs? These unusual business practices gave rise to the outcome economy, in which consumers paid more for the service or experience connected with a product than for the actual product. In this instance, the business is advertising the product's quantifiable effectiveness.

Truck tires are a more concrete example of this in action. Without knowing how long their tires will survive, logistics companies would rather not spend the money on new tires for all of their trucks up front [8]. But in the result of economy, the logistics firm simply has to pay monthly in arrears for the actual mileage and tire wear that it causes. Great for them, but how does the tire maker benefit from this? (We feel it necessary to emphasize that this is not a renting situation.) The IIoT seems to be the key to its success. This is made possible by the sensors included in each tire, which communicates back to the manufacturer over a wireless Internet connection on how many miles they have traveled and how much wear they have sustained. The tire manufacturer sends the logistics firm a monthly bill for the tires' depreciation. Everyone wins since they end up with something they desired all along, albeit in a roundabout fashion. Since the logistics firm was taking all the risk, initially they required tires but were only willing to pay the bare minimum up front. In any case, because individuals now pay in arrears and receive the services they desire, purchasing things carries less risk. The price increase is dispersed throughout the course of the tire's usable life, but the tire manufacturer still reaps the rewards and gains access to new services that it might be able to charge for.

Manufacturers may inform buyers on how their vehicles were driven by releasing data such as sensor readings for shock occurrences and high rates of speed. If the client has a logistics business, for instance, this service might help them train their drivers to reduce fuel use and costs. Jet engines made by Rolls Royce are another product of the result economy [3, 8]. A large airline in this case would rather invest in the peace of mind provided by Rolls Royce's TotalCare than in aircraft engines. The consumer forks out cash to guarantee problem-free operation of their jet engines. In exchange, Rolls Royce provides the engines and assumes complete responsibility for their upkeep and repair. Again, in this instance, Rolls Royce uses tens of thousands of sensors to continuously monitor the engines' operation and collect enormous amounts of data that may be used to forecast when a certain component may fail. Rolls Royce can use the information it collects and stores to create a "digital twin" of the actual engine. Because the digital and physical twins are virtually identical copies of one another, engineers can avoid opening the engine to service components that turn out to be fine without touching or taking the engine out of operation [3]. The industrial Internet and manufacturing depend on the "digital twin" concept because it enables big data analytics to produce recommendations that can be tested on a "digital twin" machine before being implemented in actual production [1].

13.1.13 Mixed human and machine labor force

In industrial environments, where they carry out a range of tedious and frequently hazardous tasks like heavy lifting, robots are now commonplace. Human workers are only allowed to perform the complex, delicate, and cognitive activities that call for human dexterity. This is a typical practice in the industrial sector, such in a car plant. Large object handling is handled by robots at one station, and tasks like connecting the electrical wiring loom to all the electronics are completed by a human worker at another station. Similarly, no machines are employed in the production of smartphones because of the intricate nature of the job involved in placing the many small components into the printed circuit board (at present). But experts predict that will alter in the coming decade as robots improve in intelligence and dexterity. Some academics believe that the future of industry will involve humans collaborating with robots rather than humans being displaced by them [5].

Humans and robots can work together effectively because they bring different skills to the table. We have the mental capacity to handle small objects with care and accomplish other tasks that call for dexterity and a sensitive touch. However, robots excel at repetitive tasks because of their incredible speed, strength, dependability, and efficiency. You should not get too close to an industrial robot, though. Most have sensors that can detect when people are nearby, allowing them to slow down or stop entirely if necessary.

A new sort of robot, however, that can peacefully and most crucially, safely cohabit with people at work is on the horizon. Perhaps that is not so far-fetched when you think about the reality of today's augmented reality (AR) solutions, which were science fiction just a few short years ago. According to the most recent findings in IIoT, in the future, robots and humans will coexist in the workplace [5]. When it comes to detecting, processing data, and acting, robots are essentially tiny versions of the industrial Internet. Therefore, robots are a perfect technological fit for the IIoT since they are essentially programmable machines made to take the place of humans in the workforce. For that reason, as sensor technology develops and software gets better, robots will get smarter and be able to comprehend their environment. After all, we already have self-driving cars and drones, so that day is not too far off. The supermarkets and shopping centers near you will soon be staffed by robots.

13.2 Industrial Internet use cases

Manufacturing, energy production, healthcare, logistics, aviation, transportation, and the production of oil and gas are just a few of the many productivity-related industries where the industrial Internet may have a big impact [7]. Business executives will sit up and take note of the IIoT because it has so many uses. It is true that even a modest increase in productivity can have a big influence on a company's bottom line; for example, a productivity increase of just 1% could save an organization like the airline industry a lot of money. If these potential benefits are to be realized, industry must accept and adapt to the IIoT.

Finding, recognizing, and effectively utilizing the IIoT's opportunities are more difficult than it might first appear. Therefore, creating use cases that are relevant to vertical industries is essential. The logistics sector's needs are different from the needs of the manufacturing sector, which are different from the needs of the healthcare sector. The innovation, experience, and funding that can be made available to create specific sector applications will be subject to many different limitations. Healthcare will cost vast sum of money with little to no financial return, in contrast to the oil and gas industry, which will demand enormous amounts of capital and operating expense.

The logistics sector will also have its own set of operational needs because it depends so heavily on the supply chain, product tracking, and transportation [8]. The IIoT may offer a workable solution for all vertical organizations by fusing current advancements in sensor technology, wireless communications, networking, cloud computing, and big data analysis. These tools can be used by businesses of any size and in any sector to benefit from the IIoT. The possible advantages and benefits to particular sectors are demonstrated in the following use scenarios.

13.2.1 Medical care

In this case study, we will do our best to illustrate how IIoT technology may be used to unlock and create value in the healthcare sector. Medical professionals used to visit patients' homes when they were too ill or injured to get to a clinic or hospital. Yet, doing so required a significant investment of both time and money. Therefore, doctors only visited patients at home if they were severely unable to leave their homes due to illness or injury, and everyone else had to wait in line at the clinic. Although logical, this method created substantial challenges to both patients and medical professionals, particularly for those who lived in distant areas and might have to travel a great distance while dealing with the incapacitating effects of disease or injury. It was therefore wise to have a fallback strategy. As a result, a pilot program to use smartphones as health monitors is being tested out in the UK at the Guy's and St. Thomas' National Health Service Foundation Trust [9].

The patient's equipment consists of a blood pressure cuff, scales, oxygen sensors, and a smartphone. Patients are supposed to keep a daily log of their vital signs (weight, heart rate, blood pressure, and oxygen saturation levels), sync that data to their cell phones using Bluetooth, and then transfer that data to BT's tele-health service. Services use nursing staff to examine collected information. The nurses will talk to patients about any concerns they have if the data shows anything out of the ordinary. With the help of these homecare kits, patients may take better care of themselves and better manage their chronic medical problems in the comfort of their own homes. Fifty people with heart failure are participating in a pilot program with the expectation that it will lead to improved survival rates [9]. The Scottish health officials' endeavor to automate, supervise, and communicate with distant outpatients is another example of a cutting-edge IIoT project in the modern healthcare setting.

To help dementia sufferers in the Western Isles and Shetland maintain their independence, the Giraff robot is being utilized in their homes [6]. The Giraff robots are intended to reassure loved ones by allowing them to speak with a loved one or caretaker from afar using a computer or smartphone. The Giraff video screen shows the doctor's face via the 3G audio/video channel, allowing for a Skype-like video call between the doctor and patient. Back in 2013, a test run of the Giraff was made available to the public. Giraff robots are mobile, under 5 feet tall, and have a video screen in place of a head. High-definition cameras are installed to keep an eye on things at home and to allow for remote monitoring. The Giraff paves the door for social interaction and communication between patients and their loved ones, even if they are separated by hundreds of kilometers. The caregiver can operate the robot and even drive it about the house to look for any hazards [6].

Some patients are singled out for the usage of particular assistive technologies, and the Giraff would fall under this category. Dementia patients who are elderly were once thought to have a negative reaction to the presence of a robot. They seemed to enjoy their time with the robot despite its incapacity to engage in conversation (although the likes of Siri could address that immediate problem and neither can a dog or cat). In past studies carried out in Australia, dementia sufferers showed no anxiety of the devices. They think the robots can ease the loneliness of a life in the wilderness. Baymax is a personal healthcare robot with synthetic skin that can identify medical conditions (this was an initiative based on a fictional Disney character in Big Hero 6 but it may not be far from becoming reality). Early versions of the robot teddy bear developed by the MIT Media Lab are currently being tested in a children's hospital in the United States. Two of the bear's paws on a more recent model have pressure sensors, and the bear's body has several touch sensors all over it. The smartphone gadget is housed in the robot's head, and its eyes are moving around on the screen. The robot may use the phone's built-in speaker, microphone, and camera to keep an eye on a child's health [5, 6].

13.2.2 The oil and gas sector

In order for the oil and gas business to search for untapped resources, technological innovation and scientific know-how are both crucial. For the finding and production of newly discovered oil and gas resources, cutting-edge sensors, analytics, and feedback control systems with increased connectivity, monitoring, control, and automation are important [7].

Additionally, the oil and gas industry gathers and analyzes enormous amounts of data about the functionality of machinery and processes across a field installation as well as the condition of drilling equipment. Geologists' inability to interpret the enormous volumes of data generated by a drilling rig due to storage limitations has historically been a barrier to advancements in oil and gas production technologies. Considering the sheer volume of information gathered, it seemed inevitable that as much as 90% of it would be lost because neither storage space nor processing capacity was sufficient to keep up with its rapid accumulation. However, thanks to the IIoT, that ineffective strategy is no longer required. It is more likely that drilling rigs and research stations will upload the massive amounts of raw data collected by drilling and production sensors to a cloud service for storage and later analysis. Drilling and exploring, for instance, have traditionally been time-consuming, expensive, and risky due to geologists' interpretations of sea floor maps. Even though they are unprepared, major oil and gas producers and explorers are modernizing their facilities to take advantage of the advancements driving the industrial Internet.

Technologies like high-bandwidth connections, wireless sensors, cloud-based data storage, advanced analytical tools, and intelligent networks are enabling systems that improve the predictability of field research, lower the costs of exploration, and lower the overall costs of field operations. Field operators are under increasing pressure to find effective solutions to the existing operational limits brought about by the adoption of new industry standards for well monitoring and reservoir management, among other technical requirements. Due to storage and communication limitations in the 1990s and 2000s, it was normal practice for field operators to discard nearly all of the data they had collected during drilling. Most of the information, such as drill bit temperature or rotations per second, was only relevant at the time it was collected. With the advent of downhole sensors and the consequent flood of data from downhole drilling equipment, however, the demand for novel solutions has increased, necessitating intricate analysis of both current and past data streams.

The oil and gas industry's requirement for such enormous data analytics coincided, fortunately, with the emergence of a new technology that offers the compute, data storage, and industrial scalability necessary to perform real-time data analysis. Big data mining in bulk is made possible by cloud computing and can be used for both retrospective analysis and prospective forecasting. The industrial Internet and cloud computing have made it possible to gather, store, and analyze enormous amounts of data at a reasonable cost [7]. However, the development of the industrial Internet has revolutionized the manufacturing industry in addition to bring forth affordable and scalable cloud services for computing, storing, and analyzing data. The IoT, for instance, enables organizations to create networks of cooperating intelligent objects, such as machines, devices, sensors, actuators, and even people. The inclusion of functions like self-diagnosis and self-configuration by the designers has increased the dependability and usability of these sophisticated smart products. The usage of cables and other power sources was previously required to connect devices, which was a severe problem that wireless communication has also assisted in resolving.

Thanks to the proliferation of low-power technologies and miniaturized components and advancements in wireless networking, sensors can now be placed in previously inaccessible or unwieldy locations without compromising on data collection capacity. Since the industrial Internet relies on Internet-based communications and cloud-based engagement, connectivity is essential to its operation. Consequently, the communication protocols are crucial, and this has led to the development of new protocols like 6LoWLAN and CoAP, which we shall cover in greater technical detail in the next chapters. For some industrial use cases, low-capability devices placed in end-to-end connections might be helpful.

The controller can either poll the node at predefined intervals to obtain the node's condition, or the sensor can convey data back to the controller, for instance, as an event. The publish/subscribe software design, however, provides a solution that outperforms the shortcomings of both of these methods (which will be covered in further depth below). This approach is recommended since any interested parties are notified right away of any upgrades over a common software bus. This approach is much more effective and quick than the subscriber asking the publisher about changes. However, not all publish/subscribe strategies are made equal. Although MQPP and XMPP have widespread adoption and support, they are not suited to use in industrial settings since they do not permit real-time operations.

The Data Distribution Service (DDS) can function in real time and send data at physical speeds to thousands of receivers concurrently while maintaining precise control over delivery time, data integrity, and operating system translation. These qualities are essential when applied in a professional setting, such as the oil and gas business. Oil and gas exploration and field production have been substantially altered, thanks to new IoT protocols and technology. To demonstrate how DDS is utilized in practice, let us look at how the oil and gas industry has implemented it as a publish/subscribe protocol.

The first demonstrates how automated remote control of drilling rigs has been made possible via the IoT. In a time when field professionals are growing scarce, this is not only financially prudent but also advantageous in terms of field productivity, safety, and well quality. The use of advanced sensor technology may also dramatically reduce downtime and equipment failures due to its sensors' ability to self-diagnose and self-configure. A high-speed DDS data connection connects all the sensors and actuators to a process controller as shown in Figure 13.2, which is the block diagram of a remote control topology for drilling and completion automation.

Aside from automating tasks, the layout also allows for remote data collection and analysis of operational data, equipment health, and process activity. High-speed connections between the field and the control center, and beyond that to the enterprise systems, whether wired or wireless, ensure efficient operations. Data can be broadcast from the field station and saved for use in future historical and predictive analyses, thanks to the DDS bus. By sending corrective input to the well systems, analysts and process planners on the mainland will be able to adjust and manage well operations.

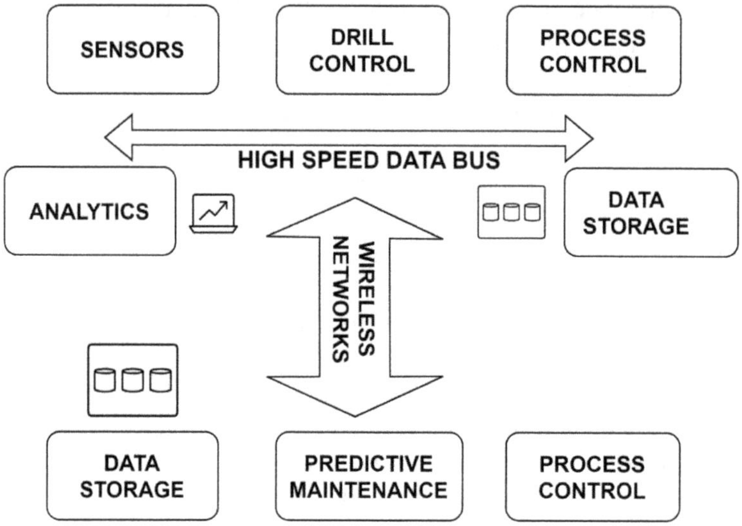

Figure 13.2: The data bus (DDS).

The IIoT also presents the chance to collect and analyze huge amounts of data. Even the wealthiest oil and gas firms could not afford to store massive amounts of data before the advent of cloud computing and widespread public access to its abundant resources. When an oil well is drilling or producing, it can produce a massive amount of data. However, this is no longer a problem because technological advancements in the realm of the industrial Internet have made it possible to store and process such enormous data quantities. A common application of this technology in intelligent well monitoring is the monitoring of vast fields of sensors and the subsequent collection of data to give a distant control center with data for historical and predictive analysis.

An additional use of IIoT in the oil and gas industry is the implementation of smart real-time reservoir management. Each component of the ecosystem needs to be linked together and add to the overall data set if analytics of any kind are to be successful. It is preferable to use a larger data set since it reduces the impact of anomalous patterns that do not accurately portray the true nature of the operation. As a very simple example, think about the chances of getting a head or tail after flipping a coin 10 times and then 10 million times. Working with real-time analytics on streaming data increases the importance of system integration due to the need for instantaneous analysis and feedback. Complexity is inherent in the design of large-scale analytical networks due to the interconnected systems and data that power them, either by way of a data bus to the cloud or via streaming analytical tools. DDS provides a single logical data highway, freeing the designer from the tangle of physical connections between computers, equipment, systems, and locations. Finally, a practical application demonstrates how IIoT protocols and technologies can simplify the development and rollout of industrial systems by decoupling software components. The development of unique, intelligent work methods made possible by the use of automated procedures, intelligent devices, and data analytics is the true promise of the IIoT.

The oil and gas industry is already making use of IIoT approaches and tools to boost productivity, security, and profitability. For the IIoT to realize its full potential, it will need to merge with the cloud, which can connect disparate local applications into more comprehensive regional and even global ones.

13.2.3 Building constructions and automation

There is a direct correlation between the energy consumption of smart office buildings and the total energy consumption in the European Union. The problem is compounded by the fact that buildings account for 36% of greenhouse gas (GHG) emissions. But keeping these numbers under control or lowering them is not simple. Improvements in building insulation, energy efficiency, and the provision of better building control systems are examples of the three-pronged plan, although progress has been excruciatingly slow.

Many factors usually contribute to this result. First, increasing the structure's insulation can significantly cut heating and cooling costs, making it a worthwhile investment for any property. In addition, it is simple to incorporate into the design and installation of brand-new buildings, and it helps cut down on both energy expenditures and carbon dioxide emissions, but it is very costly and difficult to execute in already-existing structures. This is because few older structures were constructed with sustainability in mind. The second technique, which entails modifying the building's lighting to make it more energy efficient (by replacing incandescent bulbs and fluorescent strips with LEDs, for instance), is gaining pace but is still not being fully utilized. It is possible that this is the result of a communication breakdown with landowners and company owners. However, the third option has the potential to increase building energy efficiency and minimize GHG emissions by enhancing building management through automation control systems. Sadly, adopting a building control management system (CMS) is a hard operation, both financially and in terms of business impact, similar to placing insulation into existing buildings, especially older ones. There was a time when retrofitting was necessary in order to add sensors and actuators (on things like radiators and air-conditioning units).

IoT-enabled "smart" sensors and actuators, on the other hand, considerably minimize interruption and associated costs by communicating via wireless networks. Smart buildings are made possible by the widespread adoption and further improvement of IoT-based sensors, devices, and protocols. The IoT paves the way for connected devices to exchange data with one another and with the physical environment, transforming analogue information into digital information and knowledge.

For instance, a smartphone supports communication protocols across Wi-Fi, SMS, and cellular networks, as well as acceleration and location sensors. They also have identification technologies like radio frequency identification (RFID) and near-field communication (NFC). The smartphone is therefore a crucial tool for obtaining and disseminating information. Furthermore, because of their general acceptance and availability, smartphones are ideal Human Machine Interface (HMI) for smart buildings. The IoT, however, is not without its difficulties, such as managing the enormous amounts of data produced in real time by a multitude of IoT devices dispersed throughout the structure. The inability of devices to communicate with one another is a problem, as is the market's existing proliferation of proprietary protocols and communication standards. Off-the-shelf devices might not support the protocols required to control building systems (such as heating, ventilation, and air-conditioning (HVAC) units). Before there can be widespread adoption, this needs to be corrected. Common building management systems may be difficult to retrofit into older buildings if the building was built before certain protocols were developed, such as BACnet, KNX, or LON. ZigBee, Z-Wave, and EnOcean are just a few examples of alternative protocol stacks that can be utilized to construct WSN (wireless sensor network) solutions.

Even though deployment is much simpler than with the BACnet wired bus, integration with other systems remains a difficulty. IoT6 (a working group of the European Union) built a smart office testbed in 2014 to study the potential of IPv6 and related protocols in assisting a conceptual IIoT architecture. The goals of this study were to investigate and test IPv6 to determine if it may help with the interconnection and fragmentation issues that plague IoT implementation efforts at the present time. The IOT6 team reached its conclusions by constructing a test office with commercially available sensors, devices, and protocols. Despite being the preferable choice, IPv6 was not always practical due to a lack of infrastructure. The devices were networked using a service-oriented architecture to distribute intelligence and enable Internet services, interoperability, cloud integration, mobility, and device portability.

The IOT6 Smart Office was designed with the goal of investigating IPv6's potential as a standard protocol that would make it easier to combine people and information services, like Internet and cloud-based services, with the actual building's physical structure and its different systems. It was the goal of the IOT6 team to show that conventional building automation methods could be better controlled to result in a 25% decrease in energy use. Moreover, they intended to save the time and effort required for the generally expensive refit and installation of building automation systems. They also looked into using smartphones as an HMI to better manage security and access.

The testbed would provide a setting, a fully functional smart office, to examine how well people and BIS can work together. It would have spaces for meetings and conferences, and it would also have cutting-edge interfaces (virtual assistants, etc.) installed so that building occupants can interact with their surroundings and set their own preferences for how sensors affect things like temperature, lighting, and window coverings [4]. In addition, the workplace would be fully equipped with information and services like Internet-connected computers and monitors that deliver real-time updates on global events. The smart workplace would also include a commercial coffee machine with an endless supply of hot water. The IOT6 testbed was designed to be a place where the many components of the IIoT's conceptual architecture could be tried out and their compatibility with one another verified. They were adamant about trying out as many conceivable combinations of protocols to ensure that real-world gadgets could connect to and communicate with one another (among the selected standards). They also aimed to put to the test and showcase cutting-edge IoT-related application scenarios, such as those involving business operations. They also aimed to test and validate the multi-protocol card's and IPv6 proxy's capabilities for non-IP devices, as well as evaluate the system's potential scalability. They would also deploy and evaluate the system in a true testbed environment with actual end users in order to test the various scenarios.

There were four different cases examined:

- In order to maintain a building, the first scenario required combining IPv6 with common IoT building control equipment, mobile phones, cloud services, and building management software.
- Where the office actually becomes intelligent, or "smart," is in the second scenario, which focuses on user comfort in the smart office. The CMS of the building will put the user's customized settings, such as a cozy temperature and welcoming lighting, into effect when the user enters the building and is recognized by his mobile device, NFC, or RFID [4]. When a guest comes, as identified by RFID on their cell phone, the CMS can again turn on the lights in the reception area and start playing music or a film. When the last person departs the smart office, presence of detectors will alert the CMS, informing it to dim the lights and put the HVAC system into standby.
- The third scenario places a strong emphasis on energy efficiency and outreach. This example was intended to demonstrate the usefulness of IPv6, with a focus on efficient resource management and improved end-user comprehension. Users should be able to adjust the temperature and lighting as soon as they enter a workspace using their smartphones [4]. When a modification is made, the app will show the user the effect on energy consumption after displaying the user's current settings. When the last person in the room leaves, the system is reset to its most energy-efficient setting.
- The fourth scenario, which dealt with security issues, focused mostly on detecting intruders and fires.

When magnetic switches on windows or doors are triggered or presence of detectors detect a person in a room that should be unoccupied, the security system is alerted. Similarly, fire alarms might be set off by temperature sensors or smoke detectors. In all circumstances, it will query for the IP addresses of the nearest security server and any available backups. Information is sent to the nearby data server via anycast with quality-of-service and priority routing. In the absence of a reply, it relays the same information to a second set of security nodes. The device also alerts the nearest on-duty security guard, who can then use a mobile app to remotely view the area via video feed. When using IoT devices based on IPv6-aware protocols like 6LoWPAN and CoAP over a native IPv6 network for a building CMS, the IOT6 team's technical research on the smart office demonstrated significant improvements (discussed later in the technical chapters). They noticed improvements in every element of deployment, including cost, dependability, security, flexibility, and modularity. The technical studies focused on lowering energy consumption and raising energy efficiency as success indicators.

13.2.4 Transportation and the industrial web

The logistics industry has always been at the forefront of the IoT since many IIoT opportunities and techniques are specifically designed for it. Therefore, it should not be surprising that the logistics industry has been utilizing many of the sensors and related technologies linked to the IIoT for quite some time. For many years, for instance, barcodes have been used in logistics to track packages, pallets, and containers as they enter and leave storage facilities. Compare that to the old system, when you had to open each individual delivery note and verify each item by hand, and you can see the enormous improvement. The manual scanning of barcodes was labor-intensive and, even when done carefully, could miss some pallets or products. Logistics firms sought an automated answer to these issues with stock control by implementing IIoT strategies and utilizing wireless technologies.

The solution is embedded RFID tags and the corresponding RFID readers, which can quickly scan entire rows or stacks of waiting pallets at the entry gate. The system can swiftly and precisely scan all RFID tags on all pallets within radio range, regardless of whether the tags are visible to the human eye, as opposed to reading individual barcodes. Along with the delivery itself, the RFID reader immediately enters the order ID, manufacturer, product model, type, quantity, and condition into the ERP system.

After an incoming shipment has been processed and the items have been relocated to their proper storage area, the tags can be updated to reflect the new information. Also, they can convey data about the storage environment, including readings from temperature and humidity sensors. Because of this, warehouse workers may intervene before any inventory is lost or damaged. RFID tags also facilitate speedy and precise inventory counts, which is a huge advantage. Because the RFID scanners are interfaced with an ERP application, stock levels are automatically updated whenever there is a change, and any discrepancies are flagged promptly.

When a shipment of an order is made, an RFID tag reader may scan the barcode on each pallet as it leaves the facility, updating the ERP delivery tickets and inventory levels to reflect the order's status as complete and undamaged. Thanks to the automation of stock control operations like delivery and dispatch, which has improved operational efficiency and stock control accuracy, and booking in and out products to warehouses is now a straightforward process.

Due to the competitive advantage, businesses gain from deploying sensor technology to increase operational efficiency, and logistic companies are continually ready to study new IIoT initiatives. Logistic firms are finding success in areas like improved asset utilization, in which a single system keeps track of all of their equipment and vehicles and reports on their location, status, and usage [8]. Inadvertently overusing some assets while underusing others is a common problem for warehouse managers, therefore understanding this is crucial. While other forklifts and their drivers are constantly at work, one may be sitting idle in another part of the warehouse.

Forklift productivity is a common issue in large warehouses, which adds another layer of complexity to warehouse operations. Drivers face this problem since they have to navigate the aisles in a warehouse and line up to discover the goods they need to deliver. A combination of location sensors, barcodes, RFID tags, and ERP stock information can be used to guide the driver to the locations of the stock items and provide walking directions. This method is used by Swisslog's SmartLIFT technology, which creates a visual representation of the warehouse's stock locations in relation to the forklift's current location by combining data from forklift sensors, directional barcodes on the ceiling and aisles, and warehouse stock location information. The system sends the driver to the most practical position from which to retrieve the stock using technology similar to GPS. SmartLIFT technology improves forklift productivity while reducing stock-handling errors.

Over 100,000 accidents a year are caused by forklifts in the United States alone, with over 80% involving pedestrians. As a result, the logistics sector is eager to adopt IIoT in an effort to cut down on the frequency and severity of incidents. Forklifts equipped with sensors, cameras, and radar can alert their drivers to the presence of pedestrians and other forklifts, just one example of how IIoT might be useful.

A forklift should ideally be able to communicate with other forklifts so that everyone is aware of one another and can prevent crashes by slowing down or stopping at blind intersections if another forklift is seen nearby. In wealthy nations, the human practice of choosing products from a shelf is still extensively practiced and is known as "pick-by-paper" [8].

When it comes to choosing little, intricate items out of a stock bin, forklifts, autonomous vehicles, and robots are not as useful as they are when moving massive, bulky pallets. Human labor is best suited for this kind of task. Keep in mind that the many people who were hurt when forklifts ran over them as they were walking in the warehouse. Those people walking about are probably the ones who are picked based on paper applications. Employees whose job it is to go through a list and retrieve each individual item on it. It is inefficient, and much like the forklift operators, they have trouble navigating the warehouse and finding the goods they are looking for.

AR, though, can provide assistance. Google Glass is the most well-known AR device, but it is not the only one out there. When it comes to logistics, AR, or Google

Glass for simplicity sake, is incredibly useful for human stock pickers. The pick list can be displayed on Google Glass's heads-up, hands-free display, but the wearer can also see other relevant data, such as the item's location and how to get there. In addition, it may take a picture of the item to ensure it is the right one in stock. In cases when two things are nearly indistinguishable to the naked eye, such as with a computer chip or integrated circuit, automatic barcode scanning can be used to guarantee correct item identification with no human intervention required. In addition, AR shortens the learning curve, which is especially useful given that stock pickers are typically seasonal workers. Workers are able to locate goods much more rapidly, thanks to this technology, which boosts efficiency and eradicates pick errors, resulting in higher output. Forklift drivers no longer need the fear load sheet, which specifies the sequence in which each pallet must be put onto the truck, thanks to AR glasses. Like the stock picker, the forklift operator may view the pertinent data on the glasses, speeding up loading times by reducing the need for the operator to constantly stop and consult a paper list. Scanning and verifying documents is another exciting application of the IoT and AR. In its simplest form, it allows delivery drivers to ensure that all packages or pallets have been loaded. An even more futuristic application for these glasses would be to scan international commerce documents for import/export conformity. Through the IoT integration of the AR gadget, a driver may scan the document as the software searched for the necessary keywords, phrases, and codes to verify its authenticity. This might potentially reduce the time spent at airports and border crossings fixing incorrect or missing documentation.

There are exciting possibilities for freight movement as an IIoT use case in logistics that extend beyond the warehouse. Pallets on an airplane in midair or containers on a ship at sea can both be tracked and traced by logistics firms today. Despite these features, the market is anticipating the next generation of track and trace, which will be even more efficient and secure. For instance, with more sophisticated IIoT solutions implemented, freight products may be traced meter-by-meter from dispatch to arrival, which would help combat the continuing problem of freight theft. Thanks to RFID tags on goods and sophisticated telemetric sensors in delivery vehicles, it will be possible to track the whereabouts and condition of shipments with unmatched precision and reliability. Freight items will be equipped with many sensors to track environmental factors including temperature, humidity, shock, and packaging tampering that could suggest theft. Trucks equipped with sophisticated telemetric sensors can forecast when and how repair is needed, notify the driver and maintenance staff, and even set up a service window automatically. But it is not just the trucks themselves that need to be watched out for; since they often work for long hours in dangerous conditions, driver tiredness poses a threat to both of them and other motorists. The technology to identify drowsy driving is already in use. Caterpillar, for one, employs infrared cameras to track the eyes, with the help of a computer keeping tabs on the driver's blink rate and pupil size. If it determines the drivers are drowsy, it will sound alarms and vibrate the seats to wake them up. Big data's predictive analysis approaches may also find application in the realm of supply chain management [9, 10].

Global political conditions and local weather conditions that disrupt conventional trade routes are examples of current events that the greatest logistic corporations in the world need to be aware of. Logistic corporations could cause a significant disruption that would have an impact on the stock levels of their clients if they are unaware of a planned strike by traffic controllers or crane drivers at a shipping port. Today's sensor-equipped vehicles and goods, however, allow for the global collection of this data when knowledge of potential disruptions to existing commercial routes due to social upheaval, natural disasters, and other factors is added to what is previously known. Cargo can be moved from the air to the sea in the event of a hit, enabling proactive threat management. Similar to this, if predictive analysis of all available data indicates that there is a high likelihood of civil unrest or bad weather along the route that could cause a substantial delay delivery, urgent cargo routes can be changed in real time. Over 80% of firms will implement some type of predictive analysis using big data in 2016, according to some estimates proactively mitigating possible hazards to operations and keeping freight moving to the destination is the promise of predictive analysis for global logistics.

Businesses in the retail sector, like most others, incur IT expenses and overhead costs, which eat into their earnings because they must be passed on to the consumer. Therefore, it serves retailers' interests to cut, control, or at least extract some value from their IT expenditures. A retailer's IT expenditures typically consist of the purchase and maintenance of essential IT systems used in the daily operations of the firm. The retailer can save money on IT if it can discover a more cost-effective approach to meet its needs. Retail incurs tremendous IT costs due to the replication of hardware and software licenses, break-fix callouts, and other IT support services. To reduce these high variable overheads, and to convert them into reduced fixed costs, any solution that may be implemented is preferable. There must be a system in place for retailers to accept payments, store cash and credit temporarily, track sales, returns, partial payments, and the daily cash balance, and finally provide a sales report at the end of the day. A store needs all these data to match up the day's sales statistics with the cash on hand at the end of the day. This may not seem like a big deal, but imagine how difficult it would be for a company with a thousand outlets to reconcile their end-of-day sales.

Before computers, store managers manually reconciled cash registers at the end of each day. Later, conventional cash registers were replaced by point-of-sale (POS) terminals powered by personal computer technology and outfitted with specialized peripherals like a barcode scanner and a cash box. Since this is really simply a computer running POS software, we can optimize it by shifting the brainpower and the majority of the IT workload from each store to a centralized place, the cloud. Capital (Capex) and operating expenses (Opex) are drastically reduced by moving the IT workload from individual stores to a centralized location (head office) and ultimately the cloud. As the POS application is hosted in the cloud, we are able to reduce the stores' hardware requirements to just a web browser.

In order to alleviate the stress of staffing a crowded store, sales associates can use portable devices like tablets and smartphones to do business on the go. Simply making these cuts can make a struggling shop profitable. However, the operational value of running the POS application, such as centralized administration and management, is much higher. In the past, while using legacy POS solutions, each store produced its own end-of-day report and submitted it to the retail headquarters. The retail headquarters then had to integrate all of the data into a single report. The management or owner might not have a clear view of how the day's sales performed until late the next day. However, with a cloud POS tool, both individual and aggregate information are always accurate and accessible at anytime from anywhere. Real-time access to sales data is extremely beneficial for stock control and other business operations, both on an individual and corporate level (inventory).

After all, every transaction creates a demand for stock replenishment. Due to the fact that all the data from the stores is kept in a single, multitenant database that separates the sales and inventory of each store, the cloud application can automatically manage sales and stock from both the individual and the collective perspectives, and it can do so in real time rather than the following day. With the click of a button and a cloud POS, management can quickly and effectively set standard policies across all sites and POS terminals. In addition, this facilitates the computerization of replenishment-related processes, procedures, and workflows. A significant cloud service that expedites decision making is real-time reporting and visibility.

More effective and efficient management is possible when reports and real-time status reports are available instantly. For instance, during sales promotions, the success of individual stores is monitored in real time, strategies are adjusted, and choices to restock depleting stock are made on the fly, rather than relying on yesterday's data. As we have shown, adopting a cloud POS approach for IT can help retailers cut costs, while the remaining costs can be managed and exploited to boost the company's bottom line. Savings in hardware/software and support/maintenance fees contribute directly to overall IT cost savings. A cloud-based POS system eliminates the need for these extra expenses while only charging a consistent monthly price. On its own, this lowers the price of production and boosts earnings. Further, the services offered by cloud POS allow for real-time business management and the creation of value through increased productivity and reduced waste. The retail sector has enthusiastically embraced the IIoT, and its innovations have had a far-reaching impact beyond simple efficiencies in operations. Some instances of this shift in perspective are shown below.

In-store interactive advertising, web marketing, social media, video, and AR are just a few examples of the omni-channel communication channels that the IIoT makes possible for businesses. Newspapers, magazines, and television/radio still have a large audience; therefore, we should not ignore them. However, in order to maintain their relevance, retailers must convey the appropriate message across the most widely used channels, with particular attention paid to the preferences of their target demographics.

The IoT is central to this strategy, so retail adoption of it will need to undergo significant change. The IoT has the potential to extend far beyond the confines of any particular communication channel, such as television, by linking together individuals, machines, objects, and services. Therefore, stores should make sure their marketing plans are distributed throughout omni-channels to facilitate smoother communication. This is especially crucial in the era of the Internet, when it is essential that stores cater to customers' immediate needs and improve their overall shopping experiences. Major businesses catering to youthful, trend-conscious customers, such as electronic stores and department stores, have already spent billions on the IoT.

With the help of IoT, they have been able to reap financial benefits beyond their wildest dreams through the use of cutting-edge technologies in their marketing campaigns and in-store virtual experiences for their customers. Efficiency in stock management, real-time customer-targeted promotions, and also the in-store entertainment that boosts foot traffic, builds brand awareness, and ultimately boosts sales are just a few examples of the operational and financial benefits received by retailers. These innovative IoT companies foresaw early on the need to adapt and change their procedures to meet the needs of their customers' evolving lifestyles. In addition, they saw that the IoT would eventually affect practically every facet of retail operations and customer engagement. Those stores that have made it through this shift from the old to the new have not only embraced change but also seen the value in implementing IIoT. There is a good chance that merchants are familiar with the IIoT's core ideas and the commercial and promotional possibilities it presents. Unfortunately, they find that concrete proof of the particular competitive advantage is rarely available. The IIoT is sometimes portrayed as integrating cutting-edge technology, which may not sit well with the majority of a store's current or future customers. Examples include RFID customer tracking with targeted advertising and AR. The IoT has several advantages, including better customer experiences, inventory management, digital signage, fleet management, perishable and cold chain management, and smart fulfillment centers in the back office of the shop.

Three of the biggest UK grocery chains are just one example of how the IIoT has helped some firms cut costs by up to 50%. Some merchants have to put up hardware, telematics, sensors, apps, data, devices, and cloud connectivity before using IIoT solutions. However, in addition to the obvious operational advantages, it is simple to understand how the company may profit from interactive advertising. With the aid of implanted RFID or NFC tags, this cutting-edge method of advertising can give clients extra information when the product is brought into close proximity. A digital sign underneath a product in a supermarket is activated when a customer stops in front of it to display additional information, such as nourishing information or, apparel, social media reviews and possibly a current discount. This might have an impact on the customer's purchase decision.

However, all of this might be a major turnoff to the buyer. A lot of people would rather not be subjected to such aggressive advertising. Because of this, stores are making efforts to shape the shopping experience, which they increasingly understand can be just as crucial as promoting products. Customers place a higher value on their time spent shopping than they do on the products themselves, so it makes sense to design future offerings with this trend in mind. Retailers have realized that adopting the IoT idea and technology will make their consumers' life easier and give them a more fulfilling purchasing experience, regardless of whether they use a business-to-consumer or business-to-business strategy. The deployment of IoT projects and technology aims to enhance consumer satisfaction, which in turn boosts individual sales and bottom lines.

Obviously, every client is not the same. The success of retailers using AR is evident that some people enjoy cutting-edge technology. Retailers in this era have supplied a virtual magic mirror, going above and beyond stock-tracking and NFC card payments. There has been a recent uptick in the use of AR in the retail industry since it allows customers to more actively engage with the things they are considering purchasing. The most recent IKEA catalogue is an AR example, as it is available as a mobile app that allows users to virtually place the furniture in their actual living rooms.

The consumer can virtually try out several layouts, check for size and color consistency, and make adjustments to their selections. Similarly, Uniquo, a fashion firm based in Los Angeles, has made waves with its "magic mirror," which uses audiovisual technology to recognize the customer's body shape, clothing type, and color preference before displaying a variety of complementary options. But the IoT and AR go much further than that. Yihaodian, China's largest food e-tailer, demonstrates how far virtual interfaces have the potential to go. The startup recently made headlines when it announced plans to launch the first AR grocery chain. Every one of these simulated supermarkets is empty inside and located in busy neighborhoods. Intriguingly, although those without an AR-capable gadget, such as Google Glass, will only see blank walls and floors, those with AR glasses will see shelves stocked with real food, drinks, and other items. Each virtual goods can be scanned with a customer's mobile device and added to a virtual shopping cart. The items are then sent directly to their houses for delivery.

References

[1] Digital Transformation in Manufacturing and Industrials (Accessed July 6, 2022, at https://www.ge.com/digital/industrial-internet).

[2] Industrial Internet of Things (Accessed July 6, 2022, at https://www.accenture.com/us-en/labs-insight-industrial-internet-ofthings.aspx).

[3] Data drive success at Rolls Royce (Accessed Jun 20, 2022, at (http://www.forbes.com/sites/bernardmarr/2015/06/01/how-big-datadrives-success-at-rolls-royce/#206a7e173ac0).

[4] Selling Lights, not Light bulbs (Accessed Aug 8, 2022, at http://zwcblog.org/2014/08/08/selling-light-not-light-bulbs/).

[5] Giving robots a more nimble grasp (Accessed Aug 9, 2022, at http://news.mit.edu/2015/giving-robots-more-nimble-grasp–0804).
[6] (Accessed Aug 12, 2022, at http://www.giraffplus.eu/).
[7] (Accessed Aug12, 2022, at https://www.rti.com/whitepapers/5_Ways_Oil_Gas.pdf).
[8] (Accessed at Aug 14, 2022, at http://www.dhl.com/en/about_us/logistics_insights/dhl_trend_research/Internet_of_things.html#.Vxbz49R94rgiot6.eu/sites/default/files/IoT6%20-%20D7.3.pdf).
[9] G. Sasubilli and A. Kumar, "Machine Learning and Big Data Implementation on Health Care data," 4th International Conference on Intelligent Computing and Control Systems (ICICCS), 2020, pp. 859–864, doi: 10.1109/ICICCS48265.2020.9120906.
[10] S. M. Sasubilli, A. Kumar and V. Dutt, "Machine Learning Implementation on Medical Domain to Identify Disease Insights using TMS," 2020 International Conference on Advances in Computing and Communication Engineering (ICACCE), 2020, pp. 1–4, doi: 10.1109/ICACCE49060.2020.9154960.
[11] P. K. Kotturu and A. Kumar, "Comparative Study on Machine Learning models for Early Diagnose of Alzheimer's disease: Multi-Correlation Method," 2020 5th International Conference on Communication and Electronics Systems (ICCES), 2020, pp. 778–783, doi: 10.1109/ICCES48766.2020.9137872.

K. Sasi Kala Rani, Priyadharsheni J. M., B. Karthikeyan, G. S. Pugalendhi

Chapter 14
Applications of quantum AI for healthcare

Abstract: Healthcare plays an important role in human life. Quantum artificial intelligence (AI) for healthcare is the need of the day. In order to get an insight about healthcare, certain case studies have been discussed here. Here we are going to focus on the development of healthcare within the context of the Fourth Industrial Revolution. The transition from "Healthcare 1.0" to "Healthcare 4.0" and "Industry 1.0" to "Industry 4.0." can be understood from this chapter. Better information and more individualized care are just two ways in which the Fourth Industrial Revolution can improve patients' health. The Internet of things (IoT), blockchain, big data, and AI are some of the technologies that have improved healthcare efficiency and effectiveness. Diseases can be predicted and detected at earlier stages with the help of predictive data analysis. Blockchain technology can safeguard sensitive patient information gathered from IoT-connected wearable sensors. Medical professionals can receive and analyze the data in real time. The applications covered in this chapter will be a boon to modern medicine, and the possibilities of Healthcare 4.0 seem endless. The most cutting-edge applications of Industry 4.0 medical research solutions are to compile and interpret data from high-income nations in order to assess medical sector efforts toward health value chain development 4.0. This chapter examines information on patients' symptoms, diagnoses, and ongoing care.

14.1 Introduction

The manufacturing sector, which has been around since the eighteenth century, is making rapid strides forward to automation and globalization. Production was entirely manual prior to the industrial revolution. There is significantly less need for human work these days because automated machines can do the job just as well. Industrialization has seen a number of transformations over the years, and right now, we are living through the Fourth Industrial Revolution. The healthcare industry

K. Sasi Kala Rani, Professor, Department of Computer Science and Engineering, Sri Krishna College of Engineering and Technology, e-mail: sasikalaranik@skcet.ac.in
Priyadharsheni J. M., Assistant Professor, Department of Robotics and Automation, Sri Ramakrishna Engineering College, e-mail: priya86.btech@gmail.com
B. Karthikeyan, Assistant Professor, Department of Computer Science and Engineering, KPR Institute of Engineering and Technology, e-mail: karthikeyanb@krpiet.ac.in
G. S. Pugalendhi, Assistant Professor, Department of Artificial Intelligence and Data Science, Sri Krishna College of Engineering and Technology, e-mail: pugalendhigs@skcet.ac.in

https://doi.org/10.1515/9783110791402-014

is one of the most significant areas that impact Industry 4.0. At a yearly pace of about 15%, healthcare is one of the fastest growing industries. The health and well-being of humans have been improved by and are continuing to benefit from technologically creative and advanced healthcare systems. Big data refers to a collection of very massive and complicated data that may or may not be organized in any particular way. For efficient data management, algorithms are employed to ensure that the right information is made available to the right person at the right time.

The five characteristics of big data commonly referred to as "five Vs" are volume, velocity, variety, veracity, and value. As the associated technology has improved over time, it has become progressively less daunting to deal with massive amounts of data. Blockchain is a new kind of database where information is stored in chronological blocks. Blocks are the fundamental building blocks of a blockchain and contain information such as a timestamp, transaction data, and also a cryptographic hash of the prior block. On the other hand, blockchain is a distributed, decentralized database that gets constantly updated with new data. Once a chunk of information has been input, it cannot be changed. It is important to note that blockchain technologies, like Bitcoin [1], are just one example of decentralized ledgers. Information stored in a blockchain is encrypted as it is entered. The user's private key is used to verify the public key and thus the data's authenticity. The Internet of things (IoT) is another form of network associated with big data. In 1999, the technology pioneer Kevin Ashton coined the term "Internet of things" to describe a system that used radio frequency identification tags to coordinate the flow of goods throughout a company's supply chain. The proliferation of mobile devices, wireless networks, cloud storage, and analytic tools all contribute to the importance of IoT. The IoT) extends far beyond personal computers, including automobiles, smartphones, home appliances, cameras, and even wearable health monitors.

Healthcare is only one of many industries that can benefit from the analysis and utilization of the data produced and exchanged by IoT-connected devices. Given the nascence of Industry 4.0, there is a dearth of written material on the topic's potential medical applications. Some of Industry 4.0's cutting-edge technology is out of reach for healthcare because of high prices and a lack of trained personnel. Blockchain technology, machine learning for disease detection and treatment, and the storage, security, and management of patient data have all received little academic attention. We hope to close that knowledge gap with this chapter. Section 14.1.2 analyzes the development of healthcare from its 1.0 stage (before the industrial revolution) to its 4.0 stage, while Section 14.1.1 does the same for Industry 4.0. An overview of the relevant literature is provided in Section 14.2, and the use of big data in healthcare is covered in Section 14.3. The IoT and blockchain applications in healthcare are covered in Section 14.5.

14.1.1 Differences between industries from version 1.0 to version 4.0

The industrial revolution took place between the latter half of the eighteenth century and the early part of the nineteenth century. Industry 1.0 refers to the initial stage of the industrial revolution, which occurred during this time period and marked the beginning of the shift away from human labor in favor of machines and, ultimately, higher productivity. The first stage of the industrial revolution was marked by the development of water and steam-powered machinery and was mostly driven by the iron and textile industries. Thomas Newcomen devised the steam engine, sometimes known as an "atmospheric engine," in 1712. Iron engines required coal for fuel. Craftsmen in the countryside gave way to wage workers in the cities when industry moved there. The country's economy grew as a result of all these transition. Additionally, they influenced population expansion by lowering child and infant mortality. Industry 1.0 paved the way for Industry 2.0, which developed in the United States from the late nineteenth century to early twentieth century.

Inventions like the phonograph, steamships, and electric lights all appeared during this time period. People also discovered the wireless telegraph, as well as electric railways, elevators, and motors during this time period,. It was at this time that Charles Babbage developed the first mechanical computer. As a result of Industry 2.0, many areas of social and economic life, including communication, commerce, and transportation, advanced. More than 80% of children were enrolled in some form of formal education, and significant advancements were made in healthcare, thanks to the discovery of antibiotics and the development of health insurance. This time period is notable for the rise of mechanized harvesting, which contributed to the expansion of the agricultural industry. Machines helped boost output across the board, particularly in the metals and petroleum sectors. In 1910, the first coal-fired steam plants and hydroelectric power stations went online. The development of electronic tools like computers ushered in the Third Industrial Revolution in the late twentieth century. This is when things like semiconductors, desktop PCs, and supercomputers all started to take off. Technology shifted from analogue to digital. A rise in the employment of robots, PLCs, and ICs for previously human-performed labor can be observed. Previous computers were largely theoretical, unreliable, and disproportionately bulky for their processing capacity.

Digitalization and integration through the merger of electronics with automated technologies characterize Industry 3.0, which emerged as a direct result of advances in computing and information technology. Klaus Schwab, founder and executive chairman of the World Economic Forum, named the current era the "Fourth Industrial Revolution." This revolution, also known as Industry 4.0, has ushered in numerous transitions, most of which pertain to the realms of the Internet and telecommunications. Technologies such as big data analytics, virtual systems, cloud computing, robots, IoT, and artificial intelligence (AI, 3D printing) are all examples. Healthcare 4.0 and the various technologies that improve human health are discussed in reference [2]. There is a timeline and brief explanation of each of the four stages of the industrial revolution in Table 14.1. There are three ways to categorize the integration of Industry 4.0: horizontally, vertically, and technologically. The term "horizontal integration" describes the practice of working with and competing with other businesses in the same industry to maximize production efficiency. The term "vertical integration" refers to the practice of linking together several departments and functions inside a single company. It guarantees great adaptability and facilitates straightforward production line setup. Information on product-centric value generation, customer happiness, product quality improvement, product recycling technology, production planning, and maintenance can all be gleaned from engineering integration. Data simulation, IoT, enhanced realism, cybersystems, additive production, mechanical robots, blockchain, and so on are all hallmarks of the Fourth Industrial Revolution.

Table 14.1: Industry 1.0 to Industry 4.0.

	Duration	Particulars	Applications
Industry 1.0	Eighteenth century	The machines were powered by water and steam	As manufacturing moved from the countryside to the metropolis, many artisans lost their livelihoods and became wage workers instead
Industry 2.0	Mid of the nineteenth century	Electricity was the major concern during this age	Advances in various facets of social and economic life, including communication, commerce, and transportation
Industry 3.0	Twentieth century	Computers were the major concern during this age	The integration of electronics and IT to improve manufacturing dynamics
Industry 4.0	Twenty-first century	Internet of things was the major concern during this age	Big data analytics and robotics

Through the implementation of Industry 4.0, global corporations are able to boost their competitiveness on the global market. Industrial Revolution 4.0 productions save money and time. Several potential roadblocks to implementing Healthcare 4.0 in practice were identified and addressed in reference [3]. Seven different strategies were highlighted in reference [4], all with the goal of protecting extensive electronic patient records. Sajid et al. [5] analyzed the impact of data science on the Fourth Industrial Revolution. The evolution of the sector's infrastructure is depicted in Figure 14.1.

The advent of Industry 4.0 has aided multinational corporations in becoming more competitive on the global stage. Productivity gains and cost savings are realized in Industry 4.0 manufacturing. Some potential roadblocks to implementing Healthcare 4.0 in practice were highlighted and solutions proposed in ref. [3]. Seven distinct strategies were highlighted in reference [4], with the goal of securing huge amounts

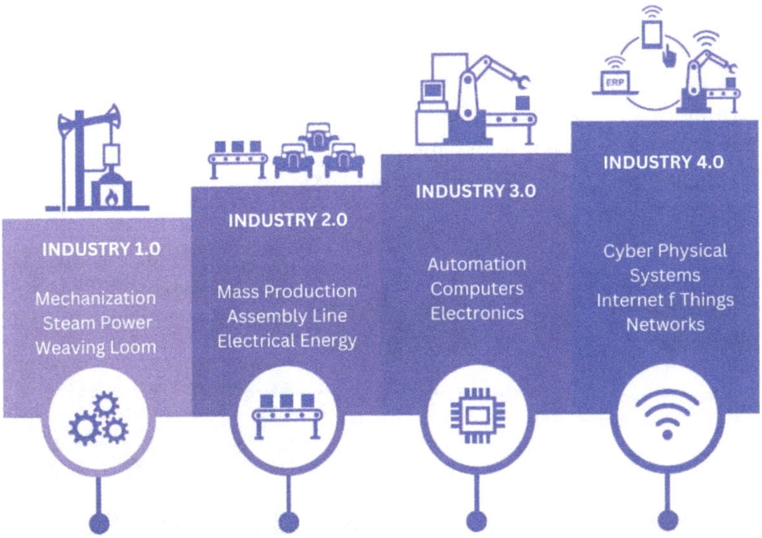

Figure 14.1: Chronological approach to industry.

of patient data that is held electronically. The significance of data science in the Fourth Industrial Revolution was analyzed in reference [5]. Figure 14.1 displays a chronological framework of the business sector.

14.1.2 Healthcare 1.0 to 4.0

Healthcare 1.0 provided answers to widespread health issues. These methods worked because they were supported by evidence [6]. With the advent of Healthcare 2.0, hospitals expanded and medical education soared, leading to a more highly educated and specialized medical staff. Both antibiotic production and responsible use skyrocketed and reached its peak during this time period. As a result of digitization, rapid pharmaceutical production became possible with the introduction of healthcare 3.0. In this era, patient data storage also improved in simplicity and effectiveness. The healthcare industry is currently undergoing a revolution known as "Healthcare 4.0," which is being fueled by the rapid advancement of new technologies. The emphasis is on the individual receiving care as opposed to the institution providing it, or the hospital. The foundations of Healthcare 4.0 are AI and IoT, which aid in the management of health and quality of life for both patients and healthcare providers. AI is helping doctors make better diagnoses and prognoses. The data collected by sensors and safely stored in the cloud can also be extracted, processed, and delivered with the help of AI. Thanks to the IoT, patients can keep tabs on their health at all times. There are many positive outcomes associated with healthcare innovations. For instance, advances in medicine have led to better care in rural areas. Insurance fraud and false claims are easier to spot, thanks to analytical technologies, which have also contributed to lowered treatment costs. In order to better handle and analyze large amounts of data, healthcare practitioners are turning to blockchain technology. Electronic health records (EHRs) save time and are safer than paper records. Healthcare 4.0 offers numerous benefits. Healthcare 1.0, 2.0, and 3.0 are depicted in Figure 14.2.

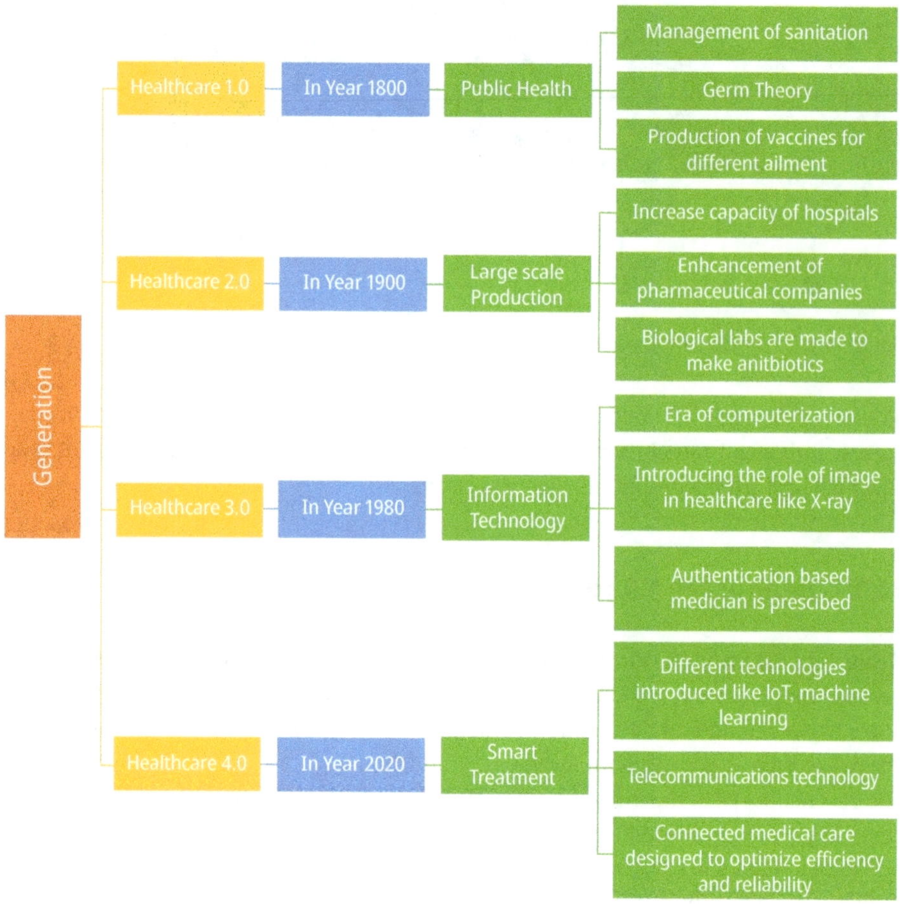

Figure 14.2: Generation hierarchy in the healthcare system.

14.2 Literature review

Data privacy, preservation, and sharing are all addressed by the GuardHealth blockchain system proposed in reference [7]. The system's goal is to control how sensitive data is handled, including its privacy, authenticity, storage, and distribution. Li and Carayon [8] stress the significance of Healthcare 4.0 and its integration with smart devices for better health. The opportunities and threats to research in this field are also discussed. The effects of Industry 4.0 on the medical field are examined in reference [9]. "Smart hospitals" are healthcare facilities that use technology to maximize efficiency across all areas of operation. The authors also make mention of "smart patients," or those who use health monitoring devices that are connected to the Internet. Based on the type of enabling technology being discussed, the authors of this study divided the scientific advancements in healthcare into eight distinct buckets. Cyber-physical systems (CPS), IoT, edge/cloud computing [28], addictive manufacturing, robotics, AI, virtual reality (VR), and augmented reality are all examples of technologies that fall into this broad category.

Kumari et al. [10] analyzed cloud computing, fog computing, and IoT. With the authors' three-tiered fog computing approach, data can be collected, processed, and transmitted in real time, improving the accuracy of diagnoses and responsiveness to treatments. Javaid and Haleem [11] conducted a comprehensive analysis of the medical applications of Industry 4.0 and found 19 critical uses for the technology, all of which serve to improve such areas as diagnostics, telemedicine, and patient-doctor interactions. The four pillars of Industry 4.0 are automated processes, automated production components, and intelligent systems. Included in this category are digital rollout, automation robots, sensing components, and cybersecurity. It allows medical facilities to electronically exchange data, which improves patient care and also helps save money and time. Patient data security is the main concern of Hathaliya et al. [12].

Secure remote access to individual patients' EHRs is a priority; thus, they suggest a biometric-based authentication system. Healthcare is one sector that has been profoundly affected by the Fourth Industrial Revolution, as described by Monterio et al. [13]. Many people highlight the IoT and IoS as crucial developments in the new Industry 4.0. Automobiles, home appliances, and other devices that can connect to the Internet via the Internet of services (IoS) are all part of the IoT. Patients can keep tabs on their health with the aid of biosensors, artificial organs, and smart devices. Furthermore, virtualization technologies allow patients to communicate with their doctors from anywhere. When resources are limited, value chain organizations can be of great assistance in facilitating efficiency and usefulness. In contrast to Healthcare 3.0, which centered on medical facilities, Healthcare 4.0 prioritizes the needs of individual patients. Industry 4.0 is characterized by its central architectures of interoperability, virtualization, decentralization, real-time capabilities, service-orientedness, and modularity. All levels of information must be able to freely flow between systems for them to be considered interoperable. One of the most important services that does this is a CPS, which combines data reading and recording. All physical objects can be replicated in a digital form, thanks to virtualization. Medical supplies and equipment are manufactured in "smart" factories that create digital duplicates and are monitored by means of CPS.

Using VR, doctors, nurses, and other medical professionals can provide more individualized care to their patients. There is increased adherence between CPSs for better healthcare. In order to ensure the safety of real-time CPS, technologies such as 5G, IoT, narrowband IoT, cloud computing, big data, network slices, and cryptography are used. It is difficult for developing countries to deal with the fact that healthcare is increasingly becoming decentralized. Smart devices and healthcare-related gadgets like wearables and bioactuators are not of guaranteed quality. The healthcare industry can benefit from decentralization with the help of mobile edge cloud computing. Chanchaichujit et al. [14] investigated core concepts in the medical services industry and provided suggestions to improve it through the use of technology. Sharma et al. [15] discussed the historical healthcare system's absence of policies.

They stress the importance of implementing a new healthcare system that is well equipped with a variety of technologies to close these gaps. Gupta et al. [16] addressed many of the concerns and difficulties of modern telesurgery. Koul and Krishna [17] explained the applications of blockchain technology in the healthcare industry, including the protection of patient records and supply chain management systems.

Table 14.2: Comparison of different healthcare technologies by various authors.

Reference	Particulars	Description
[2]	Improvements in healthcare made possible by cutting-edge technology	In what ways is the delivery of healthcare services and products being disrupted by the introduction of Industry 4.0 technologies?
[18]	Blockchain applications in medical care	Using blockchain technology, healthcare providers will be better able to share patient data while protecting their patients' confidentiality
[14]	Technology in healthcare implementation	First steps in the Healthcare System 4
[12]	Patient data security	For safe remote access to electronic health records (EHRs), a biometric authentication system is being developed
[3]	Barriers to the execution of Healthcare 4.0	The healthcare industry in India faces 15 potential roadblocks to the widespread implementation of Healthcare 4.0
[8]	Revolution of Healthcare 1.0 to Healthcare 4.0	The future of Healthcare 4.0: new frontiers in research and emerging challenges

Table 14.2 (continued)

Reference	Particulars	Description
[10]	Fog computing in healthcare	Using a three-tiered fog computing architecture, medical professionals' interpretation of patient data can be made more efficient and accurate
[7]	What's new in Healthcare 4.0	Technology based on the blockchain and machine learning algorithms
[19]	Blockchain technology in healthcare	Blockchain technology for managing patient data
[20]	Edge-based approach called BodyEdge	Applications of big data and AI

Examining the healthcare field from various authors' perspectives, Table 14.2 displays the contrast. Industry 4.0 is characterized by its central architectures of interoperability, virtualization, decentralization, real-time capabilities, service-orientedness, and modularity. All levels of information must be able to freely flow between systems for them to be considered interoperable. One of the most important services that does this is a CPS, which combines data reading and recording.

All physical objects can be replicated in a digital form, thanks to virtualization. Medical supplies and equipment are manufactured in "smart" factories that create digital duplicates and are monitored by means of CPS. Using VR, doctors, nurses, and other medical professionals can provide more individualized care to their patients. The use of 5G, IoT, narrow Band IoT, cloud computing, big data, network slices, and encryption for security in real-time CPS to enhance healthcare delivery. The cutting edge of mobile technology computing in the cloud is a tool for healthcare decentralization.

14.3 Big data in healthcare

As a result of the industry's rapid growth, hospitals and private clinics acquire an overwhelming amount of data. Because of its volume and complexity, this information is classified as "big data." It is now an integral part of every aspect of healthcare. There is a 48% annual increase in the rate at which healthcare data is being digitized. The 10 Vs of healthcare big data discussed in reference [21] are volume, velocity, variety, veracity, validity, viability, volatility, vulnerability, visualization, and value. The healthcare industry generates a growing number of data every day. Data in the healthcare industry can include a wide variety of formats, including patient information, clinical notes, medical records, scanning results, insurance information, radiology pictures, and 3D imaging. The information could be in the form of text, images, or even videos. In the past, medical records were kept on paper. Data like this is now managed in a hybrid fashion, using both human and machine labor via computers and mobile devices connected to the web. In addition, data transfer rates are evolving. Data was easier to manage when it was handled manually. The three types of healthcare data are structured, semistructured, and unstructured types. When we talk about data quality, we also talk about its truthfulness. Integrity, veracity, completeness, lack of noise, and objectivity are all desirable qualities in data quality measurements.

A study can be considered valid if the results are true and accurate. Finding the right data in a big sample is essential to making a project viable. What we mean by "volatility" is the time frame during which the information is most useful. Safety and privacy concerns are intertwined with data vulnerability. Insights can be gleaned from data with the aid of proper visualization. Analysis and management are the ones that give data its worth. The exponential growth of healthcare data makes it increasingly important to manage this information and extract the data that is most useful. The analysis of large datasets is quite useful in this context. It aids in in-depth medical examinations, which in turn improves patient care. Big data analytics can be broken down into four subfields: descriptive analytics, diagnostic analytics, predictive analytics, and prescriptive analytics. Diagnostic analytics is a subfield of data science that applies the findings of descriptive analytics to medical diagnosis.

Disease forecasting [31, 32] makes use of predictive analytics. AI, statistical modeling, data mining, and other related technologies are all brought together. Patients can be made aware of their illnesses via predictive analytics, allowing them to take preventative measures. To figure out what to do with the projected model, prescriptive analytics integrates descriptive analysis and predictive analytics. Using prescriptive analytics, doctors can quickly determine the best course of action for each patient. This aids the doctor in making a quick and precise diagnosis, which is crucial in healthcare.

One of the numerous benefits of big data in healthcare is improved patient care. It has decreased patient costs by shifting the healthcare system toward a model centered on them. Hospital readmissions have decreased as a result. In the past, patients often needed to return for repeated checkups. Big data analytics has allowed doctors to sift through health records and determine who among their patients actually require additional attention and who can safely do without it. Big data analytics can also help in optimizing their staff. As a result, the organization can better prepare for the expected volume of patients on any given day. Real-time notifications are another area where big data analytics are proven useful. As soon as a medical emergency arises, the doctor will be notified via the patient's wearable gadget.

Patients' EHR maintenance and security rely heavily on big data analytics. If a patient agrees, their health records can be stored and shared securely on a blockchain and accessed from anywhere in the globe. Analyzing hospital networks, managing public health research, improving the productivity of medical practices, introducing new safeguards, and enlisting the participation of patients are just some of the many applications of big data in healthcare. Cancer, diabetes, substance misuse, cardiovascular disease, and other illnesses can all be foreseen and treated with its help. The use of big data technology has had a profound effect on contemporary medical practice. Several uses of big data and AI to improve Industry 4.0 have been demonstrated in reference [20].

14.4 Healthcare and IoT

The benefits of wearable technology for health are discussed in reference [22]. The IoT is a cutting-edge technology that connects sensors, microcontrollers, and transceiver-equipped devices via the Internet to facilitate real-time global communication and collaboration. Heart rate monitors, continuous glucose monitors (CGMs), sleep tracking devices, and more are just a few examples of the many types of sensor-based IoT devices that can be connected to the Internet. These tools help us move closer to a healthcare system that is tailored to each individual's unique psychological, social, and biological makeup. The IoT ensures individualization of medical care by linking each patient's digital identifier to their records. One area where the IoT really shines is in remote health monitoring. It allows people to keep tabs on their health from the comfort of home, relieving some of the burden on doctors, nurses, and administrators. It is also a time-saver. People in remote places, the elderly, and those who are too busy to get regular medical attention can all benefit greatly from this. There have been enormous shifts in the treatment of chronic diseases, the effectiveness of rehabilitation, and the comfort of those who live in ambient-assisted environments, thanks to the IoT application in medicine. Many different uses of IoT have been implemented in the medical industry. As an illustration, IoT devices can be used to track movement and identify tremors in people with Parkinson's disease [31]. By monitoring the heart's rhythm, ECG sensors can foresee cardiac problems.

The SPHERE wearable device is another case in point; it is a wrist-worn sensor node designed specifically for low-maintenance health and behavior monitoring in the home. It is made to make life easier and more secure for the elderly as they continue to live at home. Sensors to track vital signs, temperature, and blood pressure are also examples of useful IoT devices in the medical field. In order for the IoT to function properly, a number of characteristics must be met. These include secrecy, trustworthiness, verifiability, verification, user-friendliness, newness of data, support, recommendation, robustness, fault tolerance, self-healing, security, interoperability, and isolation. The IoT has revolutionized healthcare by making it more efficient, cheaper, and patient-centered. Koul [23] described machine learning and deep learning applications in agriculture. Today, Industry 4.0 is demonstrating its worth not just in healthcare but also in other industries, such as agriculture.

14.5 Healthcare and blockchain

Nakamoto [1] suggest a decentralized Bitcoin-based electronic cash system for use between individuals (blockchain). The term "blockchain" was first used in reference [24], which also explores the viability of blockchain-based solutions to governance difficulties in genomic data sharing. Numerous medical fields can benefit from implementing blockchain technology. Using cryptographic methods, this technology helps to preserve and safeguard sensitive patient data. As a sophisticated data structure, a blockchain organizes information into blocks. Blockchain technology is distinguished by its decentralized nature, its traceability, its transparency, and its immutability. Because blockchain is decentralized, there is no central repository for any information exchanged among users. The blockchain ledger is immutable if information placed onto it cannot be edited or changed after it has been stored. When we talk about data being "traceable," we mean that it can be tracked back to specific locations and periods.

It is said that a network is transparent when all its participants have full access to the system's data. Data, a timestamp, hashes of the current block and the previous block, and a hash of the preceding block are the components that make up a blockchain's block. Tanwar et al. [19] outline numerous ways in which blockchain technology might be used to improve the healthcare industry. For better access to patient data, the authors devised an algorithm. Public blockchains, private blockchains, and hybrid blockchains are all described in detail in reference [18]. Any individual or organization can join a public blockchain and view its transactions as they occur. In private blockchain technology, only approved individuals have access to the ledger and may view the transactions that take place within it. A hybrid blockchain allows for both public and private data storage options. The benefits of blockchain technology extend beyond the corporation to the patient as well. It ensures the confidentiality of patient information while facilitating its usage in insurance claims management and the monitoring of clinical trials throughout the pharmaceutical distribution process.

Blockchain technology has many potential advantages but also faces numerous difficulties. These include challenges with scalability, user access and security, energy consumption, slow operations, and user ignorance of technical details. Together, blockchain and smart contracts form a powerful system. Automated transaction protocols and computer programs are used to carry out, regulate, and record activities of legal significance in accordance with the provisions of a contract or agreement. A registrar contract is an Internet-wide agreement that connects a user's unique ID to their unique Ethereum address. In this context, strings are preferred to the conventional ID form employed in cryptographic public key identification. Similarly, a summary contract and a registrar agreement map to one another. The following form is the Patient Progress Record (PPR) agreement between a patient and a healthcare professional. There are two hubs in the PPR agreement. The healthcare information for the second node is stored and managed by the first. Caregiver data is revealed by PPR's collection of data identifiers and the ease of access to permission. The next element of the agreement is the summary contract, which is used by the participants to keep track of their health information. As part of the summary contract, all involved parties will receive alerts, and the status of the connection will be recorded. Providers can initiate the relationship in the patient's summary contract and update the data to establish a new relationship. If a patient wants to keep a relationship despite its past, then that is their prerogative. The smart contract's final structure is a description of the system's nodes. Connectivity to EHR systems is built into the system node. The backend library, Ethereum client, database gatekeeper, and EHR manager are the software components built into the design of a node in a system. In our network, patient nodes are built out with the same fundamental parts as provider nodes.

The implementation of patient nodes can be run locally on a PC or remotely on a mobile phone through the Internet. Any lost information can be retrieved from the network at any time utilizing the summary contract's nodes. A patient's health status can be summarized using these databases as input. Possible consequences of blockchain technology [30, 31] on designing and implementing cyber-physical production systems in the real world are discussed by Lee et al. [25]. In order to get a comprehensive overview of blockchain technology's potential applications in the medical field, one can refer to reference [26]. Using examples from the medical field, Siyal et al. [27] discuss the current state and potential future applications of blockchain technology in the healthcare sector [35, 36].

14.6 Advanced medication and tracking of drug use

Unfortunately, forgetting to take prescriptions is a typical problem among the elderly. Assistive technology used to track patients' medication intake could help with this issue. There are a number of apps that can remind people to take their medications at the right time. In the pharmaceutical industry, "smart" drugs [29] are ones that may be purchased online and shipped directly to the customer.

14.7 Patient-centered medical care system

A healthcare system that is tailored to each individual patient's needs is called a "personalized" system. Wearable sensor technology allows patients to collect data about their health from a wide range of sources. Certain intelligent medical gadgets have been medically tested and found to be beneficial to patients.

14.8 Case studies

The case studies presented here show how Healthcare 4.0 technologies are being used to better treat patients across a range of medical specialties.

The first example in the case study is a person with diabetes. Diabetes mellitus can be either type 1, also known as "juvenile diabetes," or type 2, which affects adults. Both forms are long-lasting, so they need to be managed continuously under a doctor's supervision. Diabetic complications include an elevated danger of cardiovascular disease, kidney failure, eye disease, and more [32, 33]. Before Healthcare 4.0, people with diabetes typically had to schedule regular office visits with their doctor and take medication to control their condition. Because of this, diabetics often need to make frequent trips to their doctors, which can take up a lot of time and money.

Diabetics can now self-monitor a variety of vital signs, including blood sugar, heart rate, oxygen level, pulse rate, and blood pressure, with the help of a wide variety of modern home monitoring devices. In this way, the patient can monitor their own health and report any unusual symptoms to their doctor over the IoT. The physician can then make treatment recommendations and keep tabs on the patient's progress as necessary (e.g., weekly and monthly).

Patient's wrist can be equipped with a CGM device. Patients' glucose levels are monitored by sweat testing, and the results are transmitted to a mobile device (e.g., smartphone). The patient's doctor can access this information by downloading it from IoT. Before the advent of CGMs, people with diabetes had to perform regular finger-prick tests to monitor their blood sugar levels. CGMs are an effective and less invasive innovation in Healthcare 4.0. COVID-19 is the subject of the second case study. Early on in the COVID-19 pandemic, a diagnosis of the virus required a patient to attend a medical facility specifically equipped to make such a determination. But since doing so could potentially infect others, it was not a good idea. Consequently, COVID testing was assigned as a priority for Healthcare 4.0 to develop a safe and effective remote method. As a result, kits for performing tests at home were created. At-home kits include a sterile nasal swab, extraction tube, test card, biohazard bag, and directions. An app for reading the bar code from the smartphone is required before the self-test can be performed. After finishing the test, an individual will see the results on their mobile device. Time and money are saved, and the person can avoid potentially infecting others by staying at home, thanks to these quick tests.

One of Healthcare 4.0's novelties is the availability of quick diagnostic tests for use at home. The third example is cardiovascular disease. Machine learning algorithms like linear regression, support vector machines, decision trees, logistic regression, *k*-nearest neighbors, and Gaussian naive Bayes are used to forecast the probability of a heart attack in this case study [32]. In the past, a doctor had to have access to information about the patient's past and similar case studies, both of which were not always readily available, in order to detect and diagnose the disease. The delay in diagnosis is directly related to the lag in receiving this information. Thanks to advances in machine learning, however, today's doctors can access a patient's complete medical record and information about previous cases with the click of a mouse. Several machine learning algorithms were utilized in this investigation for the purpose of cardiac disease diagnosis. Support vector machines and logistic regression are the best algorithms for predicting the risk of heart attack.

14.9 Final thoughts

This section provided an in-depth analysis of the development of healthcare within the context of the Fourth Industrial Revolution. It documented the transition from "Healthcare 1.0" to "Healthcare 4.0" and "Industry 1.0" to "Industry 4.0." Better information and more individualized care are just two ways in which the Fourth Industrial Revolution can improve patients' health. The IoT, blockchain, big data, and AI are some of the technologies that have improved healthcare efficiency and effectiveness. Big data analytics is a perfect example of where the adage "prevention is better than cure" rings true. Diseases can be predicted and detected at earlier stages with the help of predictive data analysis. Blockchain technology safeguards sensitive patient information gathered from IoT-connected wearable sensors. Medical professionals can receive and analyze this data in real time. The applications of innovations covered in this chapter have been and will continue to be a boon to modern medicine, and the possibilities of Healthcare 4.0 seem endless. The most cutting-edge applications of Industry 4.0 medical research solutions are to compile and interpret data from high-income nations in order to assess medical sector efforts toward health value chain development 4.0. This study examines information on patients' symptoms, diagnoses, and ongoing care.

Health chain efficiency and speed will benefit from this investigation of 4.0. In the future, it will be important to verify and evaluate factors related to the growth of various items by certain firms in the medical health sector. The economic sector as a whole may be benefitted, and linked businesses may see an uptick in output as a result. To analyze the unique traits of the 4.0 sector, it is necessary to conduct a survey at the hospital by gathering information of all patients, which would serve as the case study. How forward-thinking healthcare facilities are giving staff more autonomy while also using new technologies each year to better manage patients' personal health information is determined.

References

[1] S. Nakamoto, Bitcoin: A Peer-to Peer Electronic Cash System, 9, 2008. www.Bitcoin.org.

[2] G. Aceto, V. Persico, A. Pescape, Industry 4.0 and health: Internet of things, big data, and cloud computing for Healthcare 4.0, J. Ind. Inf. Integr., 18, 2020. https://doi.org/10.1016/j.jii.2020.100129,100129.

[3] P. Ajmera, V. Jain, Modelling the barriers of Health 4.0–the fourth healthcare industrial revolution in India by TISM, Oper. Manag. Res., 12, 2019, 129–145.

[4] J. J. Hathaliyaa, S. Tanwar, R. Evans, Securing electronic healthcare records: A mobile-based biometric authentication approach, J. Inf. Sec. Appl., 53, 2020, 102528.

[5] S. Sajid, A. Haleem, S. Bahl, M. Javaid, T. Goyal, M. Mittal, Data science applications for predictive maintenance and materials science in context to Industry 4.0, Mater. Today: Proc., 45, 2021, 4898–4905. https://doi.org/10.1016/j.matpr.2021.01.357.

[6] G. L. Tortorella, F. S. Fogliatto, M. Vijaya Sunder, A. M. C. Vergara, R. Vassolo, Assessment and prioritisation of Healthcare 4.0 implementation in hospitals using quality function deployment, Int. J. Prod. Res., 2021. https://doi.org/10.1080/00207543.2021.1912429.

[7] Z. Wang, N. Luo, P. Zhou, GuardHealth: Blockchain empowered secure data management and graph convolutional network enabled anomaly detection in smart healthcare, J. Parallel Distrib. Comput., 142, 2020, 1–12. https://doi.org/10.1016/j.jpdc.2020.03.004.

[8] J. Li, P. Carayon, Health Care 4.0: A vision for smart and connected health care, IISE Trans. Healthc. Systems Eng., 2021. https://doi.org/10.1080/24725579.2021.1884627.

[9] L. Cassettari, C. Patrone, S. Saccaro, Industry 4.0 and its applications in the healthcare sector: A systematic review, XXIV Summer School "Francesco Turco"-Ind. Syst. Eng., 2019, 136–142.

[10] A. Kumari, S. Tanwar, S. Tyagi, N. Kumar, Fog computing for Healthcare 4.0 environment: Opportunities and challenges, Comput. Electr. Eng., 72, 2018, 1–13. https://doi.org/10.1016/j.compeleceng.2018.08.015.

[11] M. Javaid, A. Haleem, Industry 4.0 applications in medical field: a brief review, Curr. Med. Res. Pract., 9, 2019, 102–109. https://doi.org/10.1016/j.cmrp.2019.04.001.

[12] J. J. Hathaliya, S. Tanwar, S. Tyagi, N. Kumar, Securing electronics healthcare records in Healthcare 4.0: A biometric-based approach, Comput. Electr. Eng., 76, 2019, 398–410. https://doi.org/10.1016/j.compeleceng.2019.04.017.

[13] A. C. B. Monterio, R. P. Franca, V. V. Estrela, A. Khelassi, N. Razmjooy, Y. Iano, Health 4.0: Applications, management, technologies and review, Med. Technol. J., 2, 2018, 262–272.

[14] J. Chanchaichujit, A. Tan, F. Meng, S. Eaimkhong, An Introduction to Healthcare 4.0, in: Healthcare 4.0, Palgrave Pivot, Singapore, 2019, pp. 1–15. https://doi.org/10.1007/978-981-13-8114-0_1.

[15] D. Sharma, G. S. Aujla, R. Bajaj, Evolution from ancient medication to human-centered Healthcare 4.0: A review on health care recommender systems, Int. J. Commun. Syst., 2019. https://doi.org/10.1002/dac.4058.

[16] R. Gupta, S. Tanwar, S. Tyagi, N. Kumar, Tactile-internet-based telesurgery system for Healthcare 4.0: An architecture, research challenges, and future directions, IEEE Netw., 33(6), 2019, 22–29. https://doi.org/10.1109/MNET.001.1900063.

[17] S. Koul, T. Krishna, Introduction to Blockchain Technology and Its Role in the Healthcare Sector, in: R. S. Ambikapathy, D. Logavani (Eds.), Reinvention of Health Applications with IoT: Challenges and Solutions, 1st, CRC Press, Boca Raton, 2022, pp. 55–80. https://doi.org/10.1201/9781003166511.

[18] I. Abu-elezz, A. Hassan, A. Nazeemudeen, M. Househ, The benefits and threats of blockchain technology in healthcare: A scoping review, Int. J. Med. Inform., 142, 2020, 1–9. https://doi.org/10.1016/j.ijmedinf.2020.104246.

[19] S. Tanwar, K. Parekh, R. Evans, Blockchain-based electronic healthcare record system for Healthcare 4.0 applications, J. Inf. Sec. Appl., 50, 2020. https://doi.org/10.1016/j.jisa.2019.102407,102407.

[20] S. K. Jagatheesaperumal, M. Rahouti, K. Ahmad, A. Al-Fuqaha, M. Guizani, The duo of artificial intelligence and Big Data for Industry 4.0: Review of applications, techniques, challenges, and future research directions, IEEE Internet Things J., 2020. https://arxiv.org/abs/2104.02425.

[21] P. K. D. Pramanik, S. Pal, M. Mukhopadhyay, Healthcare Big Data: A Comprehensive Overview, in: N. Bouchemal (Ed.), Intelligent Systems for Healthcare Management and Delivery, IGI Global, 2018, pp. 72–100.

[22] S. Pirbhulal, O. W. Samuel, W. Wu, A. K. Sangaiah, G. Li, A joint resource-aware and medical data security framework for wearable healthcare systems, Futur. Gener. Comput. Syst., 95, 2019, 382–391.

[23] S. Koul, Machine Learning and Deep Learning in Agriculture, in: Patel, et al. (Ed.), Smart Agriculture: Emerging Pedagogies of Deep Learning Machine Learning and Internet of Things, CRC Press, 2021, pp. 1–19.

[24] M. Shabani, Blockchain-based platforms for genomic data sharing: A de-centralized approach in response to the governance problems?, J. Am. Med. Inform. Assoc., 26, 2019, 76–80. https://doi.org/10.1093/jamia/ocy149.

[25] J. Lee, M. Azamfar, J. Singh, A blockchain enabled cyber-physical system architecture for Industry 4.0 manufacturing systems, Manuf. Lett., 20, 2019, 34–39.

[26] C. C. Agbo, Q. H. Mahmoud, J. M. Eklund, Blockchain technology in healthcare: A systematic review, Healthcare, 7, 2019, 56. https://doi.org/10.3390/healthcare7020056.

[27] A. A. Siyal, A. Z. Junejo, M. Zawish, K. Ahmed, A. Khalil, G. Soursou, Applications of blockchain technology in medicine and healthcare: Challenges and future perspectives, Cryptography, 3(1), 2019. https://www.mdpi.com/2410-387X/3/1/3.

[28] P. Pace, G. Aloi, R. Gravina, G. Caliciuri, G. Fortino, A. Liotta, An edge-based architecture to support efficient applications for healthcare industry 4.0, IEEE Trans. Ind. Inf., 15(1), 2019, 481–489. https://doi.org/10.1109/TII.2018.2843169.

[29] J. Wan, S. Tang, D. Li, M. Imran, C. Zhang, C. Liu, Z. Pang, Reconfigurable smart factory for drug packing in healthcare Industry 4.0, IEEE Trans. Ind. Inf., 15(1), 2019, 507–516. https://doi.org/10.1109/TII.2018.2843811.

[30] R. Gupta, S. Tanwar, S. Tyagi, N. Kumar, M. S. Obaidat, B. Sadoun, "HaBiTs: blockchain-based telesurgery framework for healthcare 4.0,", 2019 International Conference on Computer, Information and Telecommunication Systems (CITS), 2019, pp. 1–5. https://doi.org/10.1109/CITS.2019.8862127.

[31] P. Bhattacharya, S. Tanwar, U. Bodke, S. Tyagi, N. Kumar, BinDaaS: Blockchain-based deep-learning as-a-service in healthcare 4.0 applications, IEEE Trans. Netw. Sci. Eng., 8, 2021, 1242–1255. https:/doi.org/10.1109/TNSE.2019.2961932.

[32] A. Janosi, W. Steinbrunn, M. Pfisterer, R. Detrano, UCI Machine Learning Repository, University of California, School of Information and Computer Science, Irvine, CA, 1988. http://archive.ics.uci.edu/ml.

[33] S. R. Swarna, A. Kumar, P. Dixit, T. V. M. Sairam, "Parkinson's Disease Prediction using Adaptive Quantum Computing," 2021 Third International Conference on Intelligent Communication Technologies and Virtual Mobile Networks (ICICV), 2021, pp. 1396–1401, doi: 10.1109/ICICV50876.2021.9388628.

[34] G. Sasubilli, A. Kumar, "Machine Learning and Big Data Implementation on Health Care data," 4th International Conference on Intelligent Computing and Control Systems (ICICCS), 2020, pp. 859–864, doi: 10.1109/ICICCS48265.2020.9120906.

[35] S. M. Sasubilli, A. Kumar, V. Dutt, "Machine Learning Implementation on Medical Domain to Identify Disease Insights using TMS," 2020 International Conference on Advances in Computing and Communication Engineering (ICACCE), 2020, pp. 1–4, doi: 10.1109/ICACCE49060.2020.9154960.
[36] S. M. Sasubilli, A. Kumar, V. Dutt, "Improving Health Care by Help of Internet of Things and Big Data Analytics and Cloud Computing," 2020 International Conference on Advances in Computing and Communication Engineering (ICACCE), 2020, pp. 1–4, doi: 10.1109/ICACCE49060.2020.9155042.

Biography

Dr. Abhishek Kumar is Assistant Director/Associate Professor in Computer Science and Engineering Department at Chandigarh University, Punjab, India. He holds a Ph.D. in computer science from University of Madras and is a Post-Doctoral Fellow in Ingenium Research Group Ingenium Research Group Lab, Universidad De Castilla-La Mancha, Ciudad Real, Spain. He has more than 11 years of academic teaching experience. He is having more than 100 publications in reputed, peer-reviewed national and international journals, books and Conferences. His research area includes Artificial Intelligence, Renewable Energy Image Processing, Computer Vision, Data Mining, and Machine Learning. He has been session chair and keynote speaker of many international conferences and webinars in India and abroad. He has been the reviewer for IEEE and Inderscience journals. He has authored/coauthored 7 books published internationally and edited 27 books (published and ongoing with IET, Elsevier, Wiley, IGI GLOBAL, Springer, Apple Academic Press, De Gruyter, CRC, etc. He also serves as Series Editor for three books series: Quantum Computing with De Gruyter, Germany; Intelligent Energy System with Elsevier; and Sustainable Energy with Nova, USA.

Dr. Ashutosh Kumar Dubey is an Associate Professor in the Department of Computer Science, Chitkara University School of Engineering and Technology, Chitkara University, Himachal Pradesh, India. He is also a Postdoc Fellow of Ingenium Research Group Lab, Universidad de Castilla-La Mancha, Ciudad Real, Spain. He received both BE and M.Tech. degrees in Computer Science and Engineering from Rajiv Gandhi Proudyogiki Vishwavidyalaya, Bhopal, India. He completed his Ph.D. in Computer Science and Engineering from JK Lakshmipat University, Jaipur, Rajasthan, India. He is also a Senior Member of IEEE and ACM. He has more than 14 years of teaching experience. He has authored/edited 10 books and published over 50 articles in the peer-reviewed international journals and conference proceedings. He serves as editor/editorial board member/reviewer of many peer-reviewed journals. His research areas are Machine Learning, Renewable Energy, Health Informatics, Nature Inspired Algorithms, Cloud Computing, and Big Data.

Dr. Surbhi Bhatia is an Associate Professor in the Department of Information Systems, College of Computer Sciences and Information Technology, King Faisal University, Saudi Arabia. She holds a Ph.D. in Computer Science and Engineering in the area of Machine Learning and Social Media Analytics. She has earned Project Management Professional Certification from a reputed project management institute in USA. She has more than 10 years of academic and teaching experience in different universities. She has published many papers in reputed journals and conferences in high indexing outlets. She is currently serving as an Academic editor, Associate Editor and Guest editor in reputed journals. She has published 10 international patents from India, Australia, and USA and also authored and edited around 12 books. She has completed many research-funded projects approved from the Ministry of Education, Saudi Arabia, and Deanship of Scientific Research. Her area of interest is Information Systems, Artificial Intelligence, and Data Analytics.

https://doi.org/10.1515/9783110791402-015

Dr. Oswalt Manoj S is an Associate Professor in the Department of Computer Science and Engineering, Sri Krishna College of Engineering and Technology, Coimbatore, Tamil Nadu, India. He obtained his Ph.D. in the field of Information and Communication Engineering from Anna University in Chennai. The title of his thesis is "Analysis and Prediction of Rainfall in Agriculture based on Deep Learning Framework." He has 13 years of experience in teaching, research, mentoring, and administration. His research focuses on Data Mining, Machine Learning, Deep Learning, Big Data Analytics, and Cloud computing. He has his publications in 24 international journals, 19 international conferences, and 21 national conferences. He is also a reviewer and editorial board member for several journals and professional associations. He has more than four patents published and one Indian patent grant. He has published two books and eight book chapters. More than six book chapters are under review. He is a member of more than 14 professional bodies. He is a Publons-certified reviewer.

Dr. Pethuru Raj is working as a chief architect at Reliance Jio Platforms Ltd. (JPL) Bangalore. Earlier, he worked with IBM Global Cloud Center of Excellence (CoE), Wipro Consulting Services (WCS), and Robert Bosch Corporate Research (CR). He has more than 22 years of IT industry experience and 8 years of research experience. He completed the Council of Scientific and Industrial Research (CSIR)-sponsored Ph.D. at Anna University, Chennai, and continued with the University Grants Commission (UGC)-sponsored postdoctoral research in the Department of Computer Science and Automation, Indian Institute of Science (IISc), Bangalore. He has received a number of international research fellowships (JSPS and JST) to work as a research scientist for 3.5 years in two leading Japanese universities.

He focuses on some of the emerging technologies such as the Internet of Things (IoT), Artificial Intelligence (AI), Model Optimization Techniques, Big, Fast and Streaming Analytics, Blockchain, Digital Twins, Cloud-Native Computing, Edge and Serverless Computing, Reliability Engineering, Microservices Architecture (MSA), Event-Driven Architecture (EDA), 5G, etc. More details are found in the following pages:

- Technology books: https://peterindia.net/MyBooks.html
- Google Scholar: https://scholar.google.co.in/citations?user=yaDflpYAAAAJ&hl=en
- The digital technologies and tools portal: www.peterindia.net/Digital.html

Index

https://doi.org/10.1515/9783110791402-016

www.ingramcontent.com/pod-product-compliance
Lightning Source LLC
LaVergne TN
LVHW081517300926
849161LV00006B/1201
9783110791259